T0262672

Carbohydrates: Integrated Research on Glycobiology and Glycotechnology

Volume II

Carbohydrates: Integrated Research on Glycobiology and Glycotechnology Volume II

Edited by **Sydney Marsh**

New York

Published by Callisto Reference,
106 Park Avenue, Suite 200,
New York, NY 10016, USA
www.callistoreference.com

**Carbohydrates: Integrated Research on Glycobiology
and Glycotechnology
Volume II**
Edited by Sydney Marsh

International Standard Book Number: 978-1-63239-108-7 (Hardback)

Contents

Preface

This book was inspired by the evolution of our times; to answer the curiosity of inquisitive minds. Many developments have occurred across the globe in the recent past which has transformed the progress in the field.

This book has been compiled for those interested in the study of carbohydrates. The book has many topics for those who are involved in glycobiology and related fields, as it encompasses the fundamentals of carbohydrates in metabolism, the influences of environment and fungi in plant carbohydrates and application of carbohydrates in microbes. This comprehensive book can serve well as an uncomplicated introduction to different disciplines of carbohydrate investigators and glycobiologists.

This book was developed from a mere concept to drafts to chapters and finally compiled together as a complete text to benefit the readers across all nations. To ensure the quality of the content we instilled two significant steps in our procedure. The first was to appoint an editorial team that would verify the data and statistics provided in the book and also select the most appropriate and valuable contributions from the plentiful contributions we received from authors worldwide. The next step was to appoint an expert of the topic as the Editor-in-Chief, who would head the project and finally make the necessary amendments and modifications to make the text reader-friendly. I was then commissioned to examine all the material to present the topics in the most comprehensible and productive format.

I would like to take this opportunity to thank all the contributing authors who were supportive enough to contribute their time and knowledge to this project. I also wish to convey my regards to my family who have been extremely supportive during the entire project.

Editor

Carbohydrate Metabolism

Food Structure and Carbohydrate Digestibility

Suman Mishra, Allan Hardacre and John Monro

Additional information is available at the end of the chapter

1. Introduction

Carbohydrate is almost universally the major dietary source of metabolic energy. Nearly all of it is obtained from plants, and nearly all of it requires digesting before it is available for metabolism. While digestion is aimed at breaking down molecular structure within carbohydrate molecules, there is a raft of further plant structural impediments to be overcome before most plant carbohydrates are available for digestion.

Starch, for instance, represents energy stored, not for animals, but for the plant that made the starch. It is a reserve available to carry a plant between seasons, to sustain it during periods when photosynthesis is limited, to prepare it for times of intense energy use such as flowering, and to support its progeny in seeds before autonomous growth. But so accessible is free starch as a form of energy that plants have taken special measures, many of them structural, to protect it physically from all sorts of opportunist consumers – animals, fungi and bacteria - and from the effects of existing in a hydrating entropic environment. All these structural barriers have to be overcome before the carbohydrate becomes available for digestion, and can be used as a source of food-derived energy.

In the human diet, lack of available carbohydrate is associated with under-nutrition and its attendant problems, while at the same time a surfeit of available carbohydrate is associated with obesity, metabolic syndrome, and diabetes – scourges of the developed world. Therefore, as food structure can have a critical role in determining the proportion of carbohydrate that is made available by food processing and digestion (Bjorck et al. 1994), it is of fundamental importance to nutrition and health.

This chapter discusses the importance of carbohydrate digestibility to human health, various forms of plant and food structure that have an impact on carbohydrate digestibility, and how food processing methods of various types overcome them.

2. The nutritional importance of carbohydrate digestion

Digestibility, energy and the glycemic response

The nutritional importance of available carbohydrate currently extends far beyond its role as a major source of sustenance for humans. Thanks to modern agriculture, transport and food technology, and to the market-driven economy in which appetite-driven food wants, rather than nutritional needs and survival, have come to determine the types of foods available to consumers, energy intakes have far exceeded energy requirements. As a result, the "developed" world is now facing an obesity crisis. Carbohydrate digestibility has gained new importance, not only because of its contribution to obesity, but also because a secondary consequence of obesity is the metabolic syndrome for which a defining feature is glucose intolerance – an impaired ability to control blood glucose concentrations after a carbohydrate meal.

It is now evident that the adipose tissue of obesity is not a passive fat storage tissue, but is physiologically active and intimately involved in glucose homeostasis. It plays a key role in glucose intolerance and Type 2 diabetes by producing factors, including free fatty acids, that induce insulin resistance (Saltiel & Kahn 2001). Resistance to insulin leads to a reduced rate of clearance of glucose from the blood, and the resulting increased concentration of glucose in the blood leads to generalized damage throughout the body, from chemical bonding (glycation) of proteins, increased oxidative stress, and damage to numerous biochemical processes (Brownlee 2001). In response to increased blood glucose and to the rate of blood glucose loading, insulin production increases, with its own damaging effects (Guigliano et al. 2008). Ultimately, exhaustion of the capacity of the pancreas to produce adequate insulin means that the insulin resistance of Type 2 diabetes evolves into the insulin insufficiency of Type 1 diabetes. The generalized, cumulative, systemic damage of prolonged and/or repeated exposure to high blood glucose concentrations manifests itself as a raft of disorders associated with long-term diabetes – kidney failure, circulatory problems, neuropathy, heart disease, blindness and so on – that are imposing enormous costs in suffering and resources (Zimmet et al. 2001).

In the context of the pandemic of obesity and glucose intolerance in the modern world, new ways of manipulating the rate and extent of digestibility of carbohydrate are being sought. The rate of starch digestion is important because the degree to which blood glucose loading exceeds blood glucose clearance determines the acuteness of the net increase in blood glucose concentrations, and consequently, the intensity of the insulin response required to remove the glucose overload and restore normal blood glucose concentrations. The rate of digestion also determines how sustained will be the supply of glucose by continued digestion in the gut, and therefore, how prolonged its contribution to delaying the urge to eat again will be.

Carbohydrate digestibility and colonic health

The extent of digestion during transit through the foregut is important because it determines the proportion of starch that is available to the colon as polysaccharide for fermentation, which has a role in colonic health (Fuentes-Zaragoza et al. 2010) and probably also in appetite control through the colonic brake feedback mechanism (Brownlee 2011). Undigested food residues, including both food structures and the carbohydrates and other nutrients that they have protected from digestion, are now recognized as being not simply gastrointestinal refuse, but a valuable feedstock for the colonic ecosystem. Through both fermentation of the residues and through the ability of a proportion of them to survive colonic transit, they play an essential part in maintaining gut health and function, as well as good health in general (Buttriss & Stokes 2008).

It is increasingly recognized that events in the colon influence the body as a whole, through products of colonic fermentation, through effects on the immune system mediated by the colonic epithelium, and through neuronal and hormonal feedback from the colon to upstream regions of the digestive tract (Wikoff et al. 2009). Short chain fatty acid products of colonic fermentation, propionic acid in particular, may play a direct role in blood glucose control by suppressing the release of plasma triglycerides, which contribute to insulin resistance. Colonic fermentation also appears to have indirect effects on hormones from the pancreas and adipose tissue that are involved in the regulation of energy metabolism (Nilsson et al. 2008).

Recent research suggests that obesity is associated with a colonic microbiota that is more effective in scavenging energy from undigested food polysaccharides than the microbiota from lean individuals (Turnbaugh et al. 2006). Although the daily increments in energy gain may be small, over time they accumulate in expanding adipose tissue. Recovering undigested energy by colonic fermentation could make the important difference between starving and surviving in an energy-depleted environment where food is scarce and of poor quality, or under the precarious conditions in which we evolved. However, in the present developed world of plenty, it may contribute to the difference between remaining trim and being overtaken by creeping obesity.

3. Forms of food structure affecting carbohydrate digestion

Food structure can take a number of forms that can affect the availability of carbohydrate in a number of different ways and at a number of different levels – molecular, cellular, plant tissue and food.

3.1. Molecular level

In the case of short chain sugars, such as the disaccharides sucrose, maltose and lactose, the structural constraint on digestion to monosaccharides lies solely within the glycosidic linkage between monosaccharide units, and is easily overcome by disaccharidases of the gut brush border (Wright et al. 2006). But even then, the rate at which the monosaccharide units traverse the gut wall, and so the extent to which absorption is completed during small intestinal transit, depends on the ability of membrane-bound transporters to recognize the structure of monosaccharides. Glucose transporters (SGLT1, GLUT 2) achieve active ATP-driven facilitated transport against a gradient, whereas the transporters that recognize fructose as a structure carry out less effective absorption by facilitated transport, which may result in overflow of fructose into the terminal ileum and colon, leading to intestinal discomfort from the resulting osmotic and fermentative effects (Gibson et al. 2004). Similarly, the structural specificity of lactase means that decline in lactase activity leads to the severe gastrointestinal problems of lactose intolerance.

Starch

Starch presents a different challenge for digestion from that of the common food disaccharides. Although it consists solely of α-D-glucose units, it may have a degree of polymerization of thousands or millions, and the glucose units may be α(1-4) linked into long linear amylose chains, or shorter amylose chains may be connected at α(1-6)-linked branch points. Most starch (~70%) is branched (amylopectin) and has a molecular weight of 50-500 million, and a degree of polymerization in the millions, depending on the plant species (French 1984; James et al. 2003; Thomas & Atwell 1998). The long regular string of glucose units in both amylose and amylopectin provides the opportunity for interactions between starch chains, leading to the buildup of pseudo-crystalline regions, which may sterically inhibit amylase access.

a.   Native starch and Starch granules (RS2)

Above the scale of amylose and amylopectin molecules, the starch is organized during growth in plants into granules that impose further restrictions on enzyme access (Ayoub et al. 2006; Gallant et al. 1997). Starch granules characteristically consist of concentric rings of alternating amorphous and pseudo-crystalline structures laid down during granule growth (**Figure 1**). The amorphous starch corresponds to regions that are rich in branches at (α(1-6) glycosidic bonds, while in the pseudo-crystalline regions the starch is highly organized as closely packed short branches, approximately 10-20 glucose subunits in length (Gallant et al. 1997; Ratnayake & Jackson 2007; Waigh et al. 2000). The high degree of organization of the pseudo-crystalline region is revealed by the typical Maltese cross birefringence pattern of

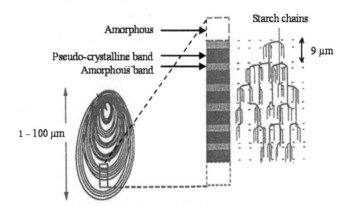

Figure 1. Schematic view of the organization of starch within a native starch granule

native starch when viewed in polarised light. The pseudo-crystalline regions are far more resistant to digestion by α-amylase than the amorphous regions (Donald 2004), and the highly organized starch granule as a whole may be relatively resistant to digestion, thanks to protein and lipid at the granule surface, which together form a coating resistant to water and digestive enzymes (Debet and Gidley, 2006).

Although covered with a resistant coating, almost all types of starch granules have been shown to bear surface pores that are entrances of channels that reach the near centre (hilum) of the granule (Huber & BeMiller 2000). The pores may be well developed in maize and nearly absent and much smaller in potato and tapioca (Juszczak et al. 2003). They may play an important role in digestion by allowing penetration of water and enzymes into the centre of the granules (Copeland et al. 2009) and leaching of glucose outwards, so the native starch granules often appear to be digested from the inside out (Gallant et al. 1997; Oates 1997; Planchot et al. 1995; Tester & Morrison 1990). However, digestion remains relatively slow while the starch is organized in its native (ungelatinized) state.

b.   Gelatinized starch

Gelatinization is the loss of the pseudocrystalline structure of the starch granules and is characterised by a loss of the maltese cross pattern in polarised light and rapid water absorption and digestion in the presence of amylase. It involves a dramatic loss of structural organization of starch granules in response to temperatures above about 60°C in conjunction with excess moisture, or by processing at temperatures above 120°C at high shear, even at low moisture levels, such as during extrusion processing (**Figure 2**).

Various techniques used to study the gelatinization process suggest that the profound change in structure during gelatinization in moderate heat and in the presence of excess water is due principally to water invasion and swelling of the amorphous regions of the starch granule (Donald 2004). Because the molecules of the amorphous regions have connecting bonds with the semi-crystalline regions, as the amorphous regions swell they force the molecules of the pseudo-crystalline regions to dissociate. As the swelling and dispersion progresses, the starch becomes increasingly accessible to digestive enzymes, and the glycemic impact of the starch rises dramatically. Starch granule pores may assist by allowing water to invade deeply into the granule interior.

Starches differ in their susceptibility to gelatinization, and have been classified as those that swell rapidly, those that have restricted swelling associated with surface lipids and proteins (Debet & Gidley 2006), and a third group of granules that contain high amounts of amylose (high semi-crystalline content), which do not swell significantly at temperatures below 100°C.

c.   Retrograded starch (RS3)

Retrogradation of starch is a form of structural change that has a large effect on digestibility. It occurs as the linear portions of starch molecules that have been dispersed during gelatinization randomly re-crystallize, without the organizing guidance of the living plant, when the gelatinized starch is cooled. Both amylose and amylopectin will retrograde.

However, amylose chains being less branched than amylopectin, will tend to re-crystallize almost irreversibly and again become nearly resistant to amylase digestion, while retrogradation of branched amylopectin is less complete and more reversible, and digestion by amylase is retarded less.

d.   Modified starches (RS4)

As starch is a long digestible polymer covered in exposed hydroxyl groups, there are many ways that it may be modified. It may be partly depolymerized by enzymes or acid, substituent groups may be added (e.g. acetylated), it may be oxidized, cross-linked, pre-gelatinized and retrograded. Most modifications to starch are designed to change its functional properties as a food ingredient by altering its rheological characteristics (Taggart 2004; Whistler & BeMiller 1997). All the chemical/processing modifications involve structural change at the molecular level and many alter the digestion characteristics of the starch. Where chemical modification of starch causes resistance to digestion, type 4 resistant starch (RS4) is formed (Sajilata et al. 2006).

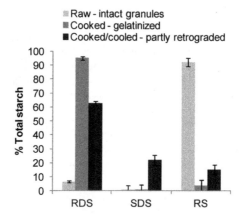

Figure 2. Effects of food structure at the molecular level - dependence of starch digestibility *in vitro* on its molecular form: rapidly digested (RDS), slowly digested (SDS) and digestion-resistant starch (RS) in potatoes digested raw (pseudo-crystalline, intact starch granules), freshly cooked (starch dispersed after gelatinizing), and cooked-cooled (starch partially recrystallized by retrogradation). (Mishra and Monro, unpublished)

e.   Occluded starch (RS1)

In the mature endosperm of most cereals, the thin cell walls are largely obliterated and the endosperm becomes a protein matrix containing embedded starch granules (Eliasson & Wahlgren 2004; White & Johnson 2003). The density and occluding effect of the protein reduces water uptake during cooking by preventing the swelling of the starch granules and as a result reduces the rate of digestion. In species with hard endosperm, such as certain wheat and maize varieties and in rice, the protein matrix is almost continuous, whereas in wheat and maize cultivars with soft endosperm, and in cereals such as oats, rye and sorghum (Earp et al. 2004), there are many discontinuities that create pathways for water and enzyme penetration into the endosperm. As a result, soft endosperm variants hydrate more quickly and present a greater internal surface area of starch for water absorption and digestion.

In pastas based on high-protein durum wheat, a relatively slow rate of digestion and low glycemic impact has been attributed to protein coating the starch granules, inhibiting both gelatinization and amylase access to starch (Colonna et al. 1990; Jenkins et al. 1987). Microscopy has revealed that protein-starch conglomerates survive in cooked pasta (Kim et al. 2007). Because of the protein occlusion of starch, carbohydrate digestion in pasta may be enhanced by cooperative protease activity (Holm & Bjorck 1988). In fatty or oily tissues such as nuts, the hydrophobic nature of the fat may also be a factor protecting the starch from hydration, gelatinization and subsequent digestion.

f.   Complexed starch

Complexing of starch with other macromolecules may involve a change in structure that is associated with reduced digestibility. Amylo-lipid complexes, formed when starch is gelatinized in the presence of lipid, are regarded as crystalline (Eliasson & Wahlgren 2004). The rate of digestion of amylose-lipid complexes is less than digestion of amylose, but greater than digestion of retrograded amylase (Holm et al. 1983).

3.2. Cell and tissue level

So far we have been discussing structural factors at the sub-cellular level that may affect carbohydrate digestibility. At the multicellular level, many sources of food carbohydrate are swallowed in the form of plant tissue fragments in which cell walls, and multiple overlying layers of cells, may act as partial barriers to both digestive enzyme penetration into, and carbohydrate diffusion out of the fragment or particle. In fruits, cereal kernels, nuts and pulses, tissue structure may influence the availability of carbohydrate and other nutrients (Mandalari et al. 2008; Palafox-Carlos et al. 2011; Tydeman et al. 2010a; Tydeman et al. 2010b).

Cereals

Seeds have evolved as dry, mechanically resistant structures that protect the embryo and the starchy endosperm from insect and animal attack until germination. In addition, in many mature grains such as rice, maize, the hard wheat varieties and some legumes, the molecular structural organization of starch and the protein that surrounds it results in a very hard endosperm that fragments into particles when crushed. Although the surface of such kernel particles is available for attack by digestive enzymes, penetration into the dense particles, especially when uncooked, is slow, and a high proportion of the starch may reach the colon. To obtain the digestible energy available in such grains, they must be subjected to processes such as grinding, flaking and cooking before being consumed. However, the dependence of digestibility on particle size provides a means by which the rate of starch availability, and the amount escaping foregut digestion to act as a substrate for colonic bacteria, may be influenced. The progressive decrease in rapidly digestible starch and increase in inaccessible (resistant) starch with increasing particle size is a very clear trend (**Figure 3**).

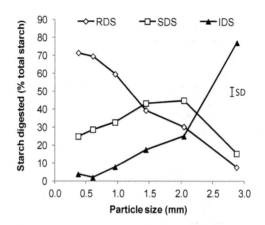

Figure 3. Effect of tissue structure on digestibility: Effect of particle size in chopped kernel fragments of the wheat cultivar 'Claire' on the *in vitro* digestibility of starch; RDS = rapidly digested (0-20 min), SDS = slowly digested (20-120 min), IDS = inaccessible digestible starch (undigested until residue homogenized) (Monro and Mishra, unpublished).

Pulses

In pulses, the starch-containing reserve tissues of the cotyledons differ in structure from the endosperm of cereals, in that the cells of the storage tissue are living and the walls retain an organized structure separating cells and contributing to tissue support in species in which the cotyledons become "seed leaves" after germination (Berg et al. 2012). In contrast, cereal endosperm cell walls are thin and usually disintegrated, and the structural integrity of the kernel is maintained by the starch/protein concretion of the endosperm, combined with the tough surrounding testa or seed coat (the bran in wheat). The differences in structure between the pulses and cereal products are reflected in the patterns of carbohydrate digestion from them (**Figure 4**).

The thick and resistant cell walls of pulses and may retard the gelatinization of starch by confining it within the cell lumen (Tovar et al. 1990; Tovar et al. 1992). When the starch is densely packed within resilient clusters of intact cells with robust cell walls, swelling is constrained. In addition, an encapsulating layer of gel from unconstrained starch in the outer cell layers of pulse fragments may create a barrier that impedes water penetration. However, partly because they are pectin rich compared with cereals, when processing is harsh or prolonged, the cell walls of pulses will degrade enough for the cells to separate. Then the starch becomes free to swell and disperse, and digestion is more rapid.

In domestic cooking, the robust cell walls of pulses are often able to survive moist heat enough to remain intact, so that cohesive plant cell clusters with encapsulated starch remain after cooking, making pulses some of the most slowly digested carbohydrate sources, with a typically low glycemic effect compared with other carbohydrate foods (Venn et al. 2006). Pulses such as kidney beans and chick peas are typical, and *in vitro* show a slow linear digestion that is usually incomplete (**Figure 4**). *In vivo* and particularly when cooking is incomplete or the food fragments are poorly comminuted before swallowing, they load the colon with fermentable starch (Type 1 resistant starch), which is largely responsible for the flatulence generated by pulses.

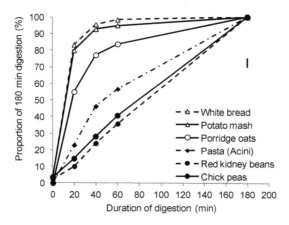

Figure 4. *In vitro* digestion patterns (%carbohydrate available after 180 min digestion) associated with different types of food structure: Porous, no intact cell walls (white bread), crushed and dispersed (mashed potato), crushed but partially intact native structure (porridge oats), dense non-porous structure (pasta; acini), robust and intact plant cell walls encapsulating starch (chick peas and red kidney beans). The bar represents the mean between duplicate ranges (5%). (Mishra and Monro, unpublished).

Fruits and vegetables

In most ripe fruits, available carbohydrates are in the form of soluble sugars – glucose, fructose and sucrose – which are highly soluble and mobile. Glucose and fructose are absorbed by specific transporters in the intestinal wall, while sucrose is hydrolyzed by a brush border invertase (Wright et al. 2006). Therefore, the only direct structural impediments to sugar availability from fruits are those that retard sugar diffusion. The parenchyma cell walls of fruits, even after mincing and digesting *in vitro*, can markedly retard sugar diffusion (**Figure 5**) and removing them from fruit puree increases its glycaemic impact (Haber et al. 1977). Such retardation can be regarded as a structural effect, as the presence of cell wall fragments with their enmeshed pectic polysaccharides increases the length of the diffusion pathways to a degree that would make a significant difference to blood glucose loading *in vivo*.

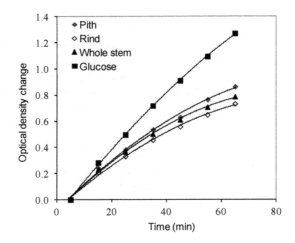

Figure 5. Effect of structure in digested plant tissue remnants on a process important to the digestive process - diffusion. Glucose diffusion was retarded about 40% by the presence of digestion-resistant remnants of broccoli tissue in an unstirred system. Pith – parenchyma cells. Rind (cortex) – parenchyma, fiber and vascular cells. The tissue remnants (cell walls) were at settled bed density, after they had been predigested *in vitro* and allowed to settle overnight by gravity. All the solutions contained 10% glucose (w/v) at the start of dialysis (Monro, unpublished). The mean between duplicate ranges was <0.1 OD units.

3.3. Food level - secondary structure established by processing

The characteristics of carbohydrate digestion in many carbohydrate foods are determined by the structure of the food matrix established during food processing involving cooking, with and without disruption of the original cellular structure of the plant source. Such secondary food structure exerts its influence largely by affecting the accessibility of the digestion medium to starch in the food. At the larger level, food particle geometry may create structures that have an enormous impact on digestibility through their influence on the surface area available for digestion (Monro et al. 2011).

a.   Open porous structures

Foods with an open porous structure include leavened products such as breads and cakes, puffed products produced by steam expansion, including many snack foods, and breakfast cereals such as puffed rice. These structures have a high internal surface area that is almost immediately available for amylase attack, and as they are precooked to eliminate native starch granule structure, they typically digest very rapidly (**Figure 4**), causing an acute blood glucose response. Such foods are typified by a high rapidly digested starch (RDS) content, little slowly digested starch (SDS) and a small proportion of retrograded resistant starch (RS Type 3) (**Figure 6, Group A**).

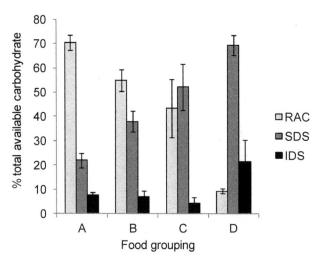

A: Little structure, starch gelatinized e.g. extruded, puffed and flaked precooked cereal products
B: Some structure, minimally processed, incomplete starch gelatinization e.g. rolled oats
C: Dense secondary structure, non-porous, surface digestion e.g. pasta
D: Intact, robust cell walls encapsulating starch in native tissue structure e.g. pulses.

Figure 6. Effects of food structure typical of various food groupings on the content of rapidly digested (RDS), slowly digested (SDS) and resistant (inaccessible; IDS) starch that they contain. Error bars are SDs of food means within groups. (Mishra and Monro, unpublished).

b.   Dense low porosity structures

Dense low porosity structures include products such as pastas, in which hydrated flour is force-molded into a dense configuration and then cooked under conditions such as boiling, and in which porosity as a result of gas formation in the food matrix does not occur. Foods such as pastas allow digestion only as fast as digestive enzymes can erode superficial layers of the food, to expose underlying carbohydrate, so their rate of digestion depends strongly on their surface area, and is therefore dependent on particle geometry. The dependence of digestion rate on particle geometry has been examined in detail using pastas of different shapes and gelatinized sago as models. As surface area of a sphere depends on the square of the radius, small increases in particle size can have a large influence on digestion rate (Monro et al. 2011).

The role of the dense structure of unexpanded starch in retarding digestion becomes clear during *in vitro* digestion of solid foods such as pasta and tapioca with and without homogenizing to eliminate the occlusive effect of the starch. After homogenizing the pasta and tapioca, starch was immediately digested, in contrast to the starch in the intact pasta and tapioca, which was gradually released as the starch was digested by superficial erosion (**Figure 7**). In dense low-porosity structures such as pastas made from durum wheat, the occlusive effects of protein will also to retard starch digestion.

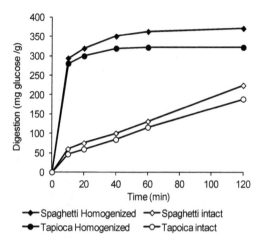

Figure 7. Effects of secondary structure on digestibility *in vitro*. Effect of occlusion by starch and dependence on surface area of food in dense non-porous foods revealed *in vitro* by the increased rate of digestion after homogenizing the foods to expose interior starch to digestive enzymes. The mean between duplicate range was < 5% (Mishra and Monro, unpublished).

4. Manipulating food structure to control carbohydrate digestibility

Many of the properties of foods that retard carbohydrate digestion discussed in the previous section are the result of the carbohydrate being stored by the plant during biosynthesis in a stable, organized, semi-crystalline and protected form, until it is required by the plant. For animals and humans to use plant carbohydrate as an energy source, it is necessary to overcome or reverse the steps the plant has taken to protect its energy reserves. Thus, before the process of enzymatic depolymerization of the individual starch molecules in the gut can provide the minimized molecular forms in which starch is absorbed by the gut border – as glucose, maltose and dextrins - there are several obstructions to be removed: firstly, the protective tissues of the plant; secondly, the barrier function of the starch granule surface; and thirdly, the obstructive molecular packing of starch within the starch granule must be overcome.

The inaccessibility of starch due to plant tissue structure has been overcome in four main ways, by external mechanical disruption, by cooking, by chewing, and in the intestine by the weak shearing and abrasion of intestinal contractions acting on digesta. In practice, with most starchy foods all four processes are used sequentially to obtain available carbohydrate from food for absorption, but the contributions that each may play in determining the amount of carbohydrate digested have been demonstrated individually.

4.1. By mechanical disruption during ingredient preparation

Mechanical disruption of food structure is one of the most effective ways of increasing carbohydrate energy availability from foods, and milling, crushing, pounding and such processes have been used for thousands of years to improve energy extraction from all types of plant tissue, but especially from the well protected form of seeds. On the other hand, reducing tissue disruption to lower carbohydrate digestibility of grain products has been found to be an effective strategy in reducing the glycemic impact of foods in populations with excessive energy intakes, obesity and diabetes (Venn & Mann 2004).

a. Milling

The most widely used procedures for milling of cereals such as wheat involve a combination of cutting and grinding, and applying strong shearing forces that break the seed coat and tear the endosperm tissue apart. More than any other process, milling can convert kernels of grain in which starch remains almost completely protected from digestion, even when cooked, to the finely divided form of flour, in which the same starch is rapidly digested because the inaccessibility of the starch to digestive enzymes caused by intact cell walls and protective plant structures has been eliminated. The effective protection of starch in cooked but intact plant tissue, and its susceptibility to digestion as soon as the tissue is disrupted by milling, is revealed by the changing distribution of starch between RDS (0-20 min), SDS (20-120 min), and RS (digestion-resistant) starch fractions measured by *in vitro* digestion (**Figure 2)**. With decreasing particle size the RDS fraction increases and the RS fraction decreases, while there is very little change in SDS; the cooked starch is either inaccessible ("resistant") or accessible, and if accessible, is rapidly digested - with the cell wall barrier removed starch is quickly degraded once gelatinized (Hallfrisch & Behall 2000).

The effects of milling do not result solely from the disruption of surrounding plant tissue, but damage to the starch granules, including cracking, fracturing, and internal changes to the granules occur that also increase their susceptibility to gelatinization and digestibility (Donald 2004).

b. Chopping/cutting

Cutting and chopping such as that occurring during the kibbling of grain does not cause the internal tissue disruption caused by the strong shearing force of grinding mills, and produce a more slowly digested product. Even simple crushing has an enormous influence on the availability of starch, because it involves forces strong enough to disrupt starch-protein conglomerates and cells containing encapsulated starch and other nutrients effectively, and creates pathways for ingress of digestive enzymes. Cutting and chopping without the shearing forces of grinding may increase digestibility much less because plant tissue damage is more restricted to the cut surfaces (**Figure 8**). Penetration of the effects of digestion through layers of intact cells underlying the cut surfaces may be a relatively slow process (Mandalari et al. 2008).

Recent detailed studies of the release of nutrients from nut fragments have shown that cell walls formed a very effective barrier against the intestinal environment (Mandalari et al. 2008; Palafox-Carlos et al. 2011; Tydeman et al. 2010a; Tydeman et al. 2010b). Mandalari et al. (2010) showed that after 3 h of simulated gastric plus duodenal digestion of almond fragments, the intracellular contents had been lost from only the first layer of the cells, at the fracture surface. After 12 h of digestion, the loss of nutrients had extended to only three to five cell layers deep. In the zone of digestion the cell walls appeared to change in structure, in that they swelled, but without any detectable change in composition of the cell wall polysaccharides. There was no evidence of cell wall fracture, so any enzyme penetration into the food particles could occur only by diffusion through the cell walls.

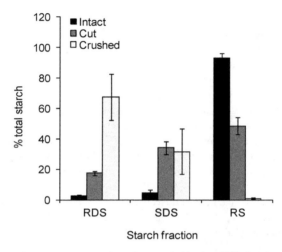

Figure 8. Influence of tissue structure on digestion revealed by the effect of cutting and crushing of cooked wheat kernels on *in vitro* digestion of starch. Individual whole hydrated kernels (n = 5/treatment) were cooked and digested either intact ("Intact"), after they had been cut transversely in half ("Cut") or crushed to 1 mm thickness ("Crushed"). RDS = rapidly digested, SDS = slowly digested, RS = resistant starch. Error bars are the standard deviations (Monro, unpublished).

4.2. By controlling gelatinization during cooking

Cooking starch under hydrating conditions brings about an often dramatic increase in starch digestibility as a result of gelatinization. In some plants, such as potato, the raw starch granules are virtually indigestible, but as soon as gelatinization occurs they, are rapidly and totally digested (**Figure 2**). Between the extremes of raw starch and total gelatinization, the degree of gelatinization may be controlled by limiting the amount of water available for hydration and by carefully controlling the cooking temperature. Because starch gelatinization requires a combination of heat and water, the availability of water during cooking can substantially modulate the effects of cooking on digestibility. In food products in which the water content is not adequate to gelatinize starch fully, the glycemic impact may be correspondingly reduced. Rolled oats, for instance, are prepared under conditions in which incomplete gelatinization of starch occurs. If consumed directly in the form of muesli, the starch digestibility is relatively low, but if further cooked and hydrated to form porridge, starch digestion is greatly increased along with the glycemic impact of the oats (**Figure 9**).

The sensitivity of starch digestion to the degree of hydration during cooking of a number of starches for 10 minutes at 95°C – maize (normal starch, 27% amylose), Hi maize (70% amylose), Mazaca (waxy maize, 2% amylose), pea, potato, rice, tapioca and wheat – increased steadily as moisture content was increased in 5 or 10% intervals from 0% moisture, reaching a maximum digestibility at about 60% moisture (**Figure 10**), which corresponds approximately to the degree of gelatinization that occurs in water-unlimited conditions at 100°C. The dependence of digestion on hydration during cooking was very similar for all the starches, even though the starches (controls) cooked with no added water differed considerably in their susceptibility to digestion in the raw (granular) state, as indicated by different Y intercepts in Figure 10. The differences in the digestibility of uncooked starch (Fig 10 A) probably reflect the importance of granule morphology, including pore size and surface chemistry, rather than any intrinsic differences in the starch molecules in determining the initial rate at which raw starch is digested. In food products containing hydrating components other than starch, such as non-starch polysaccharides in cell wall remnants, intrinsic and added gums, and sugars, competition for water may reduce the gelatinization of starch during cooking (Pomeranz et al. 1977), allowing the digestion-inhibiting effect of native starch structure to persist. Even in white bread, which contains about 60% moisture, partially intact birefringent starch granules remain in the cooked product.

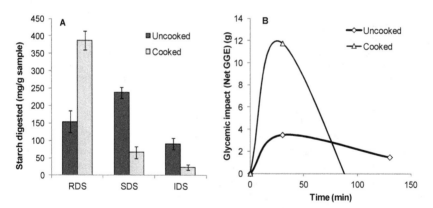

Figure 9. Effects on *in vitro* digestibility of further processing (boiling) of rolled oats (a minimally processed product) in preparing porridge, during which loss of starch structure (gelatinization) and extensive loss of oat grain structure occurs. A starch fractions of different digestibility defined in caption to Figure 3. **B** Theoretical blood glucose response curves based on RDS and SDS release with glucose disposal allowed for using the method of Monro et al. (2010). Error bars are standard deviations (Mishra and Monro, unpublished).

Two hydrothermal treatments of starch, annealing and heat-moisture treatment (HMT), cause changes in the structural and physicochemical properties and increase the digestibility of raw starch enough for it to qualify as slowly digested starch (digested between 20 and 120 min *in vitro*) (Lehmann & Robin 2007). In annealing, the starch is heated to below gelatinization temperature but long enough for some molecular rearrangement of the starch to occur. In HMT, higher temperatures are used but water is restricted so that full gelatinization does not occur. Although structural change is sufficient to increase digestibility, the granular structure and birefringence of the starch granules remain.

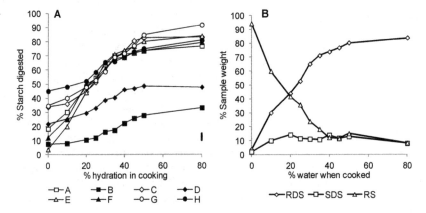

Figure 10. Effect of structural change in starch granules as a function of hydration during cooking, on *in vitro* digestion. Chart A. Rapidly digested starch as a % of total starch during cooking of seven starches at progressively increasing degrees of hydration: A. Maize; B. Hi Maize (high amylose); C. Mazaca (low amylose); D. Pea ('Sonata'); E. Potato; F. Rice; G. Tapioca; H. Wheaten cornflour. Chart B. Example of changes in starch fractions differing in digestibility: Potato starch showing that increased hydration caused the interconversion of resistant and rapidly digestible starch without substantially increasing slowly digested starch. The bar is the mean between-duplicate range.

4.3. By controlled retrogradation after cooking

During cooling of starch dispersed by gelatinization, the linear sections of amylopectin and amylose chains anneal to form hydrogen bonded alignments that limit access by digestive enzymes. The partial retrogradation of amylopectin in cooked potato to form slowly digested starch (SDS) is a good example of the effect of retrogradation (Figure 1; cooked-chilled), and has led to the suggestion that a way to reduce the glycemic impact of potatoes would be to consume them in the form of cold potato salad rather than as freshly cooked hot potato (Leeman et al. 2005; Monro & Mishra 2009).

Retrogradation is now used industrially to produce resistant starches (RS3). They have become widely used as bland low energy ingredients for which colonic benefits are claimed, in nutritionally enhanced bakery and other products.

4.4. By retaining tissue structure in whole foods – minimal processing

Tissue structure can be retained by consuming whole foods, in which the incomplete comminution achieved by chewing allows some survival and influence of food structure on carbohydrate availability. The diffusion of sugars from fruit pieces, for instance, is much slower than from fruits consumed as a puree (Haber et al. 1977).

The use of retained tissue structure in cooked food is most widely used in the baking industries, when fragments of intact kernels are included in grain breads. Chopping and cutting (kibbling) of grain kernels that would otherwise be ground to flour has found a place in food processing for populations with high rates of obesity and glucose intolerance, where there is need to reduce both the rate and extent of carbohydrate digestion. By substituting partially intact kibbled kernels for flour in bread products, the rate of digestion and the resulting glycemic impact of foods can be significantly reduced, as is revealed by the increases in digestion rates and *in vitro* glycemic index estimates when kernel-rich breads are homogenized to remove grain structure (**Figure 11**). Pumpernickel is an extreme example, as it consists largely of a conglomerate of rye grains, and in line with its low digestibility, it has a much lower glycemic index than most other breads (Jenkins et al. 1986).

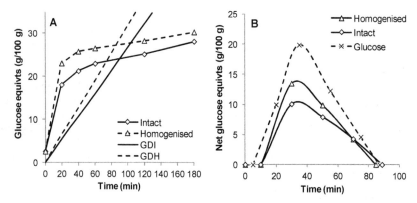

Figure 11. Effect of homogenizing to remove structure: A commercial wholegrain bread containing 25% kernel fragments > 2 mm in diameter was homogenized and the intact and homogenized breads digested *in vitro*. A. Digestion profiles (with markers) and lines of theoretical glucose disposal (GDI = glucose disposal for intact bread, GDH = glucose disposal for homogenized bread). Mean inter-duplicate range <5% at each time point. B. The curves after taking into account theoretical glucose disposal and response lag. The areas under the curves compared with the area under the glucose reference curve gave *in vitro* glycemic index values of 70 (high) for the homogenized bread compared with 55 (low) for the unhomogenized bread. Mean between-duplicate range was < 5% (Monro and Mishra, unpublished).

Minimal processing is not a term specific to any procedure, but it generally refers to processing that is the minimum required to make a product palatable, or saleable in a particular form for future cooking (Fellows 2000). The digestive advantages of minimal processing for nutrition are a reduced rate and extent of starch digestion, more resistant starch and non-starch polysaccharide for colonic function (as in bran), and greater nutrient retention than in a refined product from the same cereal source. The minimally processed category includes a range of cereal grains that have been steamed to partially precook and then rolled or flaked to eliminate hardness, such as rolled oats and barley. Depending on how thinly they are rolled they may have a relatively low glycemic impact until further processed into the form of hot porridge (Granfeldt et al. 2000) (**Figure 9**).

One of the main points supporting the argument in favor of consuming a greater proportion of dietary carbohydrate as whole grains, in which native structure is partially retained, is that "wholeness", in the sense of intactness, is associated with a reduced blood glucose loading, and increased colonic loading of resistant starch (Venn & Mann 2004). A significant association of increasing particle size with decreasing glycemic response (Fardet et al. 2006) and increased colonic fermentation (Bird et al. 2000) has been demonstrated and is consistent with results of *in vitro* analyses of grain starch digestibility.

4.5. By replacing native structure with secondary structure in food processing

a.   Formation of open textures

Extrusion cooking under shear and pressure often at moisture contents of less that 15% is a means of producing highly digestible crisp, dry food products. The extrusion process involves high temperatures, extreme shearing forces, and release of hot product under high pressure, to yield expanded dry products that are gelatinized, porous and retain almost no native tissue or starch granule structure to resist amylase activity. Leavened products such as white bread are similar, except that tissue disruption is achieved by milling before cooking, and porosity is achieved by including a leavening agent such as yeast in the product formulation. The porosity of such products coupled with the absence of any integrated plant cell wall structures ensures rapid penetration of digestive enzymes and almost immediate collapse and digestion, so that the conversion of starch to sugar during digestion rapidly increases to a plateau where digestion is complete (**Figure 4**). Accordingly, such products are often of high glycemic impact (Foster-Powell et al. 2002).

b.   Formation of dense low-porosity structures

Dense, low porosity food structures such as pasta are produced by extrusion at low temperatures with limited gelatinization and no puffing. In these foods, carbohydrate digestion is relatively slow and related to food geometry, because little enzyme penetration is possible, and any digestion is dependent on surface area (Monro et al. 2011). In such foods there is potential to use food shape to influence digestion rate as long as a proportion of the particles survive mastication. Dense foods that are soft enough to be swallowed partially intact, so that the influence of surface area on digestion rate is retained, may be useful in delivering available carbohydrate without an acute postprandial blood glucose response. Development of foods that use particle shape to determine digestion rate will therefore need also to consider the influence of food texture on the urge to chew.

4.6. By combinations of processes

Many food products are made using combinations of processes, each affecting food structure and carbohydrate digestibility in different ways. Some examples are given below (Table 1).

4.7. By changing the molecular structure of starch

Molecular modifications to starch that affect degree of branching or the ability of starch chains to interact or retrograde will affect digestibility, as already discussed. The feasibility of using starch modification at the point of biosynthesis in the plant is now being investigated, through genetic manipulation of the enzymes involved in establishing starch structure. Transgenic potato lines deficient in granule-bound starch synthase, and in two

Product type	Processes	Structure	Carbo-hydrate digestion	Relative Glycemic impact (GGE/g)[1]
Raw fruits	None	Plant tissue structure intact. Available carbohydrate as mono and disaccharides	High	Low
Pasta	Milling, cold low pressure, low shear extrusion, then time limited boiling	Dense, polymeric available carbohydrate (starch), incomplete starch gelatinization, protein occlusion of starch granules, superficial digestion	Slow but complete	Low-moderate
Biscuit	Milling then high temperature-low moisture cooking	Dense, friable, incomplete starch gelatinization. High fat levels coat starch	Moderate	Moderate
Bread, white	Milling then high temperature, high moisture cooking	Porous, more complete starch gelatinization than biscuits	High	High
Bread, kibbled grain	Milled flour plus cut grain, then high temperature, high moisture cooking	Porous gelatinized matrix containing intact kernel fragments with limited access of digestive enzymes	Moderate	Low-moderate[2]
Extrusion cooked/ puffed	Milling prior to high temperature, high shear, high pressure extrusion and puffing	All plant tissue structure and starch granule structure eliminated. Highly porous, readily accessible gelatinized starch	High	High
Muesli/ rolled oats	Crushing, steaming, limited cutting	Plant tissue structure partially intact, starch partially gelatinized	Moderate	Low[3]
Porridge	Crushing, steaming, then moist cooking	Plant tissue structure less intact than previous, starch gelatinized	High	Moderate - high

[1] Glycemic glucose equivalents/g sample
[2] Depends on inclusion rate of kibbled grain
[3] Depends on degree of crushing and chopping

Table 1. Examples of the relationship between processing combinations, food structure, carbohydrate digestibility, and relative glycemic impact per equal food weight in several food types.

starch-branching enzymes have been produced. Lines in which more linear starch was produced showed a lowered susceptibility to *in vitro* digestion than the parent lines (Karlsson et al. 2007).

Traditional plant breeding and selection of mutant types has already produced varieties with alterations in starch structure that affect starch digestibility. A high amylose maize cultivar, for instance, is used as a source of commercially available resistant starch (HiMaize®, Figure 10).

4.8. By structural breakdown during food consumption and digestion

a.   Chewing

As chewing is the natural way to increase carbohydrate availability by reducing food structure, it has been seriously suggested that a way of using structure to reduce carbohydrate digestion rate, to reduce glycemic impact, is to swallow food without chewing it (Read et al. 1986). However, chewing of food is such an important part of enjoying it and converting it into a lubricated form that can be swallowed, that not chewing food is not a practical option for controlling glycemic impact.

Evolution has equipped humans with an effective grinding mechanism in the form of teeth, and the sensitive and dexterous combination of cheek and tongue to sort and position food particles accurately between the molar grinding surfaces. The effectiveness of crushing in bringing about the conversion of inaccessible digestible ("resistant"; RS) to rapidly digested (RDS) starch **(Figure 8)** underscores the importance of mastication to the successful exploitation of starch in an omnivorous diet.

Chewing has three very important functions, all related to food structure. Firstly, chewing crushes foods to release nutrients; secondly, it reduces the size of food fragments so they may be comfortably swallowed; and thirdly, and most importantly, it churns and mixes the food with saliva to convert it into the form of a well-lubricated semi-solid bolus that may be easily swallowed. Not until the food is swallowed can effective digestion commence.

The purpose of chewing has been revealed in studies of individual differences in chewing. Although individuals differ markedly in the mechanical details of how they go about their oral comminution of foods, they all arrive at a remarkably similar endpoint in terms of the particle size reduction in the mouth **(Figure 12)**. It is apparent that the urge to stop chewing and swallow is determined more by the physical properties of the bolus that results from chewing, than by details of the mastication process that lead to the bolus. Dentition, the number of chews, the rate of chewing and so on are less important than the final result (Peyron et al. 2004).

In many processed foods produced nowadays, digestibility is not very dependent on structural degradation due to crushing by chewing, because the foods are based on ingredients, such as flour, that have been thoroughly comminuted by milling before cooking. Foods for which chewing makes a difference are usually those consisting of, or containing intact grains, such as rice and kibbled grains, and those consisting of dense, non-porous starch matrices. For many starchy processed foods the combination of mastication and salivary α-amylase activity reduces the adhesiveness of starch, and the structural cohesion of food in the form of a bolus, allowing the stomach to separate available carbohydrate from more fibrous components quite rapidly for transfer to the duodenum, which is the primary site of carbohydrate digestion in the gut.

The combination of salivary α-amylase with chewing, and the fact that starch is usually gelatinized and not intrinsically fibrous, means that most of the starch component is quickly reduced to a small particle-containing slurry that can be separated in the stomach and moved on to the ileum for digestion with little delay. The rapid dispersion of starch in most foods explains why, despite the tendency of the stomach to retain large particles, blood glucose responses to foods almost invariably commence after a lag of only about 10 min from ingestion and almost always reach a peak between 30 and 40 minutes from ingestion (Brand-Miller et al. 2009).

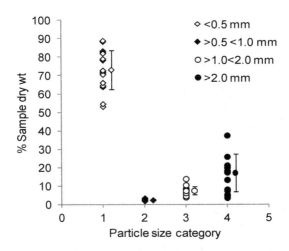

Figure 12. Effect of chewing on reduction of structure in a carbohydrate food – rice. Range of particle sizes in within size categories from cooked white rice chewed by 20 subjects. The subjects chewed intact whole rice grains in quantities they would normally consume and expectorated them when they felt the urge to swallow. For all subjects, most of the chewed sample was less than 0.5 mm in diameter. Means ± SD shown for each size category. Based on data used in Ranawana et. al. (2010).

b.   Gastric and small intestinal digestion

The stomach and intestine are not passive reservoirs, but are motile reactors in which food, while undergoing enzymatic dismemberment, is also continually subjected to shearing and abrasion from circumferential, longitudinally migrating contractions of the gut wall (Lentle & Janssen 2008). Compared with the concentrated forces exerted at the molar surfaces by jaw muscles in chewing, the forces on food particles in the stomach and intestine due to peristalsis are very small (Lentle & Janssen 2008). However, in combination with digestive enzyme action they have the important role of reducing food structure by sloughing off the hydrated and digestively weakened surface layers of food particles, to improve access of digestive enzymes to the interior.

When the mechanical and processing steps of food preparation and mastication have reduced food structure to the extent that the food can be swallowed, and the stomach has then separated a starchy slurry from the large particles remaining, the starch is ready for digestion in the ileum by α-amylase from the pancreas working in concert with enzymes in the gut wall.

The rate of starch digestion depends partly on the rate of gastric emptying (Darwiche et al. 2001), but also the structural form in which it arrives in the ileum, as intact starch granules, as disorganized or dispersed gelatinized starch, or as once-disorganized starch that has re-aggregated to form retrograded starch. Digestion of intact food particles or starch granules is relatively slow in the gut, as it is *in vitro,* and the digestion pattern depends very much on the botanical origin of the granules (Donald 2004; Oates 1997).

a.   Colonic digestion

In the colon, carbohydrate digestibility is governed by a completely different set of parameters than in the foregut, to which the discussion has referred so far. Carbohydrates entering the colon are those that were unable to be digested and/or absorbed during gastric/ileal transit. They include starch that has survived digestion for the reasons of structure and chemistry already discussed, including crystallinity in ungelatinized and retrograded starch, encapsulation by plant tissue cell walls, and occlusion by fat. However the main carbohydrate source entering the colon consists of the non-starch polysaccharides that constitute the plant cell wall and in the colonic ecosystem into which they pass they are exposed to a myriad of bacterial enzymes that are absent from the foregut. The colonic bacteria disassemble the cell walls to provide carbohydrate substrate for bacterial fermentation. The products of colonic fermentation are short chain fatty acids that are absorbed and enter intermediary metabolism, where they may provide as much as 10% of dietary energy requirements for humans.

The physical/structural constraints that modulate colonic fermentation of polysaccharide residues involve molecular structure, occlusion and particle size, which may all affect availability of substrate for bacterial enzymes and the ability of bacteria to colonize and invade fragments of plant tissue and cell walls. In the bacterial ecosystem that consists of thousands of species of bacteria that can adapt rapidly to changes in available substrates, there are few natural plant polysaccharides that on their own can resist the multipronged and coordinated attack of the diverse colonic microflora. However, some, such as psyllium gum, which is a complex and highly branched polysaccharide, are fermented so slowly that much of their molecular structure, and the hydration capacity that depends on it, remains intact after passage through the colon. Such polysaccharides make very effective faecal bulking agents but may lead to problems such as reduced mixing and fermentation in the colon when present at high concentrations. Cellulose, which exists as highly crystalline fibrils, is also slowly fermented, and is a major component of the dietary fiber that survives and contributes faecal bulk in plant-containing diets (Monro & Mishra 2010) .

Plant residues are often non-fermentable because of the occlusive effects of secondary thickening of the tissues, initially by cellulose but followed in many cases by the deposition of lignin, resulting from phenolic condensation within pre-existing cell walls, usually when they have already undergone secondary thickening by cellulose deposition (Esau 1967). The combined effects of increasing amounts of crystalline cellulose and lignin is clearly seen in the contrast between fermentation of parenchymatous pith cells and of secondarily thickened rind (cortex) cells of broccoli stem in the hind gut. Although derived from the same parenchymatous ground tissue, the pith remains as parenchyma, while the rind differentiates to contain a high proportion of secondarily thickened and lignified xylem tissue. The parenchyma cells are almost completely consumed by the bacterial flora in the hind gut, while the cells of the rind remain apparently intact and recognizable in the feces (Monro & Mishra 2010).

Even within tissues consisting entirely of parenchyma cells, the rate of fermentation can be modulated by structure. Paradoxically, multicellular clumps of carrot cells were found to be more rapidly fermented than cell wall fragments, probably because the intercellular spaces and angles between cells in multicellular clusters provided colonization sites for colonic bacteria (Day et al. 2012). There is, however, likely to be an optimal size at which increased colonization sites are counterbalanced by inaccessibility in large particles.

5. Conclusion

Food structure can affect carbohydrate digestibility in a range of ways. From the level of molecular conformation to plant anatomy, structure plays an important role in carbohydrate digestion. Attempts to manipulate structure as a means of controlling nutritional attributes of foods related to carbohydrate digestion have been in progress for thousands of years. They remain an important focus of modern food technology aimed at addressing the health problems associated with both inadequate energy intakes in developing countries, and excessive energy consumption in the developed world.

Author details

Suman Mishra and John Monro
Food Industry Science Centre, The New Zealand Institute for Plant & Food Research Limited, Palmerston North, New Zealand

Allan Hardacre
Institute of Food Nutrition and Human Health, Massey University, Palmerston North, New Zealand

6. References

Ayoub A, Ohtani T, Sugiyama S 2006. Atomic force microscopy investigation of disorder process on rice starch granule surface. Starch-Starke 58(9): 475-479.

Berg T, Singh J, Hardacre A, Boland MJ 2012. The role of cotyledon cell structure during in vitro digestion of starch in navy beans. Carbohydrate Polymers 87(2): 1678-1688.

Bird AR, Hayakawa T, Marsono Y, Gooden JM, Record IR, Carrol RL, Topping DL 2000. Coarse brown rice increases faecal and large bowel short chain fatty acids and starch but lowers calcium in the large bowel. J Nutr 130: 1780-1787.

Bjorck I, Granfeldt Y, Liljeberg H, Tovar J, Asp NG 1994. Food Properties Affecting the Digestion and Absorption of Carbohydrates. American Journal of Clinical Nutrition 59(3): S699-S705.

Brand-Miller JC, Stockmann K, Atkinson F, Petocz P, Denyer G 2009. Glycemic index, postprandial glycemia, and the shape of the curve in healthy subjects: analysis of a database of more than 1000 foods. American Journal of Clinical Nutrition 89(1): 97-105.

Brownlee IA 2011. The physiological roles of dietary fibre. Food Hydrocolloids 25(2): 238-250.

Brownlee M 2001. Biochemistry oand molecular biology of diabetic complications. Nature 414: 813-820.

Buttriss JL, Stokes CS 2008. Dietary fibre and health: an overview. Nutr Bull 33: 186-200.

Colonna P, Barry JL, Cloarec D, Bornet F, Gouilloud S, Galmiche JP 1990. Enzymic susceptibility of starch from pasta. Journal of Cereal Science 11(1): 59-70.

Copeland L, Blazek J, Salman H, Tang MCM 2009. Form and functionality of starch. Food Hydrocolloids 23(6): 1527-1534.

Darwiche G, Ostman EM, Liljeberg HGM, Kallinen N, Bjorgell O, Bjorck IME, Almer LO 2001. Measurements of the gastric emptying rate by use of ultrasonography: studies in humans using bread with added sodium propionate. American Journal of Clinical Nutrition 74(2): 254-258.

Day L, Gomez J, Oiseth SK, Gidley MJ, Williams BA 2012. Faster Fermentation of Cooked Carrot Cell Clusters Compared to Cell Wall Fragments in Vitro by Porcine Feces. Journal of Agricultural and Food Chemistry 60(12): 3282-3290.

Debet MR, Gidley MJ 2006. Three classes of starch granule swelling: Influence of surface proteins and lipids. Carbohydrate Polymers 64(3): 452-465.

Donald AM 2004. Understanding starch structure and functionality. In: A-C E ed. Starch in Food. Cambridge, UK, Woodhead Publishing Limited.

Earp CF, McDonough CM, Rooney LW 2004. Microscopy of pericarp development in the caryopsis of Sorghum bicolor (L.) Moench. Journal of Cereal Science 39(1): 21-27.

Eliasson AC, Wahlgren M 2004. Starch-lipid interactions and their relevance in food products. In: Eliasson AC ed. Starch in food. Cambridge, Woodhead Publishing Limited. Pp. 441-460.

Esau K 1967. Plant Anatomy. New York, John Wiley & Sons Inc.

Fardet A, Leenhardt F, Lioger D, Scalbert A, Remesy C 2006. Parameters controlling the glycaemic response to breads. Nutrition Research Reviews 19(1): 18-25.

Fellows P 2000. Food processing technology: principles and practice.

Foster-Powell K, Holt SHA, Brand-Miller JC 2002. International table of glycemic index and glycemic load values: 2002. American Journal of Clinical Nutrition 76(1): 5-56.

French D 1984. Organisation of starch granules. In: Whistler RL, BeMiller JN, Paschall EF eds. Starch chemistry and technology. New York, Academic Press. Pp. 184-242.

Fuentes-Zaragoza E, Riquelme-Navarrete MJ, Sanchez-Zapata E, Perez-Alvarez JA 2010. Resistant starch as functional ingredient: A review. Food Research International 43(4): 931-942.

Gallant DJ, Bouchet B, Baldwin PM 1997. Microscopy of starch: Evidence of a new level of granule organization. Carbohydrate Polymers 32(3-4): 177-191.

Gibson GR, Probert HM, Van Loo G, Rastall B, Roberfroid MB 2004. Dietary modulation of the human colonic microbiota: updating the concept of prebiotics. Nutrition Research Reviews 17: 259-75.

Granfeldt Y, Eliasson AC, Bjorck I 2000. An examination of the possibility of lowering the glycemic index of oat and barley flakes by minimal processing. Journal of Nutrition 130(9): 2207-2214.

Guigliano D, Ceriello A, Esposito K 2008. Glucose metabolism and hyperglycemia. Amer J Clin Nutr 87 (suppl): 217S-222S.

Haber GB, Heaton KW, Murphy D 1977. Depletion and disruption of dietary fibre. Effects on satiety plasma glucose and serum-insulin. Lancet ii: 679-682.

Hallfrisch J, Behall KM 2000. Mechanisms of the effects of grain on insulin and glucose response J Amer Coll Nutr 19: 320S-325S.

Holm J, Bjorck I 1988. Effects of Thermal-Processing of Wheat on Starch .2. Enzymic Availability. Journal of Cereal Science 8(3): 261-268.

Holm J, Bjorck I, Ostrowska S, Eliasson AC, Asp NG, Larsson K, Lundquist I 1983. Digestibility of Amylose-Lipid Complexes Invitro and Invivo. Starke 35(9): 294-297.

Huber KC, BeMiller JM 2000. Channels of maize and sorghum starch granules. Carbohydrate Polymers 41: 269-276.

James MG, K. , Denyer KaAM, Myers AM 2003. Starch synthesis in the cereal endosperm. Current Opinions in Plant Biology 6: 215-222.

Jenkins DJ, Wolever TM, Jenkins A, Giordano C, Giudici S, Thompson LU, Kalmusky J, Josse RG, Wong GS 1986. Low glycemic response to traditionally processed wheat and rye products: Bulgar and pumpernickel bread. American Journal of Clinical Nutrition 43: 516-520.

Jenkins DJA, Thorne MJ, Wolever TMS, Jenkins AL, Rao AV, Thompson LU 1987. The Effect of Starch-Protein Interaction in Wheat on the Glycemic Response and Rate of Invitro Digestion. American Journal of Clinical Nutrition 45(5): 946-951.

Juszczak L, Fortuna T, Krok F 2003. Non-contact Atomic Force Microscopy of Starch Granules Surface. Part I. Potato and Tapioca Starches. Starch/Staerke 55: 1-7.

Karlsson ME, Leeman AM, Bjorck IME, Eliasson A-C 2007. Some physical and nutritional characteristics of genetically modified potatoes varying in amylose/amylopectin ratios. Food Chemistry 100(1): 136-146.

Kim E, Petrie J, Motoi L, Sutton K, Morgenstern M, Mishra S, Simmons L 2007. Effect of structural and physicochemical characteristics of the protein matrix in past on in-vitro starch digestibility. Food Biophysics 3(2): 229-234.

Leeman M, Ostman E, Bjorck I 2005. Vinegar dressing and cold storage of potatoes lowers postprandial glycaemic and insulinaemic responses in healthy subjects. European Journal of Clinical Nutrition 59(11): 1266-1271.

Lehmann U, Robin F 2007. Slowly digestible starch - its structure and health implications: a review. Trends in Food Science & Technology 18(7): 346-355.

Lentle RG, Janssen PWM 2008. Physical characteristics of digesta and their influence on flow and mixing in the mammalian intestine: a review. Journal of Comparative Physiology B-Biochemical Systemic and Environmental Physiology 178(6): 673-690.

Mandalari G, Faulks RM, Rich GT, Lo Turco V, Picout DR, Lo Curto RB, Bisignano G, Dugo P, Dugo G, Waldron KW, Ellis PR, Wickham MSJ 2008. Release of protein, lipid, and vitamin E from almond seeds during digestion. Journal of Agricultural and Food Chemistry 56(9): 3409-3416.

Monro J, Mishra S 2009. Nutritional Value of Potatoes: Digestibility, Glycemic Index, and Glycemic Impact. In: Singh. J, Kaur. L eds. Advances in Potato Chemistry and Technology Elsevier Inc. Pp. 371-394.

Monro JA, Mishra S 2010. Digestion-resistant remnants of vegetable vascular and parenchyma tissues differ in their effects in the large bowel of rats. Food Digestion 1(1-2): 1 (1-2) 47-56.

Monro JA, Mishra SS, Hardacre A 2011. Glycaemic impact regulation based on progressive geometric changes in solid starch-based food particles during digestion. Food Digestion 2(1--3): 1-12.

Nilsson AC, Ostman EM, Hoist JJ, Bjorck IME 2008. Including indigestible carbohydrates in the evening meal of healthy subjects improves glucose tolerance, lowers inflammatory markers, and increases satiety after a subsequent standardized breakfast. Journal of Nutrition 138(4): 732-739.

Oates CG 1997. Towards understanding of starch granule structure and hydrolysis. Trends Food Sci Tech 8: 375-382.

Palafox-Carlos H, Ayala-Zavala JF, Gonzalez-Aguilar GA 2011. The Role of Dietary Fiber in the Bioaccessibility and Bioavailability of Fruit and Vegetable Antioxidants. Journal of Food Science 76(1): R6-R15.

Peyron MA, Mishellany A, Woda A 2004. Particle size distribution of food boluses after mastication of six natural foods. Journal of Dental Research 83(7): 578-582.

Planchot V, Colonna P, Gallant DJ, Bouchet B 1995. Extensive degradation of native starch granules by alpha-amylase from aspergillus fumigatus. Journal of Cereal Science 21(2): 163-171.

Pomeranz Y, Shogren MD, Finney KF 1977. Fiber in breadmaking – Effects on functional properties. Cereal Chem 54: 25-41.

Ranawana V, Monro JA, Mishra S, Henry CJK 2010. Degree of particle size breakdown during mastication may be a possible cause of interindividual glycemic variability. Nutrition Research 30(4): 246-254.

Ratnayake WS, Jackson DS 2007. A new insight into the gelatinization process of native starches. Carbohydrate Polymers 67(4): 511-529.

Read NW, Welch IM, Austen CJ, Barnish C, Bartlett CE, Baxter AJ, Brown G, Compton ME, Hume KE, Storie I, Worlding J 1986. Swallowing Food without Chewing - a Simple Way to Reduce Postprandial Glycemia. British Journal of Nutrition 55(1): 43-47.

Sajilata MG, Singhal RS, Kulkarni PR 2006. Resistant starch – a review. Comprehensive Review in Food Science Safety 5: 1-17.

Saltiel AR, Kahn CR 2001. Insulin signalling and the regulation of glucose and lipid metabolism. Nature 414(6865): 799-806.

Taggart P 2004. Starch as an ingredient: manufacture and applications. In: Eliasson AC ed. Starch in food. Cambridge, Woodhead Publishing Limited. Pp. 363-392.

Tester RF, Morrison WR 1990. Swelling and gelatinization of cereal starches .1. Effects of amylopectin, amylose, and lipids. Cereal Chemistry 67(6): 551-557.

Thomas DJ, Atwell WA eds. 1998. Starches. 2 ed. St Paul, Minnesota USA, Eagen Press. 100 p.

Tovar J, Bjorck IM, Asp NG 1990. Analytical and nutritional implications of limited enzymic availability of starch in cooked red kidney beans. Journal of Agricultural and Food Chemistry 38(2): 488-493.

Tovar J, Granfeldt Y, Bjorck IM 1992. Effect of processing on blood-glucose and insulin responses to starch in legumes. Journal of Agricultural and Food Chemistry 40(10): 1846-1851.

Turnbaugh PJ, Ley RE, Mahowald MA, Magrini V, Mardis ER, Gordon JI 2006. An obesity-associated gut microbiome with increased capacity for energy harvest. Nature 444(7122): 1027-1031.

Tydeman EA, Parker ML, Faulks RM, Cross KL, Fillery-Travis A, Gidley MJ, Rich GT, Waldron KW 2010a. Effect of Carrot (Daucus carota) Microstructure on Carotene Bioaccessibility in the Upper Gastrointestinal Tract. 2. In Vivo Digestions. Journal of Agricultural and Food Chemistry 58(17): 9855-9860.

Tydeman EA, Parker ML, Wickham MSJ, Rich GT, Faulks RM, Gidley MJ, Fillery-Travis A, Waldron KW 2010b. Effect of Carrot (Daucus carota) Microstructure on Carotene Bioaccessibilty in the Upper Gastrointestinal Tract. 1. In Vitro Simulations of Carrot Digestion. Journal of Agricultural and Food Chemistry 58(17): 9847-9854.

Venn BJ, Mann JI 2004. Cereal grains, legumes and diabetes. European Journal of Clinical Nutrition 58(11): 1443-1461.

Venn BJ, Wallace AJ, Monro JA, Perry T, Brown R, Frampton C, Green TJ 2006. The glycemic load estimated from the glycemic index does not differ greatly from that measured using a standard curve in healthy volunteers. Journal of Nutrition 136(5): 1377-1381.

Waigh TA, Kato KL, Donald AM, Gidley MJ, Clarke CJ, Riekel C 2000. Side-chain liquid-crystalline model for starch. Starch/Starke 52: 450-460.

Whistler RL, BeMiller JM 1997. Carbohydrate Chemistry for Food Scientists. Minnesota, USA, Eagan Press.

White PJ, Johnson LA 2003. Corn Chemistry and Technology, 2nd Edition, American Association of Cereal Chemists.

Wikoff WR, Anfora AT, Liu J, Schultz PG, Lesley SA, Peters EC, Siuzdak G 2009. Metabolomics analysis reveals large effects of gut microflora on mammalian blood metabolites. Proceedings of the National Academy of Sciences of the United States of America 106(10): 3698-3703.

Wright EM, Loo DDF, Hirayama BA, Turk E 2006. Sugar Absorption Physiology of the Gastrointestinal Tract. In: Johnson LR ed. Physiology of the gastrointestinal tract. 4th ed. New York, Elsevier Academic Press. Pp. 1653-1666.

Zimmet P, Alberti K, Shaw J 2001. Global and societal implications of the diabetes epidemic. Nature 414(6865): 782-787.

Chapter 2

Resistant Dextrins as Prebiotic

Katarzyna Śliżewska, Janusz Kapuśniak, Renata Barczyńska and Kamila Jochym

Additional information is available at the end of the chapter

1. Introduction

In current days, the way each of human beings live indicates his or her health in the future time. Many factors determine the risk of illnesses, or reversibly, the possibility of being healthy. Being physically active and consumption of appropriate diet are examples of daily routines that may influence the condition of an organism. Lack of physical activity, particularly if associated with over consumption, increases the risk of development of nutrition related chronic diseases, such as obesity, hypertension, cardiovascular diseases, osteoporosis, type II diabetes, and several cancers. Over the last decade, drastic changes have taken place in the image and assessment of the importance of the daily diet. Foods are no longer judged in terms of taste and immediate nutritional needs, but also in terms of their ability to improve the health and well-being of consumers. The role of diet in human health has led to the recent development of the so-called functional food concept. A functional food is dietary ingredient, that has cellular or physiological effects above the normal nutritional value. Functional food can contain probiotics and/or prebiotics.

2. The prebiotic concept

A number of different strategies can be applied to modify microbial intestinal populations. Antibiotics can be effective in eliminating pathogenic organisms within the intestinal microbiota. However, they carry the risk of side effects and cannot be routinely used for longer periods or prophylactically [17, 33].

The consumption of probiotics aims to directly supplement the intestinal microbiota with live beneficial organisms. Lactobacilli and bifidobacteria are numerically common members of the human intestinal microbiota, and are nonpathogenic, nonputrefactive, nontoxigenic, saccharolytic organisms that appear from available knowledge to provide little opportunity for deleterious activity in the intestinal tract. As such, they are reasonable candidates to target in terms of restoring a favorable balance of intestinal species [18, 85].

Prebiotics represent a third strategy to manipulate the intestinal microbiota. Rather than supplying an exogenous source of live bacteria, prebiotics are nondigestible food ingredients that selectively stimulate the proliferation and/or activity of desirable bacterial populations already resident in the consumer's intestinal tract. Most prebiotics identified so far are nondigestible, fermentable carbohydrates. Intestinal populations of bifidobacteria, in particular, are stimulated to proliferate upon consumption of a range of prebiotics, increasing in numbers by as much as 10–100-fold in faeces [9, 17].

3. Advantages and disadvantages of the prebiotic

The prebiotic strategy offers a number of advantages over modifying the intestinal microbiota using probiotics or antibiotics.

Advantages over probiotics [17]:

- Stable in long shelf life foods and beverages;
- Heat and pH stable and can be used in a wide range of processed foods and beverages;
- Have physicochemical properties useful to food taste and texture;
- Resistant to acid, protease, and bile during intestinal passage;
- Stimulate organisms already resident in the host, and so avoid host/strain compatibilities, and the need to compete with an already established microbiota;
- Stimulate fermentative activity of the microbiota and health benefits from SCFA (short chain fatty acids);
- Lower intestinal pH and provide osmotic water retention in the gut.

Advantages over antibiotics [18]:

- Safe for long-term consumption and prophylactic approaches;
- Do not stimulate side effects such as antibiotic-associated diarrhea, sensitivity to UV radiation, or liver damage;
- Do not stimulate antimicrobial resistance genes;
- Not allergenic;

Disadvantages of prebiotics [17]:

- Unlike probiotics, overdose can cause intestinal bloating, pain, flatulence, or diarrhea.
- Not as potent as antibiotics in eliminating specific pathogens.
- May exacerbate side effects of simple sugar absorption during active diarrhea.

A consumed probiotic strain must compete with an already established microbiota, and in most cases they persist only transiently in the intestine. Individuals also harbor their own specific combination of species and unique strains within their intestinal bacteria suggesting that certain host–microbiota compatibilities exist. By targeting those strains that are already resident in the intestinal tract of an individual, the prebiotic strategy overcomes the need for probiotic bacteria to compete with intestinal bacteria that are well established in their niche [12, 17, 89, 103].

4. Definition of term "prebiotic"

The term prebiotics was first introduced in 1995 by Gibson and Roberfroid, defining "non-digestible food ingredient that beneficially affects the host by selectively stimulating the growth and/or activity of one or a limited number of bacteria in the colon, and thus improves host health" [33]. Definition brought up to date by Gibson specified prebiotic as "selectively fermented ingredient that allows specific changes, both in the composition and/or activity in the gastrointestinal microflora that confers benefits upon host well-being and health" [32]. Current definition of prebiotics was suggested during ISAPP experts' meeting in 2008 and it states that prebiotic is "dietary prebiotic is a selectively fermented ingredient that results in specific changes, in the composition and/or activity of the gastrointestinal microbiota, thus conferring benefit(s) upon host health" [24, 44].

Substances with prebiotic properties have to possess following properties [42, 74, 114, 116]:

- selectively stimulate growth and activity of chosen bacterial strains that have positive influence on health,
- lower pH of the bowel content,
- show positive for human spot action in the intestinal tract,
- be resistant to hydrolysis, action of intestinal tract enzymes and gastric acids,
- should not get soaked up in the upper part of the intestinal tract,
- should act as a selective substrate for one or for determined amount of beneficial species of microorganisms in the colon,
- should be stable in the process of food processing.

In order to evaluate and reason, if the given product is a prebiotic, the source of the substance should be given, as well as its purity, chemical composition and structure. It is very important to specify the carrier, concentration and amount in which it should be given to the host. Relating to the newest definition of the prebiotic, it was decided to type out three main criteria that have to be fulfilled by the substance in order to include it to the group of prebiotics [24].

1. Substance (component) – it is neither an organism, nor a medicine; substance that may be characterized chemically; in most cases this is a nutrient component.
2. Health benefits – calculable, exceeding any adverse effects.
3. Modulation – represents, that the presence of the substance and the preparatory, in which it is handled, changes the composition or activity of host microflora.

Prebiotics, similarly to other nutrient elements, have to fulfill certain safety parameters established in a given county. In the assessment of the final product following points should be taken into account [24, 39], (Figure 1):

1. If according to the legislation in the country, the history of safe use of the product in host is known (GRAS or its equivalent). If yes, the conductance of the following toxicological tests on animals and humans may not be necessary.
2. Safe, allowable norms for the consumption with minimal symptoms and adverse effects.

3. Product must not be infected and it should not contain any impurities.
4. Prebiotic cannot change the microflora in such a way, to cause a long-lasting harmful effect on host.

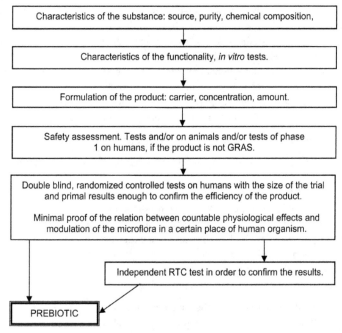

Figure 1. Guidelines to the assessment and proof of the action of prebiotics

5. Criteria of prebiotics classification

According to Wang [114] there are 5 most important basic criteria for classification of prebiotics (Figure 2). The first one assumes, that prebiotics are undigested in upper parts of intestinal tract and thus they are able to get to the large bowel where they can be fermented by potentially beneficial bacteria, which in the meantime meets the second criterion [61]. This fermentation may lead to increase in expression of short chain fatty acids, enlargement of faecal mass, some small reduction of large bowel pH, reduction of end nitrogen compounds and decrease of faecal enzymes, as well as general improvement of immunological system of host organism [17]. All these features contribute to improvement of consumer's health, which is the third criterion for prebiotics. The following one, that has to be fulfilled for the product to be recognized as prebiotic, is selective stimulation of growth of bacteria potentially thought to be connected with health improvement [32]. In order to assess the ability of prebiotic to selectively stimulate positive bacteria of species *Bifidobacterium* and *Lactobacillus*, so-called Prebiotic Index (PI) was introduced, and it can be calculated from the following formula [71]:

$$PI = (Bif/Total) - (Bac/Total) + (Lac/Total) - (Clos/Total)$$

Where:

Bif – *Bifidobacterium*

Bac – *Bacteroides*

Lac – *Lactobacillus*

Clos – *Clostridium*

Total – total bacteria

This PI allows to track the changes in the population in the given time in vitro conditions. At last, but not least, prebiotic should be able to survive the conditions in which the food would be stored, remain unchanged chemically and be accessible for bacteria metabolism [71].

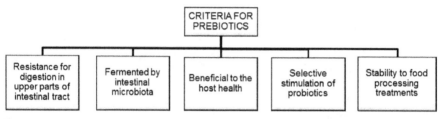

Figure 2. Criteria for prebiotics classification

6. Prebiotic mechanisms

The impact of prebiotics on the organism is indirect, because prebiotics do not do anything healthy for it, but they improve microorganisms that are beneficial [43]. The mechanism of how prebiotics influence the human health is presented on the Figure 3.

It is thought, that molecular structure of prebiotics is important taking into consideration the physiological effects, and that it determines which microorganisms are actually able to use that prebiotics. However, the way and progress of the stimulation of bacterial growth still remains unknown.

The most important function of prebiotic action is its influence on the microorganisms' growth and number in the large bowel [55, 90]. Going further, the tests have been conducted in order to investigate the potential ant pathogenic and anticancer action of prebiotics, their ability to decrease the presence of large bowel diseases [57]. A lot of different potential beneficial influences on human organisms are being sought, and those are, among the others: increase of the volume and improvement of stool moisture, lowering of the cholesterol level, decrease of the amount long chain fatty acids in bowels, decrease of pH in bowels, increase of mineral compounds absorption and raised short chain fatty acids production [1, 21, 90, 114], and the mechanism can be observed on the Figure 4.

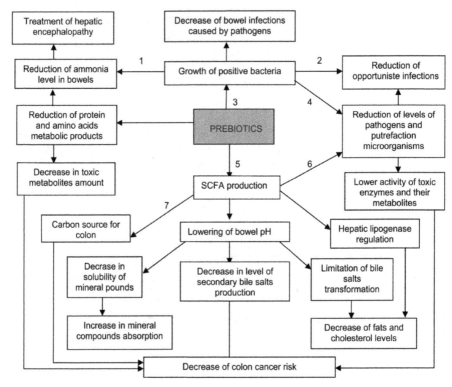

1 - bacterial proteins synthesis
2 - colonization counteraction
3 - selective carbon and energy source
4 - contention
5 - fermentation
6 - inhibition
7 - butyl acid

Figure 3. Proposed mechanism of prebiotic action

7. Production of prebiotics

Some prebiotics can be extracted from plant sources, but most are synthesized commercially using enzymatic or chemical methods. Overall, prebiotics are manufactured by four major routes (Table 1). Food-grade oligosaccharides are not pure products, but are mixtures containing oligosaccharides of different degrees of polymerization (DP), the parent polysaccharide or disaccharide, and monomer sugars. Oligosaccharide products are sold at this level of purity, often as syrups. Chromatographic purification processes are used to remove contaminating mono- and disaccharides to produce higher purity oligosaccharide products containing between 85 and 99% oligosaccharides, which are often dried to powders [17].

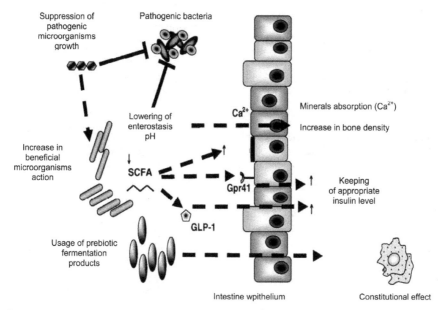

Figure 4. Mechanism of prebiotic action

Approach	Process	Prebiotic Examples
Direct extraction	Extraction from raw plant materials	Resistant starch from maize Inulin from chicory Soybean oligosaccharides from soybean whey
Controlled hydrolysis	Controlled enzymatic hydrolysis of polysaccharides; may be followed chromatography to purify the prebiotics	Fructooligosaccharides from inulin Xylooligosaccharides from arabinoxylan
Transglycosylation	Enzymatic process to build up oligosaccharides from disaccharides; may be followed by chromatography to purify the prebiotics	Fructooligosaccharides from sucrose Galactooligosaccharides from lactose Lactosucrose from lactose + sucrose
Chemical processes	Catalytic conversion of carbohydrates	Lactitol from hydrogenation of lactose Lactulose from alkaline isomerization of lactose

Table 1. Production of prebiotic carbohydrates

Carbohydrate	Chemical structure	Degree of polymerisation	Method of manufacture
Inulin	β(2-1)-Fructans	2 - 65	Extraction from chicory root and *Agave tequilana*.
Fructooligosaccharides (FOS)	β(2-1)-Fructans	2 - 9	Transfructosylation from sucrose or hydrolysis of chicory inulin.
Galactooligosaccharides (GOS)	Galactose oligomers and some glucose/ lactose/ galactose units	2 - 5	Produced from lactose by β-galactosidase.
Soya-oligosaccharides	Mixture of raffinose and stachyose	3 - 4	Extracted from soya bean whey.
Xylooligosaccharides (XOS)	β(1–4)-Linked xylose	2 - 4	Enzymatic hydrolysis of xylan. Enzyme treatments of native lignocellulosic materials. Hydrolytic degradation of xylan by steam, water or dilute solutions of mineral acids.
Isomaltooligosaccharides (IMO)	α(1–4)-glucose and branched α(1–6)-glucose	2 - 8	Microbial or enzymatic transgalactosylation of maltose. Enzymatic synthesis from sucrose.
Dextrins	Mixture of glucose-containing oligosaccharides	Various	Chemical modification of starch.

Table 2. Main candidates for prebiotic status

Different manufacturing processes also produce slightly different oligosaccharide mixtures. For example, FOS mixtures produced by transfructosylation of sucrose contain oligosaccharides between three and five monomer units, with the proportion of each oligosaccharide decreasing with increasing molecular size. These oligosaccharides contain a terminal glucose with β-1→2 linked fructose moieties. FOS produced by the controlled hydrolysis of inulin contain a wider range of β-1→2 fructooligosaccharide sizes (DP 2–9), relatively few of which possess a terminal glucose residue. Even different b-galactosidases used in the production of GOS will produce oligosaccharide mixtures with different proportions of β-1→4 and β-1→6 linkages. Hence, there can be some diversity between the structures of oligosaccharides produced by different manufacturers. The precise impact of these differences in their health effects remains to be determined [17, 100].

There are many oligosaccharides under investigation for their prebiotic potential. Fermentation of some oligosaccharides is not as selective as that of FOS, and their prebiotic status therefore remains in doubt. The main candidates for prebiotic status is provided in Table 2 [26]. There is, therefore, a need for new prebiotic substances of distinct, selective stimulation of growth of lactic acid bacteria, and non-fermented or slightly fermented by other, sometimes pathogenic intestinal bacteria. The search for functional food or functional food ingredients is beyond any doubt one of the leading trends in today's food industry.

8. Resistant dextrins as prebiotics

8.1. Resistant starch

Resistant starch (RS) includes the portion of starch that can resist digestion by human pancreatic amylase in the small intestine and thus, reach the colon. The general behaviour of RS is physiologically similar to that of soluble, fermentable fibre, like guar gum. The most common results include increased faecal bulk and lower colonic pH and improvements in glycaemic control, bowel health, and cardiovascular disease risk factors, so it has shown to behave more like compounds traditionally referred to as dietary fibre [31, 60, 97, 126].

Resistant starch is found in many common foods, including grains, cereals, vegetables (especially potatoes), legumes, seeds, and some nuts [31, 35].

Resistant starch may not be digested for four reasons [30, 38, 60]:

- this compact molecular structure limits the accessibility of digestive enzymes, various amylases, and explains the resistant nature of raw starch granules. The starch may not be physically bio accessible to the digestive enzymes such as in grains, seeds or tubers,
- the starch granules themselves are structured in a way which prevents the digestive enzymes from breaking them down (e.g. raw potatoes, unripe bananas and high-amylose maize starch),
- starch granules are disrupted by heating in an excess of water in a process commonly known as gelatinization, which renders the molecules fully accessible to digestive enzymes. Some sort of hydrated cooking operation is typical in the preparation of starchy foods for consumption, rendering the starch rapidly digestible. However, if these
- starch gels are then cooled, they form starch crystals that are resistant to enzymes digestion. This form of "retrograded" starch is found in small quantities (approximately 5%) in foods such as "corn-flakes" or cooked and cooled potatoes, as used in a potato salad.
- selected starches that have been chemically modified by etherisation, esterisation or cross-bonding, cannot be etherisation, esterisation or cross-bonding, cannot be broken down by digestive enzymes.

Resistance starch is the sum of starch itself and products of her decomposition, that are neither being digested nor absorbed in the small bowel of healthy human [23]. Resistant starch is the difference between amount of the starch exposed to the action of amylolytic enzyme complex and the amount of starch decomposed to glucose during hydrolysis performed by those enzymes [83, 123].

$$RS = TS - (RDS + SDS)$$

Where:
RS – resistant starch
TS – total starch
RDS – rapidly digestible starch
SDS – slowly digestible starch
Few types of digestible starch are recognized nowadays [36, 52, 113].

Resistant starch of type 1 – RS1 covers the starch in plant cells with undestroyed cell walls. This starch is unavailable for digestive enzymes present in human intestinal tract, and thus together with fragments of plant tissues passes through the small bowel getting to the large bowel untouched, and there it can undergo fermentation [70, 104]. RS1 is heat stable in most normal cooking operations, which enables its use as an ingredient in a wide variety of conventional foods [31].

Resistant starch type 2 – RS2 is composed of native starch granules from certain plants containing uncooked starch or starch that was gelatinized poorly and hydrolyzed slowly by R-amylases (e.g., high-AM corn starches). RS2 covers scoops of raw starch of some plant species, especially high-amylase corn, potato and banana [70, 104]. Huge size of raw potato flour scoops and hence combined with it limited area of the access for enzymes was considered as the cause of its resistance [73]. But the main reason for the resistance of the raw starch of some plant species on amylolytic enzymes is the structure of its scoops and crystallization type B that us present within them (in the scoop of potato and corn starch). Also other elements of the scoop structure have an impact on the resistance of the starch – such as the shape of the area, size of pores or susceptibility of the starch to germinate. A particular type of RS2 is unique as it retains its structure and resistance even during the processing and preparation of many foods; this RS2 is called high-AM maize starch [31, 117].

Resistant starch type 3 – RS3 covers the substance precipitated from pap or starch gel during the process of retrograding. During the germination of the starch in the lowered temperature and with the proper concentration (1.5% amylose, 10% amylopectin) colloidal solution is formed. Stable starch phase existing as double helix forms reticular structure binding water phase in its 'eyes'. During the storage of the gel (few hours in lowered temperature) helixes undergo aggregation forming thermally stable crystal structures. Such structures show the resistance to amylolytic enzymes [66, 104]. RS3 is of particular interest, because of its thermal stability. This allows it to be stable in most normal cooking operations, and enables its use as an ingredient in a wide variety of conventional foods [15].

During food processing, in most cases in which heat and moisture are involved, RS1 and RS2 can be destroyed, but RS3 can be formed. Storey et al. [99], classified a soluble polysaccharide called 'retrograded resistant maltodextrins' as type 3 RS. They are derived from starch that is processed to purposefully rearrange or hydrolyze starch molecules, and subsequent retrogradation, to render them soluble and resistant to digestion. This process results in the formation of indigestible crystallites that have a molecular similarity to type 3 RS but with a smaller degree of polymerization as well as a lower MW, converting a portion of the normal α- 1,4-glucose linkages to random 1,2-, 1,3-, and 1,4-α or β linkages [25, 31,64].

The definition of presented forms of resistant starch may be presented according to the formulas [83]:

$$RS1 = TS - (RDS + SDS) - RS2 - RS3$$

$$RS2 = TS - (RDS + SDS) - RS1 - RS3$$

$$RS3 = TS - (RDS + SDS) - RS2 - RS1$$

Where:
RS1 – resistant starch type 1
RS2 – resistant starch type 2
RS3 – resistant starch type 3
TS – total starch
RDS – rapidly digestible starch
SDS – slowly digestible starch

Resistant starch type 4 – RS4 covers the starch chemically or physically modified and achieved by combination of these two processes. During chemical modification new functional groups are brought into the starch chain, and they bind to glucose residues. Presence of the substituents and spatial changes in the chain prevent proper functioning of human digestive enzymes. In physical method, during warming of starch in high temperature process of dextrinization occurs, and it may also occur in the presence of acid as a catalyst. One of the products of dextrinization is free glucose, which binds to the chains randomly. As a result of such process between glucose residues, bonds typical for starch and those normally not existing in its chains, arise [13, 70, 124].

Resistant starch type 5 – RS5 is an AM-lipid complexed starch [31, 46], which is formed from high AM starches that require higher temperatures for gelatinization and are more susceptible to retrograde [20, 31]. In general, the structure and amount of starch-lipid in foods depend on their botanical sources. Also, Frohberg and Quanz [29] defined as RS5 a polysaccharide that consists of water-insoluble linear polyα- 1,4-glucan that is not susceptible to degradation by alpha-amylases. They also found that the poly-α-1,4-D-glucans promote the formation of short-chain fatty acids (SCFA), particularly butyrate, in the colon and are thus suitable for use as nutritional supplements for the prevention of colorectal diseases [31].

RS is the fraction of starch which is not hydrolyzed to D-glucose in the small intestine within 120 min of being consumed, but which is fermented in the colon. Many studies have shown that RS is a linear molecule of a-1,4-D-glucan, essentially derived from the retrograded AM fraction, and has a relatively low MW (1.2 x 105 Da) [31].

Resistant starch obtained during chemical or physical modification is being investigated nowadays, due to the fact that it possesses some specific physical properties, as well as because of its health benefits [49, 86, 87, 96]. During chemical modification functional groups are introduced to the starch molecule, which then leads to the changes of physical and chemical properties of obtained product, and also it lowers the availability of the starch to amylolytic enzymes, because new functional groups prevent occurring of the enzyme-substrate complex [7]. Chemical modification was found to be advantageous method of decrease of starch digestion, and therefore starch modified chemically may be the source of resistance starch RS4 [15, 36, 70].

8.2. Resistant dextrin

Resistant dextrins are defined as short chain glucose polymers, without sweet taste and performing strong resistance to hydrolytic action of human digestive enzymes [68]. In accessible throughout the whole products (resistant dextrin Nutriose, Fibersol) bigger percentage presence of (1→2)-, (1→3)-, (1→6)- α and β-glycoside bonds than in native starch which is the source of getting them [62, 115].

During warming of starch in high temperature, with or without addition of catalyst (usually acidic) dextrinization of starch is observed. Dextrinization is a complex process taking chemical side of it into account. It covers depolymerization, transglucolyzation and repolymerization [122].

During warming of wet starch random bonds (1→4) and rarely (1→6) hydrolytically break. Intermediate form in this reaction is either oxycarbonic ion, or free radicals [105].

Most probably, dextrinization undergoes the mixed mechanism. In case of warming of dry starch (or with a low moisture content), bonds (1→6) are made between two starch chains and intramolecular dehydration coincides, which results in development of 1,6-anhydro-β-D-glucose. In such a way only extreme glucose units with free hydroxyl group within anomeric carbon atom may react. In both cases exuded water has hydrolytic character. With the temperature equal to about 290°C α-(1→4)-glycosidic bonds begin to break. However during dextrinization not only bond breaking is observed, but also the isomerization (e.g. through mutarotaion) or formation of new bonds. Hydroxyl groups at C-2, C-3 or C-6 glucose unit act on oxycarbonic ions or free radicals and transglycolization, which is based on formation of (1→2), (1→3) and (1→6) bonds, undergoes. This process leads to formation of branched dextrins. Because of spherical considerations, and maybe also thermodynamic ones, formation of (1→6) bonds is privileged. 1,6-anhydro-β-D-glucose is formed, which easily forms polymers, leads to formation of 4-O-α-D-, 4-O-β-D-, 2-O-α-D- and 2-O-β-D-glukopiranozylo-1,6-dihydro-β-D-glucopyranose [105, 122].

In the next stage following reactions have to be taken into account:

- reversion, that is the reaction between glucose units leading to formation of $(1\rightarrow6)$-glycosidic bonds,
- reaction between $(1\rightarrow6)$-anhydro-β-D-glucopyranose and free radicals with formation of 1,6-glycosidic bonds,
- recombination.

From all presented reactions the most characteristic and dominating one is transglycolisation. Formation of bonds other than typical for starch $(1\rightarrow4)$ and $(1\rightarrow6)$ causes that the received product becomes unavailable for human digestive enzymes and shows properties of resistant starch.

In the presence of acidic catalyst dextrinization process progresses a bit differently. First the hydrolysis undergoes, and in the result of it $(1\rightarrow6)$ bonds break, but $(1\rightarrow4)$ bonds stay untouched. In this way, white dextrins are formed. Because $(1\rightarrow6)$ bonds are more resistant to hydrolysis than $(1\rightarrow4)$ bonds, these last ones undergo transformation into $(1\rightarrow6)$ bonds [105].

The role of basic catalyst in dextrinization process is not well known. The only thing know is that in this process deprotonating of hydroxyl groups at C-2 and C-3 is the first stage. In the presence of oxidating agents atom C-1 of terminal glucose units are being oxidized to carboxyl group [105]. Summing up, as a result of hydrolysis the reductive ends of the starch become glucose cations, which undergo intramolecular dehydration forming $(1\rightarrow6)$-anhydro-β-D-glucopyranose unit or they take part in formation of intermolecular bonds (transglycolization). As a result of this process random glycosidic $(1\rightarrow2)$, $(1\rightarrow3)$ bonds are formed [68]. Formation of bonds other than typical for starch, i.e. $(1\rightarrow4)$ and $(1\rightarrow6)$ causes that end product doesn't undergo hydrolysis through the human digestive enzymes [115].

In the process of formation of resistant dextrin, piroconversion is the first stage, and it covers following steps: thermolysis, transglucolysis, regrouping and repolymerisation. Starch thermolysis leads to breaking of α-D-$(1\rightarrow4)$ and α-D-$(1\rightarrow6)$ glycosidic bonds, which then leads to the formation of of products with lower molecular mass and higher viscosity and reducing sugars content. After transglucolysis recombination of hydrolyzed starch fragments with free hydroxyl groups happens, and formation of strongly branched structures. Repolymerisation of glucose and oligosaccharides with formation of high molecular compounds is done in high temperature and presence of acidic catalyst (e.g. hydrochloric acid) [68]. Achieved pirodextrins are the mixture of poli- and oligosaccharides with a different degree of polimerisation (DP), and simultaneously with different molecular mass. Pirodextrins are subjected to enzymatic hydrolysis or chromatography – stages, which aim is to reduce the fractions other than typical for the starch (i.e. containing bonds other than α-D-$(1\rightarrow4)$ and α-D-$(1\rightarrow6)$ glycosidic ones [6, 81, 118].

Chemical modification has long been known to inhibit in vitro digestibility of starch, the extent of which is related to the type and degree of modification, the extent of gelatinization, and the choice of enzyme [117, 119]. Starch phosphates [45], hydroxypropyl starches [51,

121], starch acetates [120], phosphorylated starch [86, 94, 119], and citrate starches [117, 124] have been tested for enzymatic degradation previously. In previous studies have been suggested that the substituted groups hindered enzymatic attack and thus also made neighboring bonds resistant to degradation. Chemical substitution of starch reduces its enzyme digestibility, probably because the bulky derivatizing groups sterically hinder formation of the enzyme-substrate complex [7]. The largest change in digestibility has been achieved through cross-linking of starch [119]. The application of organic acids, as citric acid and tartaric acid as derivatizing agents seemed to be profoundly safe. These acids are nutritionally harmless compared to other substances used for chemical modification [123]. When citrate starches were fed to rats no pathological changes could be found in comparison to native wheat and corn starches [117].

Potato starch in a natural form has limited possibilities to be used, and its chemical structure and physical properties give possibilities to many modifications, also those leading to formation of resistant substances to amylolytic enzymes. Big hopes are laid on usage of products with modified starch, especially resistant starch and resistant dextrin as substances with prebiotic properties.

Kapusniak et al. [47] resistant dextrin was receive by simultaneous pyroconversion and chemical modification (esterification/ cross-linking) of potato starch in the presence of hydrochloric acid as catalyst of dextrinization process, and citric acid as derivatizing agent. Potato starch was modified by thermolysis in the presence of acid catalyst in a sealed container at 130°C for 180 min. The effect of addition of multifunctional polycarboxylic acids (citric and tartaric) on the progress of dextrinization process, structure and properties of resulting products was investigated [47]. It seems likely that probiotic activity will be exhibited by dextrin obtained by simultaneous thermolysis and chemical modification of potato starch in the presence of a volatile inorganic acid (hydrochloric acid) as a catalyst of the dextrinization process and an excess amount of an organic acid (tartaric acid) as a modifying factor. Kapuśniak et al. [47, 48] analyzed this dextrin in terms of the solubility and pH of its 1% aqueous solution, the content of reducing sugars, molecular mass distribution, weight average molecular mass using high performance size-exclusion chromatography (HPSEC), average chain length using high performance anion exchange chromatography with pulsed amperometric detection (HPAEC-PAD), and the content of the resistant fraction using the enzymatic-gravimetric method AOAC 991.43, the enzymatic-gravimetric-chromatographic method AOAC 2001.03 [69], the enzymatic-spectrophotometric method [23] and the pancreatin-gravimetric method [94]. It was shown that the use of tartaric acid in the process of starch thermolysis yielded acidic dextrin characterized by high water solubility (about 68%) and a high content of reducing sugars (about 29%). The studies showed that dextrin modified with tartaric acid did not contain any traces of unreacted starch, and the percentage share of the main fraction (having a weight average molecular mass of about 1.800 g/mol) was 80%. The average length of the carbohydrate chain in dextrin obtained with tartaric acid was 8.2 as determined by means of HPAEC. A study by Kapuśniak et al. [47] revealed that the content of the resistant fraction in dextrin modified with tartaric acid, determined by means of the AOAC 991.43, amounted to 44.5%. However, results obtained by the Engllyst [43] enzymatic-spectrophotometric method showed that the actual content of the resistant fraction was above 68%. Kapuśniak et al. [47, 48] used the official AOAC 2001.03 method to determine the content of the resistant fraction in dextrin modified with tartaric acid. This method is the latest approved method for determining total content of dietary fiber in foods containing resistant maltodextrins. Apart from measuring the content of insoluble dietary fiber and the high molecular weight fractions of soluble fiber, this method makes it possible to determine resistant oligosaccharides (by using high-performance liquid chromatography, HPLC). The total content of dietary fiber in dextrin modified with tartaric acid was about 50% [47, 48]. In the Engllyst method, fractions undigested after 120 min are considered resistant. In the pancreatin-gravimetric method, similarly as in the Engllyst method, samples are digested with pancreatin, but resistant fractions are determined gravimetrically only after 16 h. In the case of dextrin modified with tartaric acid, the results of determination by the pancreatin-gravimetric method (67%) were similar to those obtained in the previous studies using the Engllyst method (68%), but much higher than those obtained using the AOAC 2001.03 method (50%) [43, 47]. The observed differences among the various methods in terms of the measured content of the resistant fraction in dextrin modified with tartaric acid was caused by the fact that, according to the latest reports, enzymatic-gravimetric methods (including AOAC 2001.03) using thermostable α-amylase can determine only part of resistant starch type 4 [124]. Based on the enzymatic tests, it can be argued that dextrin obtained using an excessive amount of tartaric acid may be classified as resistant starch type 4.

8.3. Prebiotic effects of resistant starch and resistant dextrin

Resistant starch has a long history of safe consumption by humans and is a natural component of some foods. Intakes vary but are generally low, particularly in Western diets. Similar to soluble fibre, a minimum intake of resistant starch (5 - 6 g) appears to be needed in order for beneficial reductions in insulin response to be observed. Estimates of daily intake of resistant starch range from 3 to 6 g/day (averaging 4.1 g/day) [4, 31]. As a food ingredient, resistant starch has a lower calorific (8 kJ/g) value compared with fully digestible starch (15 kJ/g) [31, 78], therefore it can be a substitutive of digestible carbohydrates, lowering the energy content of the final formulation. Some resistant starch products are also measured as total dietary fibre in standard assays, potentially allowing high-fibre claims [31, 35].

Resistant starch can be fermented by human gut microbiota, providing a source of carbon and energy for the 400 - 500 bacteria species present in this anaerobic environment and thus potentially altering the composition of the microbiota and its metabolic activities. The fermentation of carbohydrates by anaerobic bacteria yields SCFA, primarily composed of acetic, propionic, and butyric acids, which can lower the lumen pH, creating an environment less prone to the formation of cancerous tumours [31, 126].

RS consumption has also been related to reduced post-prandial glycemic and insulinemic responses, which may have beneficial implications in the management of diabetes, and is associated with a decrease in the levels of cholesterol and triglycerides. Other effects of resistant starch consumption are increased excretion frequency and faecal bulk, prevention of constipation and hemorrhoids, decreased production of toxic and mutagenic compounds, lower colonic pH, and ammonia levels. Considering that nowadays several diseases result from inadequate feeding, and that some may be related to insufficient fibre intake, it is reasonable to assume that an increased consumption of indigestible components would be important [31].

Resistant starch enhance the ileal absorption of a number of minerals in rats and humans. Lopez et al. [59] and Younes et al. [128] reported an increased absorption of calcium, magnesium, zinc, iron and copper in rats fed RS-rich diets. In humans, these effects appear to be limited to calcium [16, 30, 107]. Resistant starch could have a positive effect on intestinal calcium and iron absorption. A study to compare the apparent intestinal absorption of calcium, phosphorus, iron, and zinc in the presence of either resistant or digestible starch showed that a meal containing 16.4% RS resulted in a greater apparent absorption of calcium and iron compared with completely digestible starch [30, 65].

Liu and Xu [58] showed that resistant starch dose-dependently suppressed the formation of colonic aberrant crypt foci only when it was present during the promotion phase to a genotoxic carcinogen in the middle and distal colon, suggesting that administration of resistant starch may retard growth and/or the development of neoplastic lesions in the colon. Therefore, colon tumorigenesis may be highly sensitive to dietary intervention. Adults with preneoplastic lesions in their colon may therefore benefit from dietary resistant starch. This suggests the usefulness of resistant starch as a preventive agent for individuals at high risk for colon cancer development [30, 58].

Short chain fructooligosaccharides (FOS) and resistant starch (RS) may act synergistically (by combining, and thus increasing, their prebiotic effects) [31, 79], the administration of the combination of FOS and RS induced changes in the intestinal microbiota, by increasing lactobacilli and bifidobacteria in caecum and colonic contents. Several types of prebiotic fibres can be distinguished considering their rate of fermentability. Such role depends on the carbohydrate chain length as it has been demonstrated *in vitro* in a fermentation system, showing that FOS are rapidly fermented whereas long chain prebiotic, like inulin, are steadily fermented. These observations have been confirmed *in vivo* once the different prebiotics reach the large intestine: FOS are rapidly fermented, whereas RS is slowly degraded. In consequence, the particular kinetics would determine the region of the intestine where the effects will be clearer. Thus, FOS would be more active in the first parts of the large bowel whereas RS would reach the distal part of the colon. In fact, Le Blay et al. [31, 50] have reported that administration of FOS or raw potato starch induces different changes in bacterial populations and metabolites in the caecum, proximal, and distal colon, as well as in faeces. As compared with RS FOS doubled the pool of faecal fermentation products, like lactate, while the situation was just the opposite distally. These observations confirm that each prebiotic shows particular properties, which should be considered before their application for intestinal diseases; thus, rapidly fermentable prebiotics are particularly useful in those affecting the proximal part of the large intestine, while slowly fermentable prebiotics should be chosen for more distal intestinal conditions. Moreover, an association with different prebiotics with complementary kinetics should be considered when a health-promoting effect throughout the entire colon is required. So, functional foods based on the combination of two different dietary fibres, with different rate of fermentability along the large intestine, may result in a synergistic effect, and thus, in a more evident prebiotic effect that may confer a greater health benefit to the host [31, 127].

Younes et al. [127] was to examine the potential synergistic effect of a combination of these two fermentable carbohydrates (inulin and resistant starch). For this purpose, thirty-two adult male Wistar rats weighing 200 g were used in the present study. The rats were distributed into four groups, and fed for 21 d a fibre-free basal purified diet or diet containing 100 g inulin, or 150 g resistant starch (raw potato starch)/kg diet or a blend of 50 g inulin and 75 g resistant starch/kg diet. After an adaptation period of 14 d, the rats were then transferred to metabolic cages and dietary intake, faeces and urine were monitored for 5 d. The animals were then anaesthetized and faecal Ca and Mg absorption were measured. Finally, the rats were killed and blood, caecum and tissues were sampled. Ca and Mg levels were assessed in diets, faeces, urine, caecum and plasma by atomic absorption spectrometry. The inulin and resistant starch ingestion led to considerable faecal fermentation in the three experimental groups compared with the control group diet. Moreover, both carbohydrates significantly increased the intestinal absorption and balance of Ca and Mg, without altering the plasma level of these two minerals. Interestingly, the combination of the studied carbohydrates increased significantly the faecal soluble Ca and Mg concentrations, the apparent intestinal absorption and balance of Ca, and non-significantly the plasma Mg level. The combination of different carbohydrates showed synergistic effects on intestinal Ca absorption and balance in rats [127].

The example of commercially available resistant dextrin is Nutriose. It is a non-viscous soluble fiber made from starch using a highly controlled process of dextrinization. It is mostly resistant to digestion in the small intestine and largely fermented in the colon. A process of dextrinization includes a degree of hydrolysis followed by repolymerization that converts the starch into fiber by forming no digestible glycosidic bonds. Nutriose is totally soluble in cold water without inducing viscosity [37, 53]. It is produced from wheat or maize starch using a highly controlled process of dextrinization followed by chromatographic fractionation step [28]. Nutriose FB®06, produced from wheat starch, contains approx. 13% of 1,2- and 14% of 1,3- glycosidic linkages [81]. The weight average molecular weight (M_w) and the number average molecular weight (M_n) for that dextrin were nearly 5000 and 2800 g/mole respectively. The residual content of sugars (DP1-2) of Nutriose FB®06 was below 0.5% and it could be considered as sugar free [80]. The enzyme-resistant fraction content, determined according to AOAC official method 2001.03 for total dietary fiber in foods containing resistant maltodextrins, was nearly 85% for Nutriose®06 and nearly 70% for Nutriose®10 [80, 81]. Nutriose has found wide application in food and pharmaceutical industries, as components of fiber-enriched drinks [92], components of a fiber-enriched composition for enteral nutrition [88], granulation binders [93], in the preparation of low-calorie food [11], and sugar-free confectionery [91]. Very well tolerated, Nutriose may be 20-25% of a product's composition without causing discomfort or bloating. About 15% of Nutriose are absorbed in the small intestine, about 75% fermented in the large intestine, while the remainder (about 10%) is excreted in the faeces [72, 110]. Nutriose induced an increase of the colonic saccharolytic flora and decrease in potentially harmful *Clostridium perfringens* in human faeces [54, 72]. Nutriose induced a decrease in the faecal pH of human volunteers, increased production of short chain fatty acids (SCFAs) in rats [54], induced changes in faecal bacterial enzyme concentration [72, 110]. It was also shown that learning (respectively physical) performances are improved in rats 180 minutes (respectively 150 minutes) after the consumption of Nutriose, compared to dextrose. The glycaemic kinetics is not sufficient to predict this effect: even though the glycaemic peak was lower with the resistant dextrin than with dextrose, the glycaemia was the same between the two groups at 150 and 180 minutes after ingestion. These preliminary results are very encouraging [82].

Guérin-Deremaux et al. [37] found that the non-viscous soluble dietary fiber may influence satiety. The randomized, double-blind, placebo-controlled clinical study in 100 overweight healthy adults in China investigated the effect of different dosages of dietary supplementation with a dextrin, Nutriose, on short-term satiety over time. Subjects were randomized by body mass index and energy intake and then assigned to receive either placebo or 8, 14, 18, or 24 g/d of Nutriose mixed with orange juice (n = 20 volunteers per group). On days - 2, 0, 2, 5, 7, 14, and 21, short-term satiety was evaluated with a visual analog scale, and hunger feeling status was assessed with Likert scale. Nutriose exhibits a progressive and significant impact on short-term satiety, which is time and dosage correlated. Some statistical differences appear for the group 8 g/d from day 5, and from day 0 for the groups 14, 18, and 24 g/d. The hunger feeling status decreases significantly from day 5 to the end of the evaluation for the group 24 g and from day 7 for the groups 14 and 18 g. By day 5, the group 24 g showed significantly longer time to hunger between meals compared with placebo. These results suggest that dietary supplementation with a soluble fiber can decrease hunger feeling and increase short-term satiety over time when added to a beverage from 8 to 24 g/d with time- and dose-responses relationship [37].

Human clinical trials on healthy subjects have shown that Nutriose is well tolerated and able to stimulate the growth of acid-resistant bacteria. We particularly observed a beneficial shift in the bacterial microbiota profile to butyrogenic genera such as *Peptostreptococcus*, *Fusobacterium* and *Bifidobacterium* [37].

Resistant maltodextrins made from starch are commercially available. Fibersol-2 is well-known soluble, non-digestible, starch-derived resistant maltodextrin [67]. Fibersol-2 is produced from corn-starch by pyrolysis and subsequent enzymatic treatment (similar to the process to manufacture conventional maltodextrins) to convert a portion of the normal α-1,4 glucose linkages to random 1,2-, 1,3- α or β linkages [69]. Solubility of Fibersol-2 in water reaches 70% (w/w) at 20°C. It is readily dispersible in water and highly compatible with dry drink mix applications. At typical use levels, it yields clear, transparent solutions that are near water-like in performance. Fibersol-2 adds no flavor or odor. It has essentially no sweetness of its own. It shows stability to acid and heat/retort processing, including stability in high acid, hot filled, aseptic, or retorted products like juices, sauces, puddings, fluid milks, and sports. Fibersol-2 shows superior freeze-thaw stability. It shows precise and extremely low viscosity and very low hygroscopicity. It does not actively participate in non-enzymatic Maillard-type browning [69]. Fibersol-2 exhibited very important physiological properties. It was fermented slowly, producing less acid and gas than most soluble dietary fiber [27]. Studies indicated that Fibersol-2 could effectively reduce postprandial levels of blood glucose and insulin [109]. Fibersol-2 significantly reduced levels of blood triglycerides and serum cholesterol [125]. By adding stool volume, moisture, and reducing transit time, Fibersol-2 helped maintain good colon health, potentially reducing the incidence of various types of colon diseases and cancers [102]. Fibersol-2 effectively promoted the growth of a variety of beneficial bacteria (naturally occurring or ingested as probiotics) in the colon. In promoting the growth of beneficial bacteria, Fibersol-2 indirectly reduced the presence of undesirable bacterial species [41, 63].

Bodinhan et al. [8] found that a non-viscous resistant starch significantly lowered energy intake after intake of the supplement compared with placebo during both an ad libitum test meal (P = 0.033) and over 24 hours (P = 0.044). Cani et al [14] found that treatment with the fermentable dietary fiber oligofructose increased satiety after breakfast and dinner and reduced hunger and prospective food consumption after dinner, suggesting a role for the use oligofructose supplements in the management of food intake in overweight and obese patients [37].

Kapusniak et al. [47] and Śliżewska et al. [98] enzyme-resistant chemically modified dextrins resulting from heating of potato starch with hydrochloric acid as catalyst and additionally polycarboxylic acids (citric and tartaric acids) were tested as the source of carbon for probiotic bacteria (Lactobacillus and Bifidobacterium) and bacteria isolated from the human feces (Escherichia coli, Enterococcus, Clostridium and Bacteroides). It was shown that all of the tested bacteria, both probiotics and those isolated from human feces, were able to grow and utilize dextrin as a source of carbon, albeit to varying degrees. After 24 h the highest growth was recorded for the probiotic bacteria Lactobacillus and Bifidobacterium, the weakest for Clostridium and Escherichia coli bacteria. After prolonging culture time to 72–168 h, which corresponds to retarded or pathological passage of large intestine contents, the viability of intestinal bacteria in a medium with resistant dextrin was found to be lower by one or two orders of magnitude as compared to the viability of probiotic bacteria. The number of probiotics and bacteria isolated from fecal samples grown in media containing 1% glucose was lower by two or three orders of magnitude than that of corresponding bacteria grown in a medium containing dextrin. This may have been caused by lower pH values of the controls, in which the culture environment became unfavorable to preserving high viability by the studied bacteria. This may have also been caused by the protective effects of dextrin on the bacteria. After the completion of incubation, that is, at 168 h, lactobacilli and bifidobacteria were found to be highly viable. Their counts were higher by one or two orders of magnitude than those of the intestinal bacteria E. coli, Enterococcus, Clostridium and Bacteroides isolated from fecal samples. At 168 h of incubation, the probiotic bacteria amounted to over 44% of the whole population. The *Clostridium* strain showed the weakest growth, with a 9.5% share in the entire population, while *Enterococcus, Escherichia coli,* and *Bacteroides* amounted to from 15.3% to 15.8% of the population [48].

9. Conclusions

In today's world, life style is an important determinant of health in later life. Lack of physical activity, particularly if associated with over consumption, increases the risk of development of nutrition related chronic diseases, such as obesity, hypertension, cardiovascular diseases, osteoporosis, type II diabetes, and several cancers. Over the last decade, drastic changes have taken place in the image and assessment of the importance of the daily diet. Foods are no longer judged in terms of taste and immediate nutritional needs, but also in terms of their ability to improve the health and well-being of consumers. The role of diet in human health has led to the recent development of the so-called functional food concept. A functional food is dietary ingredient, that has cellular or physiological effects above the normal nutritional value. Functional food can contain probiotics and/or prebiotics [33, 40].

The studies suggests that soluble fibers help to regulate the digestive system, may increase micronutrient absorption, stabilize blood glucose and lower serum lipids, may prevent several gastrointestinal disorders, and have an accepted role in the prevention of cardiovascular disease. It is concluded that supplementation with soluble fibers (e.g. wheat dextrin) may be useful in individuals at risk of a lower than recommended dietary fiber intake.

A prebiotic is a nondigestible food ingredient that beneficially affects the host by selectively stimulating the growth and/or activity of one or, a limited number, of bacteria in the colon that can improve host health [19]. Some carbohydrates, such as fructooligosaccharides (FOS) [10, 34, 75, 101], inulin [75, 76, 77, 112] and galactooligosaccharides (GOS) [84, 108] are well-accepted prebiotics.

Promising sources of prebiotics are starch products, especially resistant starch (RS) [3, 19, 106] and products of partial degradation of starch [56].

The commercial degraded starches are known as converted starches and comprise the "thin-boiling" acid-converted starches, oxidized starches, and dextrins. There are four major groups of dextrins: maltodextrins produced by hydrolysis of dispersed starch by action of liquifying enzymes such as amylase, degradation products by acid hydrolysis of dispersed starch, cyclodextrins, and pyrodextrins produced by the action of heat alone or in a combination with acid on dry granular starch. On the market pyrodextrins are available in three major varieties: British gums, white dextrins and yellow dextrins [5, 105, 122].

In particular, almost every food oligosaccharide and polysaccharide has been claimed to have prebiotic activity, but not all dietary carbohydrates are prebiotics. Conventional fibers, like pectins, cellulose, etc. are not selectively metabolized by gut bacteria. Resistant maltodextrins, being a mixture of fractions of different molecular weight (different degree of polymerization), are dietary fibre, but there are not necessarily selective for desirable bacteria in the gut. Hence, research showing the effect of prebiotic should continue to be performed, both in the *in vito* and *in vivo*.

Author details

Katarzyna Śliżewska

Institute of Fermentation Technology and Microbiology, Faculty of Biotechnology and Food Sciences, Technical University of Lodz, Lodz, Poland

Janusz Kapuśniak, Renata Barczyńska and Kamila Jochym

Institute of Chemistry, Environmental Protection and Biotechnology, Jan Dlugosz University in Czestochowa, Czestochowa, Poland

10. References

[1] Abrams S.A., Griffin I.J., Hawthorne K.M., Ellis K.J. (2007) Effect of prebiotic supplementation and calcium intake on body mass index. J. Pediatr. 151: 293-298.

[2] Annison G., Illman R., Topping D. (2003) Acetylated, propionylated or butyrylated starches raise large bowel short-chain fatty acids preferentially when fed to rats. J. Nutr. 133: 3523-3528.

[3] Asp N.G., Van Amelsvoort J.M.M., Hautvast J.G.A.J. (1996) Nutritional implications of resistant starch. Nutr. Res. Rev. 9: 1-31.

[4] Behall K.M., Daniel J., Scholfield D.J., Hallfrisch J.G., Liljeberg-Elmståhl H.G.M. (2006) Consumption of both resistant starch and b-glucan improves postprandial plasma glucose and insulin in women. Diabetes Care 29, 976-981.

[5] BeMiller J.N. (1993) Starch-based gums. In: Industrial Gums. Polysaccharides and Their Derivatives. Whistler R.L. and BeMiller, J.N., editors. Academic Press, San Diego, USA, pp. 579-601.

[6] Berenal M.J., Periago M.J., Ros G. (2002) Effects of processing on dextrin, total starch, dietary fiber and starch digestibility in infant. J. Food Sci. 67: 1249-1254.

[7] Björck, L., Gunnarsson A., Østergård,K. (1989) A study of native and chemically modified potato starch. Part II. Digestibility in the rat intestinal tract. Starch/Stärke 41: 128-134.

[8] Bodinham B.L., Frost G.S., Robertson M.D. (2010) Acute ingestion of resistant starch reduces food intake in healthy adults. Br. J. Nutr. 103: 917-22.

[9] Boehm G., Stahl B. Oligosaccharides. In: Mattila-Sandholm and Saarela M, editors. Functional Dairy Products, Woodhead Publishing, CRC Press, England, 2003, pp. 203–243.

[10] Bouhnik Y., Vahedi K., Achour L., Attar A., Salfati J., Pochart P., Marteau P., Flourie B., Bornet F., Rambaud J.C. (1999) Short-chain fructo-oligosaccharide administration dose-dependently increases fecal bifidobacteria in healthy humans. J. Nutr. 129: 113-116.

[11] Brendel R., Boursier B., Leroux P. (2002) U.S. Patent 2002/0192344 A1.

[12] Brigidi P., Swennen E., Vitali B., Rossi M., Matteuzzi D. (2003) PCR detection of *Bifidobacterium* strains and Streptococcus thermophilus in feces of human subjects after oral bacteriotherapy and yogurt consumption. Int. J. Food Microbiol. 81: 203–209.

[13] Brown I.L., (2004) Applications and uses of resistant starch. J. AOAC Int. 87: 727-732.

[14] Cani P.D., Joly E., Horsmans Y., Delzenne N.M. (2006) Oligofructose promotes satiety in healthy human: a pilot study. Eur. J. Clin. Nutr. 60: 567-72.

[15] Champ M., Langkilde A.M., Brouns F., Kettlitz B. (2003) Advances in dietary fiber characterization. 2. Consumption, chemistry, physiology and measurement of resistant starch; implications for health and food labeling. Nutr. Res. Rev. 16: 143–161.

[16] Coudray C., Bellanger J., Castiglia-Delavaud C., Rémésy C., Vermorel M.,Rayssignuier Y. (1997) Effect of soluble or partly soluble dietary fibres supplementation on absorption and balance of calcium, magnesium, iron and zinc in healthy young men. Eur. J. Clin. Nutr., 51: 375-380.

[17] Crittenden R.G., Playne M.J. Prebiotics. In: Lee Y.K. and Salminen S., edtitors. Handbook of probiotics and prebiotics, John Wiley & Sons, New Jersey, Canada, 2009, pp. 535-584.

[18] Crittenden R.G., Prebiotics. In: Tannock GW, edtitor. Probiotics: A Critical review. Horizon Scientific Press, Wymondham, United Kingdom, 1999, pp. 141–156.

[19] Cummings J.H., Beatty E.R., Kingman S.M., Bingham S.A., Englyst H.N. (1996) Digestion and physical properties of resistant starch in the human large bowel. Br. J. Nutr. 75: 733-747.

[20] Cummings J.H., Stephen A.M. (2007) Carbohydrate terminology and classification. Eur. J. Clin. Nutr. 61, 5-18.

[21] Douglas L.C., Sandres M.E. (2008) Probiotics and prebiotics in dietetics practice. J. Am. Diet. Associat. 108: 510-521.

[22] Duncan S.H., Holtrop G., Lobley G.E., Calder A.G., Stewart C.S., Flint H.J. (2004) Contribution of acetate to butyrate formation by human fecal bacteria. Br. J. Nutr. 91: 915-923.

[23] Engllyst H.N., Kingman S.M., Cummings J.H. (1992) Classification and measurement of nutritionally important starch fractions. Eur. Clin. Nutr. 46: 33–50.

[24] FAO Technical Meeting on Prebiotics. (2007) Food Quality and Standards Service (AGNS), Food and Agriculture Organization of the United Nations (FAO). FAO Technical meeting Report, September 15-16

[25] Faraj A., Vasanthan T., Hoover R. (2004) The effect of extrusion cooking on resistant starch formation in waxy and regular barley flours. Food Res. Int. 37, 517-525.

[26] Figueroa-González I., Quijano G., Ramírez G., Cruz-Guerrero A. (2011) Probiotics and prebiotics – perspectives and challenges. J. Sci. Food Agric. 91, 1341-1348.

[27] Flickinger E.A., Wolf B.W., Garleb K.A., Chow J., Leyer G.J., Johns P.W., Fahey G.C. (2000) Glucose-based oligosaccharides exhibit different *in vitro* fermentation patterns and affect in vivo apparent nutrient digestibility and microbial populations in dogs. J. Nutr. 130: 1267-1273.

[28] Fouache C., Duflot P., Looten P. (2003) U.S. Patent 2003/6630586.

[29] Frohberg C., Quanz M. (2008) Use of Linear Poly-Alpha-1,4-Glucans as Resistant Starch, United States Patent Application 20080249297, 2008, Available on line at: http://www.wipo.int/pctdb/en/wo.jsp?WO=2005040223.

[30] Fuentes-Zaragoza E., Riquelme-Navarrete M.J., Sánchez-Zapata E., Pérez-Álvarez J.A. (2010) Resistant starch as functional ingredient: A review. Food Res. Int. 43: 931-942.

[31] Fuentes-Zaragoza E., Sánchez-Zapata E., Sendra E., Sayas E., Navarro C., Fernández-López J., Pérez-Alvarez J.A. (2011) Resistant starch as prebiotic: A review. Starch/Stärke 63: 406-415.

[32] Gibson G.R. (2004) Fibre and effects on probiotics (the prebiotic concept). Clin. Nutr. Suppl. 1: 25-31

[33] Gibson R., Roberfroid M. (1995) Dietary modulation of the human colonic microbiota: introducing the concept of prebiotics. J. Nutr. 125: 140-1412

[34] Gibson G. (1999) Dietary modulation of the human gut microflora using the prebiotics oligofructose and inulin. J. Nutr. 129 (7Suppl): 1438S-1441S.

[35] Goldring J.M. (2004) Resistant starch: Safe intakes and legal status. J. AOAC Int. 87, 733–739.

[36] González-Soto R.A., Sánchez-Hernández L., Solorza-Feria J. (2006) Resistant starch production from non-conventional starch sources by extrusion. Food Sci. Tech. Int. 12: 5-11.

[37] Guérin-Deremaux L., Pochat M., Reifer C., Wils D., Cho S., Miller L.E. (2011) The soluble fiber NUTRIOSE induces a dose-dependent beneficial impact on satiety over time in humans. Nutr. Res. 31: 665-672.

[38] Haralampu S.G. (2000) Resistant starch: A review of the physical properties and biological impact of RS3. Carbohydrate Polymers, 41, 285–292.

[39] Hautvast J.G.A.J., Meyer D. (2007) The science behind inulin and satiety. Sensus: 6-30.

[40] Holzapfel W.H., Schillinger U. (2002) Introduction to pre- and probiotics. Food Res. Int. 35: 109-116.

[41] Hopkins M.J., Cummings J.H., Macfarlane G.T. (1998) Inter-species differences in maximum specific growth rates and cell yields of bifidobacteria cultured on oligosaccharides and other simple carbohydrate sources. J. Appl. Microbiol. 85: 381-386.

[42] Huebner J., Wehling R.L., Parkhurst A., Hutkins R.W. (2008) Effect of processing conditions on the prebiotics activity of commercial prebiotics. Int. Dairy J. 18: 287-293.

[43] Imaizumi K., Nakatsu Y., Sato M., Sedamawati Y., Sugano M. (1991) Effects of xylooligosaccharides on blood glucose, serum and liver lipids and caecum short-chain fatty acids in diabetic rats. Agric. Biol. Biochem., 55: 199-205.

[44] ISAPP (2008) 6th Meeting of the International Scientific Association of Probiotics and Prebiotics, London, Ontario.

[45] Janzen G.J. (1969) Verdaulichkeit von Stärken und phosphatierten Stärken mittels Pankreatin. Stärke 21: 231-237.

[46] Jiang H., Jane J.L., Acevedo D., Green A. (2010) Variations in starch physicochemical properties from a generationmeans analysis study using amylomaize V and VII parents. J. Agric. Food Chem. 58, 5633-5639.

[47] Kapusniak J., Barczynska R. Slizewska K., Libudzisz Z. (2008a) Utilization of enzyme-resistant chemically modified dextrins from potato starch by Lactobacillus bacteria. Zeszyty Problemowe Postępów Nauk Rolniczych 530: 445-457. (in polish)

[48] Kapusniak J., Jochym K., Barczynska R. Slizewska, K., Libudzisz Z. (2008b) Preparation and characteristics of novel enzyme-resistant chemically modified dextrins from potato starch. Zeszyty Problemowe Postępów Nauk Rolniczych 530: 427-444. (in polish)

[49] Kayode O., Adebowale T., Adeniyi Afolabi B., Iromidayo Olu-Owolabi. (2006) Functional, physicochemical and retrogradation properties of sword bean (Canavalia gladiata) acetylated and oxidized starches. Carbohydr. Polym. 65: 93–101.

[50] Le Blay G.M., Michel C.D., Blottière, H.M., Cherbut, C.J. (2003) Raw potato starch and short-chain fructo-oligosaccharides affect the composition and metabolic activity of rat intestinal microbiota differently depending on the caeco-colonic segment involved. J. Appl. Microbiol. 94, 312-320.

[51] Leegwater D.C., Luten J.B (1971) A study on the in vitro digestibility of hydroxypropyl starches by pancreatin. Stärke 23: 430-432.

[52] Leeman A.M., Karlsson M.E., Eliasson A.Ch., Björck I.M.E. (2006) Resistant starch formation in temperature treated potato starches varying in amylose/amylopectin ratio. Carbochydr. Polym. 65: 306-313.

[53] Lefranc-Millot C. (2008) NUTRIOSE 06: a useful soluble fibre for added nutritional value. Nutr. Bull. 33: 234-239.

[54] Lefranc-Millot C., Wils D., Neut C., Saniez-Degrave M.H. (2006) Effects of a soluble fiber, with excellent tolerance, NUTRIOSE®06, on the gut ecosystem: a review. In: Proceedings of The Dietary Fibre Conference. June, 2006. Helsinki, Finland.

[55] Lenoir-Wijnkoop I., Sanders M.E., Cabana M.D. (2007) Probiotic and prebiotic influence beyond the intestinal tract. Nutr. Rev. 65: 469-489.

[56] Leszczyński W. (2004) Resistant Starch – classification, structure, production. Pol. J. Food Nutr. Sci. 13/54: 37-50.

[57] Lim C.C., Ferguson L.R., Tannock G.W. (2005) Dietary fibres as "prebiotics": implications for colorectal cancer. Mol. Nutr. Food Res. 49: 609–619.

[58] Liu R., Xu G. (2008) Effects of resistant starch on colonic preneoplastic aberrant crypt foci in rats. Food Chem. Toxicol. 46: 2672-2679.

[59] Lopez H.W., Levrat-Verny M.A., Coudray C., Besson C., Krespine V., Messager A., Demigné C, Rémésy C. (2001) Class 2 resistant starches lower plasma and liver lipids and improve mineral retention in rats. J. Nutr. 131: 1283-1289.

[60] Lunn J., Buttriss J. L. (2007) Carbohydrates and dietary fibre. Nutr. Bull. 32: 21-64.

[61] Maccfarlane G.T., Steed H., Maccfarlane S. (2008) Bacterial metabolism and health-related effects of galacto-oligosaccharides and other prebiotics. J. Appl. Microbiol. 104: 305-344.

[62] Mana K.K. (2003) Enzyme resistant dextrins from high amylose corn mutant starches. Starch/Stärke. 53: 21–26.

[63] Matsuda I., Satouchi M. (1997) Agent for promoting the proliferation of Bifidobacterium. U.S. Patent 5698437.

[64] Mermelstein N.H. (2009) Analyzing for resistant starch. Food Technol. 4: 80-84.

[65] Morais M.B., Feste A., Miller R.G., Lifichitz C.H. (1996) Effect of resistant starch and digestible starch on intestinal absorption of calcium, iron and zinc in infant pigs. Paediatr. Res. 39(5): 872-876.

[66] Morell M.K., Konik- Rose Ch., Ahmed R., Li Z., Rahman S. (2004) Synthesis of resistant starches in plants. J. AOAC Int. 87: 740–748.

[67] Ohkuma K., Hanno Y., Inada K., Matsuda I., Katta Y. (1997) U.S. Patent 5620873.

[68] Ohkuma K., Matsuda I., Katta Y., Hanno Y. (1999) Pyrolysis of starch and its digestibility by enzymes – Characterization of indegestible dextrin. Denpun Kagaku. 37: 107-114.

[69] Ohkuma K., Wakabayashi S. (2001) Fibersol-2: A soluble, non-digestible, starch-derived dietary fibre. In: McCleary B.V., Prosky L., editors. Advanced Dietary Fibre Technology. Blackwell Science Ltd., Oxford, pp. 509-523.

[70] Onyango C., Bley T., Jacob A. (2006) Infuence of incubation temperature and time on resistant starch type III formation from autoclaved and acid-hydrolysed cassava starch. Carbohydr. Polym. 66: 497-499.

[71] Palframan R., Gibson G.R., Rastall R.A. (2003) Development of a quantitive tool for comparison of the prebiotic effect of dietary oligosaccharides. Lett. Appl. Microbiol., 37: 281-284.

[72] Pasman W.J., Wils D., Saniez M.H., Kardinaal A.F. (2006) Long term gastro-intestinal tolerance of NUTRIOSE® FB in healthy men. Eur. J. Clin. Nutr. 60(8): 1024-1034.

[73] Ring S.G., Gee J.M., Whittam M., Orford P., Johnson I.T. (1988) Resistant starch: its chemical from in foodstuffs and effect on digestibility in vitro. Food Chem. 28: 97-109.

[74] Roberfroid M., Gibson G.R., Hoyles L., McCartney A.L., Rastall R., Rowland I., Wolvers D., Watzl B, Szajewska H., Stahl B., Guarner F., Respondek F., Whelan K., Coxam V., Davicco M-J., Léotoing L., Wittrant Y., Delzenne N.M., Cani P.D., Neyrick A.M., Meheust A. (2010) Prebiotic effects: metabolic and health benefits. Br. J. Nutr. 104: S1-S61.

[75] Roberfroid M.B. (1998) Prebiotics and synbiotics: concepts and nutritional properties. Br. J. Nutr. 80: S197-S202.

[76] Roberfroid, M.B. (2005) Introducing inulin-type fructans. Br. J. Nutr. 93: S13-S25.

[77] Roberfroid, M.B. (2007) Inulin-type fructans: Functional food ingredients. J. Nutr. 137: 2493S-2502S.

[78] Rochfort S., Panozzo J. (2007) Phytochemicals for health, the role of pulses. J. Agric. Food Chem. 55, 7981-7994.

[79] Rodríguez-Cabezas M.E., Camuesco D., Arribas B., Garrido-Mesa, N. (2010) The combination of fructooligosaccharides and resistant starch shows prebiotic additive effects in rats. Clin. Nutr. 29, 832-839.

[80] Roturier J.M., Looten P.H., Osterman E. (2003) Dietary fiber measurement in food containing NUTRIOSE®FB by enzymatic-gravimetric-HPLC method. Proc. Dietary Fiber Conference, Helsinki, Finland. May, 2003.

[81] Roturier J.M., Looten P.H. (2006) Nutroise: Analytical Aspects. Proc. The Dietary Fibre Conference, Helsinki, Finland, June 2006.

[82] Rozan P., Deremaux L., Wils D., Nejdi A., Messaoudi M., Saniez M.H. (2008) Impact of sugar replacers on cognitive performance and function in rats. Brit. J. Nutr. 100: 1004-1010.

[83] Sajilata M.G., Singhal R.S., Kulkarni P.R. (2006) Resistant starch- a review, Comp. Rev. Food. Sci. Food Safety. 5: 1-17.

[84] Sako T., Matsumoto K., Tanaka, R. (1999) Recent progress on research and applications of non-digestible galacto-oligosaccharides. Int. Dairy J. 9: 69-80.

[85] Salminen S., Gorbach S., Lee Y.K., Benno Y. (2004) Human Studies on Probiotics: What is Scientifically Proven Today? In: Salminen S, von Wright A and Ouwerhand A, editors Lactic Acid Bacteria: Microbiological and Functional Aspects. Marcel Dekker, New York, 2004, pp. 515–530.

[86] Sang S., Chang J.L. (2007) Formation, characterization, and glucose response in mice to rice starch with low digestibility produced by citric acid treatment. J. Cereal Sci. 45: 24-33

[87] Sang Y., Seib P. (2006) Resistant starches from amylose mutants of corn by simultaneous heat-moisture treatment and phosphorylation. Carbohydr.Polym. 63: 167-175.

[88] Saniez M.H. (2004) U.S. Patent 2004/6737414 B2.

[89] Satokari R.M., Vaughan E.E., Akkermans A.D.L., Saarela M., de Vos W.M. (2001) Polymerase chain reaction and denaturing gradient gel electrophoresis monitoring of fecal *Bifidobacterium* populations in a prebiotic and probiotic feeding trial. Syst. Appl. Microbiol. 2001, 24: 227–231.

[90] Saulnier D.M., Spinler J.K., Gibson G.R., Versalovic J. (2009) Mechanisms of probiosis and prebiosis: considerations for enhanced functional foods. Curr. Opinion Biotechnol. 20: 135-141.

[91] Serpelloni M. (2002) U.S. Patent 2002/0192343 A1.

[92] Serpelloni M. (2003) U.S. Patent 2003/0077368 A1.

[93] Serpelloni M. (2006) U.S. Patent 2006/0112956 A1.

[94] Shin M., Song J., Seib P. (2004) In vitro digestibility of cross-linked starches – RS4. Starch/ Stärke. 56: 478-483.

[95] Shin S.I., Lee Ch.J. (2007) Formation, characterization, and glucose response in mice to rice starch with low digestibility produced by citric acid treatment. J. Cereal Sci. 45: 24-33

[96] Singha J., Kaurb L. (2007) Review factors influencing the physico-chemical, morphological, thermal and rheological properties of some chemically modified starches for food applications—A review. Food Hydr. 21: 1–22.

[97] Slavin J., Stewart M., Timm D., Hospattankar A. (2009) International Association for Cereal Science and Technology (ICC), 1–3 July 2009, van der Kamp J.W., Vienna, Austria, p. 35.

[98] Śliżewska K., Kapuśniak J., Barczyńska R., Jochym K. (2010) The preparation of prebiotic properties. The patent application RP P-392895

[99] Storey D., Lee A., Bornet F., Brouns F. (2007) Gastrointestinal responses following acute and medium term intake of retrograded resistant maltodextrins, classified as type 3 resistant starch. Eur. J. Clin. Nutr. 61, 1262–1270.

[100] Suzuki N., Aiba Y., Takeda H., Fukumori Y., Koga Y. Superiority of 1-kestose, the smallest fructo-oligosaccharide, to a synthetic mixture of fructo-oligosaccharides in the selective stimulating activity on Bifidobacteria. Biosci. Microflora. 25: 109–116.

[101] Swennen K., Courtin CH.M., Delcour, J.A., 2006. Non-digestible oligosaccharides with prebiotic properties. Crit. Rev. Food Sci. Nutr. 46: 459-471.

[102] Takagak K., Ikeguchi M., Artura Y., Fujinaga N., Ishibashi Y., Sugawa-Katayama Y. (2001) The effect of AOJIRU drink powder containing indigestible dextrin on defecation frequency and faecal characteristics. J. Nutr. Food 4(4): 29-35.

[103] Tannock G.W., Munro K., Bibiloni R., Simon M.A., Hargreaves P., Gopal P., Harmsen H., Welling G. (2004) Impact of consumption of oligosaccharide-containing biscuits on the fecal microbiota of humans. Appl. Environ. Microbiol. 70: 2129–2136.

[104] Themeier H., Hollmann J., Neese U., Lindhauer M.G. (2005) Structural and morphological factors influencing the quantification of resistant starch II in starches of different botanical origin. Carbohydr. Polym. 61: 72-79.

[105] Tomasik P., Wiejak S., Pałasiński M. (1989) The thermal decomposition of carbohydrates. Part II. The decomposition of starch. Adv. Carbohydr. Chem. Biochem. 47: 279-344.

[106] Topping D.L, Clifton, P.M. (2001) Short-chain fatty acids and human colonic function: roles of resistant starch and nonstarch polysaccharides. Physiol. Rev. 81(3): 1031-1064.

[107] Trinidad T.P., Wolever T.M.S., Thompson L.U. (1996) Effect of acetate and propionate on calcium absorption from the rectum and distal colon of humans. Am. J. Clin. Nutr. 63: 574-578.

[108] Tzortzis G., Goulas A.K., Gibson G.R. (2005) Synthesis of prebiotic galactooligosaccharides using whole cells of a novel strain, Bifidobacterium bifidum NCIMB 41171. Appl. Microbiol. Biotechnol. 68: 412-416.

[109] Unno T., Nagata K., Horiguchi T. (2002) Effects of green tea supplemented with indigestible dextrin on postprandial levels of blood glucose and insulin in human subjects. Journal of Nutritional Food 5(2): 31-39.

[110] van den Heuvel E.G., Wils S.D., Pasman W.J., Bakker M., Saniez M.H., Kardinaal A.F. (2004) Short-term digestive tolerance of different doses of NUTRIOSE®FB, a food dextrin, in adult men. Eur. J. Clin. Nutr. 58(7): 1046-1055.

[111] Van Loo J. (2006) Inulin-type fructans as prebiotics. In: Prebiotics: Development and Application. Gibson G.R. and Rastall RA, editors. John Wiley and Sons, Chichester, UK, pp. 57–100.

[112] Van Loo J., Cummings J., Delzenne N., Englyst H., Franck A., Hopkins M., Kok N., Macfarlane G., Newton D., Quigley M., Roberfroid M., van Vliet T., van den Heuvel, E. (1999) Functional food properties of non-digestible oligosaccharides: a consensus report from the ENDO project (DGXII AIRII-CT94-1095). Br. J. Nutr. 81(2): 121-132.

[113] Wang J., Jin Z., Yuan X. (2006) Preparation of resistant starch from starch-guar gum extrudates and their properties. Food Chem. 101: 20-25.

[114] Wang Y. (2009) Prebiotics: Present and future in food science and technology. Food Res. Int. 42: 8-12.

[115] Wang Y., Kozlowski R., Delgado G.A. (2001) Enzyme resistant dextrins from high amylose corn mutant starches. Starch-Starke. 53: 21-26.

[116] Weese J., Schrezenmeir J. (2008) Probiotics, prebiotics and synbiotics. Adv. Biochem. Eng. Biotechnol. 111: 1-6.

[117] Wepner B., Berghofer E., Miesenberger E., Tiefenbacher K. (1999) Citrate starch: Application as resistant starch in different food systems. Starch/Stärke 51: 354-361.

[118] Wolf B.W., Wolever T., Bolognesib C. (2001) Glycemic response to a rapidly digested starch is not affected by the addition of an indigestible dextrin in humans. Nutr. Res. 21: 1099–1106.

[119] Woo K. S., Seib P. A. (2002) Cross-linked resistant starch: preparation and properties. Am. Ass. Cereal Chem. 79: 819-826.

[120] Wootton M., Chaundry M.A. (1979) Enzymic digestibility of modified starches. Starch/Stärke 31: 224-228.

[121] Wootton M., Chaundry M.A. (1981) In vitro digestion of hydroxypropyl derivatives of wheat starch. I. Digestibility and action pattern using porcine pancreatic α-amylase. Starch /Stärke. 33: 135-137.

[122] Wurzburg O.B. (1986) Converted starches. In: Modified starches: properties and uses. Wurzburg O.B., editor. CRC Press, Boca Raton Florida, USA, pp. 17-40.

[123] Xie X., Liu Q. (2004) Development and physicochemical characterization of new resistant citrate starch from different corn. Starches/ Starch. 56: 364 – 370.

[124] Xie X., Liu Q., Cui S.W. (2006) Studies on the granular structure of resistant starches (type 4) from normal, high amylose and waxy corn starch citrates. Food Res. Int. 39: 332–341.

[125] Yamamoto T. (2007) Effect of indigestible dextrin on visceral fat accumulation. J. Jpn. Soc. Study. Obes. 13: 34-41.

[126] Yao N., Paez A.V., White P.J. (2009) Structure and function of starch and resistant starch from corn with different doses of mutant amylose-extender and floury-1 alleles. J. Agric. Food Chem. 57: 2040-2048.

[127] Younes H., Coudray C., Bellanger J., Demigné C., Rayssiquier Y., Rémésy C. (2001) Effects of two fermentable carbohydrates (inulin and resistant starch) and their combination on calcium and magnesium balance in rats. Brit. J. Nutr. 86, 479-485.

[128] Younes H., Levrat M.A., Demige C., Remesy C. (1995) Resistant starch is more effective than cholestyramine as a lipid-lowering agent in the rat. Lipids, 30: 847-853.

Chapter 3

Carbohydrate Metabolism in Drosophila: Reliance on the Disaccharide Trehalose

Alejandro Reyes-DelaTorre, María Teresa Peña-Rangel
and Juan Rafael Riesgo-Escovar

Additional information is available at the end of the chapter

1. Introduction

Work on Drosophila has been so influential that it has even impinged on human metabolism and health. The reason for this is obvious: There is a strong evolutionary conservation between biological processes of flies and humans. Also, cloning and functional analyses of genes in Drosophila allowing study of many cellular processes have greatly aided in the identification of homologous genes and processes in humans and other organisms, as exemplified by the study of homeotic genes and mutations [1]. A great number of genes required for a myriad different functions (from transcription factors, structural proteins, ion channels and signaling molecules, to those required for behavior and sleep) is common between the two species [2]; in fact, [3] showed that from a pool of 287 genes known to be implicated in humans in cancer and malformations, and neurological, cardiovascular, hematological, immune, endocrine, renal and metabolic diseases, 178 (62%) had their counterparts in flies. These findings underscore the potential impact of Drosophila as a powerful system for gene discovery, but also for study of diseases' symptoms and complications, and the discovery of therapeutic treatments.

Drosophila melanogaster was originally introduced as a research organism and genetic model at the beginning of the twentieth century by T. H. Morgan and co-workers. From their pioneering studies on heredity and the chromosome theory of inheritance [4], work with this model organism has been the source of biological insights in many areas, including genetics and development [5]. Clearly, Drosophila offers an array of advantages: its cultures in the laboratory are easy and economical, its life cycle is short (approximately ten days at room temperature), its embryonic development lasts only twenty four hours, there is no meiotic recombination in males, it has clearly defined and easily identified structures in the cytoskeleton (subject to mutational change, and, correspondingly, easy to score), has a group of marked balancer chromosomes that impede meiotic recombination in females thus allowing the recovery and maintenance of other mutations of interest, has only four pairs of chromosomes, etc. These characteristics all greatly aid in genetic analyses [6], and since the sequencing of its genome (first released in 2000 [2, 7]), has arguably become the best characterized multicellular eukaryote model system. Critically, there are also large collections of mutant strains, with the one housed at the Bloomington Drosophila Stock Center at Indiana University [8] culturing over 49,000 different mutant stocks. Finally, since the publication of the fly genome, and its refinements with time, emphasis has also been placed on analyses and studies of physiology and signaling pathways involved, rather than just concentrating on 'gene hunts' or 'fishing expeditions' for candidate genes [9].

Since the discovery and characterization of abnormal phenotypes of insulin pathway mutants in flies [10], one of recent focuses of work in Drosophila have been metabolic studies, especially those of carbohydrates. One of the underlying reasons for this is, no doubt, the recent surge in obesity and diabetes mellitus type 2 among the world's population in both developing and developed countries [11, 12]. Just as it happens in vertebrates, *Drosophila melanogaster* regulates its levels of circulating carbohydrates, and stores excesses in the forms of glycogen and lipids, especially of triglycerides, in the fat body [13]. Another fruit of this recent surge in carbohydrate studies has been the elucidation of evolutionarily conserved signaling pathways, like the insulin/IGF/Target of Rapamycin (TOR) pathway [14].

2. The insulin pathway

Glucose homeostasis in mammals is maintained by feedback mechanisms balancing glucose cellular import and replenishment to the blood, so as to maintain a constant 5.5 mM level [15]. This level is the result of glucose cellular entry rates, glucose removal by the liver and new glucose coming from digestion of food glucose sources to the blood. Members of the Glut family of facilitated transporters, of which mammals possess several isoforms and genes, carry out glucose cellular import. It was recognized in the middle nineties that insulin stimulation in adipocytes in culture resulted in activation of a phosphatidyl-inositol-3-kinase (PI3K) isoform, and more glucose cellular import, resulting from Glut4 plasma membrane translocation [16]. Nowadays, Glut4 plasma membrane translocation from an internal vesicle pool is recognized as the rate-limiting step in glucose import by muscle (skeletal and cardiac) cells, and vertebrate adipocytes [17].

In contrast, the relation between insulin signaling and glucose cellular import in insect cells is not that clear. In Drosophila, there is a glucose transport system with similar kinetics to vertebrate Glut transporters [18]. A Drosophila Glut1 homolog cDNA has been cloned and described [19]. Nonetheless, experiments have shown that there is no increase in cellular import of 2-deoxyglucose labeled with ^{3}H in Drosophila Kc cells in culture after activation of the insulin pathway [20], and neither manipulation of activation levels of PI3K or protein kinase B (also called AKT or PKB), both molecules in the insulin pathway, in Drosophila S2 cells augments glucose cellular import [21].

These results could lead one to think that insulin pathway function in vertebrates and insects is not well conserved, and that insect insulin-like-peptides (ILPs) have no role in glucose and / or carbohydrate homeostasis, yet there is ample evidence to the contrary. The Drosophila genome possesses eight insulin-like–peptides (ILPs) [22-25] at last count, and these have been shown to interact with the Drosophila insulin receptor homolog (InR). Moreover, the Drosophila insulin receptor homolog can be functionally exchanged with the vertebrate insulin receptor [26], arguing that functional conservation must be significant.

Furthermore, ablation of the neurosecretory cells (NSC) that secrete DILPs 2, 3 and 5 in larvae gives rise to adults with developmental delays, reduced body size, and high levels of carbohydrates and lipids. Ablation of these same cells in adult flies lead to hypertrehalosemia without growth phenotypes. Knockdown of DILP2 alone by RNAi in NSC results in higher corporal trehalose levels [27, 28]. These effects, due to lower levels of circulating DILPs, causing lipid and carbohydrate accumulation, are analogous to effects observed in diabetic patients or diabetic mice when there is generalized insulin resistance, like in the insulin receptor knockout mice model [13].

Perhaps the clearest demonstration of the insulin pathway's role in carbohydrate and lipid metabolism comes from studies of Drosophila insulin pathway mutants: From the pioneering studies on Chico, a Drosophila insulin-receptor-substrate (IRS) homolog [10], it has now been demonstrated that practically all viable mutant combinations, that create partial loss-of-function or hypomorphic conditions for the insulin pathway, course with altered lipid and carbohydrate levels [29]. This is also akin to having congenital diabetes mellitus type 2, as these flies are born with a dysfunctional insulin pathway. Besides this metabolic disarray, these mutants have developmental delays, fewer and smaller cells, and concomitantly, smaller sizes, these last being cell independent [10]. In this regard, fly mutant growth phenotypes are similar to IRS1 loss-of-function mice models [30, 31] and other murine diabetic models [32] where growth phenotypes also exist; besides, in flies, the insulin and IGF (insulin growth factor) pathways are mediated by a common route. In addition, fly mutants also have nervous system abnormalities [29].

The insulin and insulin growth factor (IGF) pathways, as stated above, are united in Drosophila (see figure 1). Eight ILPs function as ligands for the sole InR in the fly genome [22, 23]. ILPs 1, 2, 3, 5 and 7 are secreted from NSC in the brain, whereas ILPs 4 and 6 are expressed in the medial intestine and the fat body. Binding of any of these ILPs to the InR causes receptor oligomerization and InR auto-phosphorylation, as the InR is a receptor tyrosine kinase. The InR itself is translated as a single polypeptide, but during its maturation within the cell and before insertion in the plasma membrane, it is proteolytically cleaved, both pieces held together by disulphur bridges. Once phosphorylated, the InRs recruit the IRS-like adaptor proteins, Chico and Lnk [10, 33]. Besides Chico and Lnk, the Drosophila InR has a long cytoplasmic tail that also acts as an IRS, in effect acting as a third IRS [26] (vertebrates have four IRS genes). InR phosphorylates the IRS proteins, like Chico, generating docking sites for other proteins in the pathway.

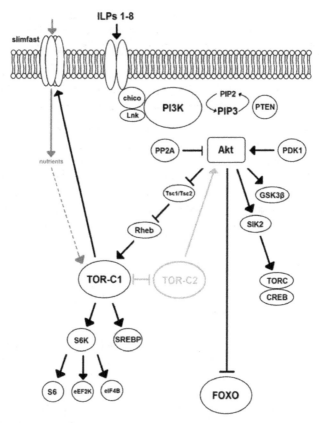

Figure 1. The insulin-signaling pathway in Drosophila. See text for details. Modified from [13].

The activated InR-IRS complex then recruits a PI3K complex (called Dp110 and Dp60, catalytic and regulatory subunits, respectively) to the membrane [14, 34]. Activated PI3K then generates phosphatidyl-inositol 3, 4, 5 trisphosphate (PIP3) from membrane phosphatidyl-inositol 4, 5 biphosphate (PIP2) and ATP. PTEN, a tumor suppressor gene, counteracts this enzymatic activity of PI3K, regulating PI3K output [35]. Plasma membrane accumulation of PIP3 then leads to membrane recruitment and activation by phosphorylation of two kinases, PDK1 and PKB (also called AKT). PKB in particular then phosphorylates and regulates several target proteins involved in metabolic control, including glycogen synthase kinase 3β (GSK-3β, or *shaggy* (*sgg*), as the gene is called in Drosophila), the transcription factor Foxo, the Tuberous sclerosis complex 2 protein, or Tsc2 (called *gigas* (*gig*) in Drosophila) and the salt-inducible kinase 2, or SIK2. PKB's activity is negatively regulated by a protein phosphatase, of class 2A (PP2A), which de-phosphorylates PKB and inactivates it. PKB can also be phosphorylated and activated by the Target of Rapamycin-complex 2 (TOR-C2) [13].

Activated *shaggy* then reduces glycogen synthesis by phosphorylation of glycogen synthase, besides participating in several other signaling networks, most notably the Wnt pathway [10, 13]. SIK2 phosphorylates and inactivates the transcriptional coactivator TORC (CRTC2 in mammals), precluding it from acting as a CREB co-activator. Foxo acts as a central catabolic regulatory point in cells; its phosphorylation by PKB enables its cytoplasmic retention, thereby inhibiting its nuclear localization, and transcriptional activity. Part of insulin's (or ILPs) anabolic effects relies on counteracting Foxo activity, which promotes energy conservation. Foxo also activates InR transcription, as part of a feedback loop.

Gigas' phosphorylation by PKB stimulates the activity of the Target of Rapamycin Complex 1 (TOR-C1) mediated by the Rheb GTPase. The Target of Rapamycin (TOR) kinase is a central anabolic kinase, regulating many aspects of general metabolism. Besides receiving input from the insulin pathway, TOR also receives input from nutrient sensors like Slimfast, an amino acid plasma membrane transporter. TOR regulates lipid, carbohydrate and autophagy levels. TOR itself forms part of two different complexes, TOR-C1 and TOR-C2. TOR-C1 is anabolic and regulates cellular growth and size, controlling cells' commitment to growth and energy utilization. Among its phosphorylation targets are the transcription factor SREBP (sterol-regulatory-element-binding-protein), and the ribosomal S6 kinase (S6K). TOR-C2 acts antagonistically to TOR-C1.

SREBP, a target of TOR-C1, regulates expression of genes involved in sterol biosynthesis. S6K, another TOR-C1 target, phosphorylates and regulates several proteins, chief of which is the ribosomal protein S6, besides the initiation and elongation factors eIF4B and eEF2K. S6K promotes growth and protein synthesis [13]. In summary, Drosophila and vertebrates have great similarities in the insulin pathway, both at the structural and functional levels.

In vertebrates, glucose homeostasis also depends on another hormone that has opposite effects from those of insulin: Glucagon. Both are secreted from the pancreas. Insulin facilitates glucose translocation from the general circulation to the cell interior via plasma membrane translocation of Glut4 glucose transporters (acting as an anabolic hormone), besides activating growth. In contrast, glucagon activates glycogen hydrolysis (acting as a catabolic hormone), helping to maintain constant glucose levels in the blood. In insects, these basic homeostatic mechanisms are evolutionarily conserved, with a family of eight peptides akin to insulin acting as anabolic hormones in glucose homeostasis, as stated, and the adipokinetic peptides, a subfamily of which constitutes the so-called hypertrehalosemic hormones, playing the role of glucagon.

Despite the close homology throughout the insulin pathway and sugar metabolism, there is one clear and fundamental difference between carbohydrate metabolism of insects and vertebrates: Insects have trehalose (or α1-D-glucopyranosyl-α1-D-glucopyranoside; see figure 2). In fact, glucagon's functions are partially diverted to trehalose synthesis by means of the afore-mentioned hypertrehalosemic hormones. In having trehalose, insects are not alone: Many other species also use trehalose.

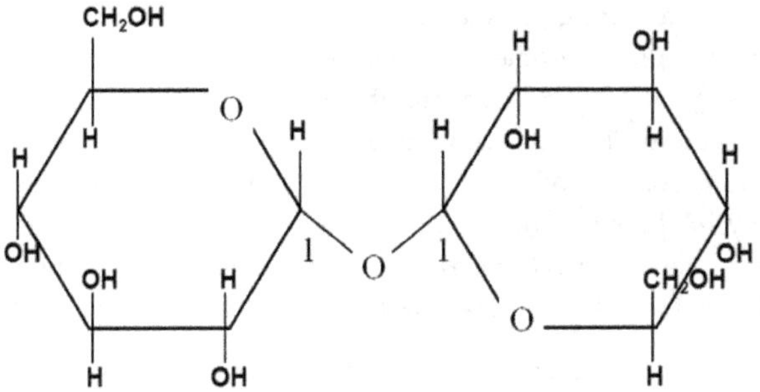

Figure 2. Structure of trehalose.

3. Why trehalose? Physicochemical properties of trehalose

Trehalose was discovered in the mid-nineteen century in rye ergot and then as a component of the cocoons of beetle species of the genus *Larinus* [36]. Many organisms use trehalose as a go-between between carbohydrate storage as glycogen and the cellular availability of glucose for energy needs (besides the other cellular roles trehalose plays in the cells' economy as an anti-stress factor): Why is this the case? Vertebrates have foregone altogether the use of trehalose, but still possess unique trehalose-degrading enzymes (trehalases; for example, human trehalase, used to digest trehalose from the diet and garner the glucose moieties within [37]). What advantages do trehalose imparts that has assured its selection in many diverse organisms (bacteria, algae, plants, fungi, invertebrates including insects), yet its loss in the vertebrate lineage? Part of the reason lies in the unique physicochemical properties of trehalose.

Trehalose is a disaccharide formed from two glucose moieties: It is synthesized by the union of two glucopyranose rings at the reducing end of the glycosyl residues. As a consequence, it is a non-reducing sugar. It demonstrates exceptional stability (with a high melting temperature, above 200°C, a high glass transition temperature, above 100°C, a slow rate of hydrolysis, and very high stability in extreme pHs; more than 99% of trehalose is still present in a pH range from 3.5 to 10 after 24 h at 100°C). It is not easily hydrolysed by acid (less so than sucrose, another non-reducing disaccharide found mainly in plant tissues), and its glycosidic bond is not cleaved by α-glycosidase, either. It also has low hygroscopicity, normally appearing as a dihydrated form. This form forms hydrogen bonds with the two water molecules, albeit at relatively short distances, making it an unusually strong interaction. In solution, all of the trehalose hydroxide groups make hydrogen bonds with water molecules. These bonds are easier to form than those with sucrose, leading to trehalose solutions with higher viscosity at equivalent concentrations, and less diffusion [36, 38].

The fact that trehalose is so resistant to acid hydrolysis may be key to understanding the use of trehalose in insect blood or hemolymph: Insect hemolymph has high amounts of peptides, free amino acids, and proteins, all of which have α amine and amino functional groups. A less resistant sugar might then react with these and be subject to condensation reactions resulting in imines and enamines, altering the structure of proteins and peptides. Also, it is less likely, from the foregoing, to cause 'browning', where the sugar's reactivity ultimately leads to the formation of pigments, and protein degradation, something sucrose is more prone to do. Sucrose circulates in plant sap without these problems mainly because these liquids have much lower levels of proteins, amino acids, and lipids. The slow rate of hydrolysis of trehalose also makes it a very stable compound, and one that is likely to aid in stabilizing other components of mixtures where trehalose is an ingredient [38].

Trehalose is highly soluble in water (up to about 50 g/100 g water at 20°C), and even if it is less water-soluble than sucrose (200 g/100g water), solubility is still high enough to allow it to accumulate to the 100 mM range in bodily fluids, like insect hemolymph. This allows trehalose to be readily transported and to accumulate in various extracellular and intracellular milieus. Also, it is a sweet compound, although, again, less so than sucrose. In solution, trehalose does not normally form intra-molecular hydrogen bonds [38].

On the other hand, trehalose also functions as a stabilizer in solutions for other organic molecules, like lipids and proteins. It seems to do so stemming from some of the physicochemical properties discussed: higher hydration capacity, higher viscosity and hydrogen bonding with proteins and other organic molecules, displacing water, than the disaccharide sucrose.

3.1. Hemolymph trehalose concentrations: Enantiostasis

Contrary to what happens with glucose in insect hemolymph (or for that matter, with glucose in vertebrates in their blood), where human glucose concentration is held constant at about 0.1% independently of nutritional status, ambient temperature, water availability, and / or oxidative stress, among other factors), trehalose levels vary normally within the hemolymph by an order of magnitude [15]. Since 1956 it was known that trehalose is a very abundant component of insect hemolymph [39], and also since then, various reports have shown that trehalose levels can vary greatly [36, 39]. As stated before, there are several reasons for this: Trehalose levels respond to a variety of environmental conditions including changes in ambient and concomitant bodily temperature, nutritional status, developmental stage, oxidative stress, salinity, etc., such that the term 'enantiostasis' was coined to describe this condition. Enantiostasis signifies that organisms' adaptations to various environmental variables require corresponding co-variance of some cellular and organismal parameters in order for the organisms to cope with the environmental challenge. In other words, trehalose variations are not only normal, but are actually adaptive, and a way of addressing changing environmental conditions. Trehalose levels can rise as much as up to 2% in hemolymph [36].

One condition that allows this compound to be used as enantiostatic is that it is non-toxic, even when in large amounts and / or high concentration. Again, the physical and chemical properties of trehalose are favorable: it is a very non-reactive sugar, much less likely than glucose or other reactive sugars to form adducts or derived metabolites. Also, a given concentration of trehalose exerts only half the osmotic effect than an equivalent concentration of glucose. Besides, as stated before, it is exceptionally stable within a large range of pH, salinity, temperature and osmotic values.

3.2. Trehalose functions

These properties lead to some of trehalose's functions in eukaryotes in general: a) as an energy source, b) as a cryoprotector, allowing hemolymph to remain liquid at normally freezing temperatures in insects, c) as a participant in a mechanism for anoxia tolerance, d) as a protein stabilizer during stress (osmotic, oxidative, thermal), and e) as a structural component for plant and bacterial cell wall polysaccharides as well as in chitin synthesis. In general, trehalose's roles can be viewed as double: on one side, as a source of energy and carbon, and on the other as a preservative and protectant against various forms of cellular and organismal stresses. These characteristics may explain in part why trehalose has been evolutionarily retained in so many organisms. Besides these roles, modern pharmacology and cosmetology makes use of trehalose in many formulations as a protein stabilizer, and for pill fabrication as a general preservative and excipient [38].

3.3. Trehalose's role as energy source

One of trehalose's most important functions, as stated before, is as a source of energy and carbon. This has been extensively reviewed in studies of metabolism in insect flight, since insect indirect flight muscles are the most energy-demanding in the animal kingdom, and are known to sustain the highest metabolic rates of all animal tissues (oxygen consumption when starting from the resting state to flight increases up to 70-fold, compared to a maximum of 20-fold increase in human muscles due to exercise). Besides, flight muscles account for about 20% of a flying insect's body mass (18% in the locust) [40, 41]. It becomes obvious that insects need "high octane" fuels to maintain these powerful motors running.

In flying insect species, flight muscles make use and oxidize different energy-rich molecules (glucose directly, glycogen, trehalose, proline, phosphoarginine and lipids) with a species-specific characteristic pattern of consumption and with time of intensive use, and according to its availability in hemolymph. The cockroach (*Periplaneta americana*) which uses glucose as its primary flying fuel [42], opts for trehalose as a high energy substrate for prolonged flights. Administration of an inhibitor of trehalase (the enzyme involved in trehalose breakdown) significantly decreases cockroach flight time [43]. Many other long distance fliers, like locusts, also use trehalose [36].

3.4. Trehalose anabolism

The ultimate source of trehalose comes from diet carbohydrates, but their use for trehalose synthesis is not necessarily direct. Under conditions where carbohydrates are abundant in the diet, hexoses are transported in the intestine by means of facilitated diffusion. A Drosophila Glut1 homolog was cloned and sequenced, with a 68% homology to human Glut1 [19]. Glut1 transporters are not specific for glucose, being able to transport other carbohydrates, like mannose, galactose and glucosamine. Insects with protein-rich diets utilize amino acids to synthesize *de novo* glucose, amino acids being also an indirect source for trehalose synthesis. When not required for energy, glucose from the diet can be stored as glycogen in the fat body. Glycogen is generally a more proximate source of glucose for trehalose synthesis.

Trehalose is synthesized in a condensation reaction of two glucose molecules, generally derived from glycogen (in effect, one of the constituent glucose substrates is in the form of UDP-glucose, derived from glucose-1P, whereas the other glucose substrate is glucose-6P). Since glucose is at the crux of trehalose anabolism, glucose regulation in the fat body impinges directly on trehalose regulation. Both vertebrates and invertebrates, in general, have very similar glucose regulatory mechanisms: insulin signaling and glucagon / hypertrehalosemic (HTH) signaling peptides, as stated above. Hypertrehalosemic hormones in particular are a family of adipokinetic (AKH) neuropeptides whose function is to derive glucose into trehalose synthesis in the fat body.

3.4.1. Insect fat body

Hemolymph circulating trehalose is synthesized in insects' fat body, a very important and extended organ analogous to both vertebrate liver and adipose tissue in insects [44]. Fat body is distributed throughout the insect body. It is also the principal organ for hemolymph circulating proteins and metabolites syntheses. It performs key roles in energy metabolism, as it both stores lipids and carbohydrates (these last in the form of glycogen), and responds to energy demands liberating both glucose and trehalose to the hemolymph. A *Drosophila melanogaster* fat body transcriptome analysis has shown that over 2200 genes are expressed in fat body cells, 290 of which attain high levels of expression. Among these are genes required for the insulin pathway, and lipid regulators. Many of the genes identified remain uncharacterized, and so, more work is required to define in greater detail the fat body transcriptome [45].

Several cell types constitute the fat body, the main one being the adipocyte (see figure 3). The adipocyte is a very dynamic cell type, with an active metabolism, regulating organismal carbohydrates and lipids, and also being an important endocrine cell [44].

The fat body is the main insect organ storing energy reserves. Most insect reserves are in the form of lipids, and these, mainly as triacylglycerols (TAGs). TAGs are stored in lipid droplets within adipocytes (figure 3). Lipid droplets have an outer coat of a monolayer of phospholipids and cholesterol molecules enclosing a core of TAGs and cholesterol molecules. In this outer coat some proteins are embedded, especially PATs (perilipin and adipocyte differentiation-related proteins). In insects two PAT proteins have been identified: Lsd1 and Lsd2 [46]. Together, they control synthesis and lipolisis of lipid stores.

Trehalose synthesis in response to energy needs normally courses through glycogen breakdown by glycogen phosphorylases in the fat body (see figure 4). Since the glucose thus obtained can also vie for glycolysis, this last path has to be inhibited. This is achieved in the fat body of cockroaches, where it has been studied, by reduction of fructose-2, 6-bisphosphate, an allosteric regulator of phosphofructokinase, the key regulatory enzyme in glycolysis [15, 47]. The adipokinetic hormones in turn, stimulate this decrease.

Figure 3. 2-D projection of a confocal stack of an abdominal adipocyte isolated from an Akt (PKB) *Drosophila melanogaster* mutant. The adipocyte was stained with Nile Red to image neutral lipids in order to evidence and quantify lipid droplets within the cell.

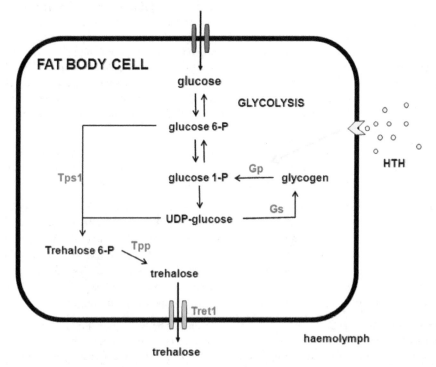

Figure 4. Trehalose synthesis in an insect fat body cell. See text for details. Abbreviations: HTH hypertrehalosemic hormones; Tret1, trehalose transporter 1; Tps1, trehalose phosphate synthase 1; Tpp, trehalose phosphate phosphatase; Gs, glycogen synthase; Gp, glycogen phosphorylase.

3.4.2. Adipokinetic hormones (Hypertrehalosemic hormones)

AKH were originally shown to induce lipid catabolism. In insects, as stated, one subgroup of these hormones produces an increase in circulating trehalose in the hemolymph [48]. About fifty HTH have been reported to date, and it is known that they participate in different functions. They are produced by neurosecretory cells of the *corpora cardiaca*, part of the ring gland of insects together with the prothoracic gland and the *corpus allatum*. The ring gland is located surrounding the anterior proventriculus in the thorax [49].

HTHs are synthesized from longer precursor peptides that are proteolitically processed to 8-10 amino acid long peptides. When the HTH neurosecretory cells are ablated, there is a significant drop in circulating trehalose levels in Drosophila larvae; over-expression of these HTHs leads to higher trehalose levels [48]. These hormones are secreted in response to various stimuli that would necessitate high circulating levels of trehalose, like lack of food, dehydration or thermal stress, or during flight. Some insects, like locusts, initiate flight utilizing trehalose, but switch to lipids later, whereas others, like some cockroaches, depend solely on trehalose supply. Still others use proline as an energy source, like some beetle species [44].

HTHs have membrane receptors. The Drosophila and homologous *Bombyx mori* HTH receptors have been cloned; they are G-protein-coupled-receptors (GPCRs) and are sequence-wise related to the gonadotrophin releasing hormone receptors of mammals [50]. Mutations in the Drosophila receptor show that it indeed mediates the HTH response: Mutants had glycogen and lipid accumulation in the fat body, irrespective of nutritional status, whereas rescue experiments returned glycogen and lipids to wild type levels [51].

There are scant data on the signaling pathways activated by HTHs. In cockroaches, HTH signaling increases inositol trisphosphate levels, leading to intracellular high Ca^{+2} concentrations and higher activity of protein kinase C (PKC). This produces an increase in the activity of glycogen phosphorylase, and trehalose synthesis. Inhibitors of PKC reduce trehalose production [52]. On the other hand, in locusts, AKH stimulation leads to cAMP production [53].

3.4.3. Final reactions of trehalose synthesis

Trehalose phosphate synthase catalyses condensation of glucose 6-P and UDP-glucose to generate trehalose 6-P in the fat body. Trehalose 6-P is then de-phosphorylated by the protein phosphatase coded by the *trehalose phosphate phosphatase (tpp)* gene to generate trehalose [54]. Both trehalose phosphate synthase and glycogen synthase use UDP-glucose as substrate, so besides competition for glucose between trehalose synthesis and glycolysis, the opposing trehalose phosphate synthase and glycogen synthase enzymes must be regulated too in order to shunt glucose to one or the other metabolic outcome [55]. Km values for trehalose phosphate synthase vary in different species where it has been studied (the moth *Hyalophora cecropia*, the cockroach *Periplaneta americana*, the blowfly *Phormia regina*, the sleeping chironomid *Polypedilum vanderplanki*, and the fruit fly *Drosophila melanogaster*). *trehalose phosphate synthase 1 (tps1)* overexpression in flies results in elevated trehalose levels, whereas mutations in this gene are lethal in the first larval instar. Trehalose phosphate synthases have also been studied in other organisms, like *Escherichia coli*, the yeast *Saccharomyces cerevisiae*, and plants like *Selaginella lepidophylla*, *Myrothamus spp.* and *Arabipdosis thaliana*. In *A. thaliana*, whose genome has been sequenced, there are eleven *tps* and ten *tpp* genes, suggesting that trehalose metabolism in plants is also very important [55, 56].

The aforementioned synthesis pathway is the one most commonly found; but there are, nevertheless, four other biosynthetic routes, three of them only present in prokaryotes. Some prokaryotes can use up to three different synthesis routes. The genes encoding trehalose synthase (*treS*), maltooligosyl-trehalose synthase (*treY*), maltooligosyl-trehalohydrolase (*treZ*) and trehalose glycosyltransfering synthase (*treT*) are prokaryotic, whereas trehalose phosphorylase (*treP*) is present in both [56] (see figure 5).

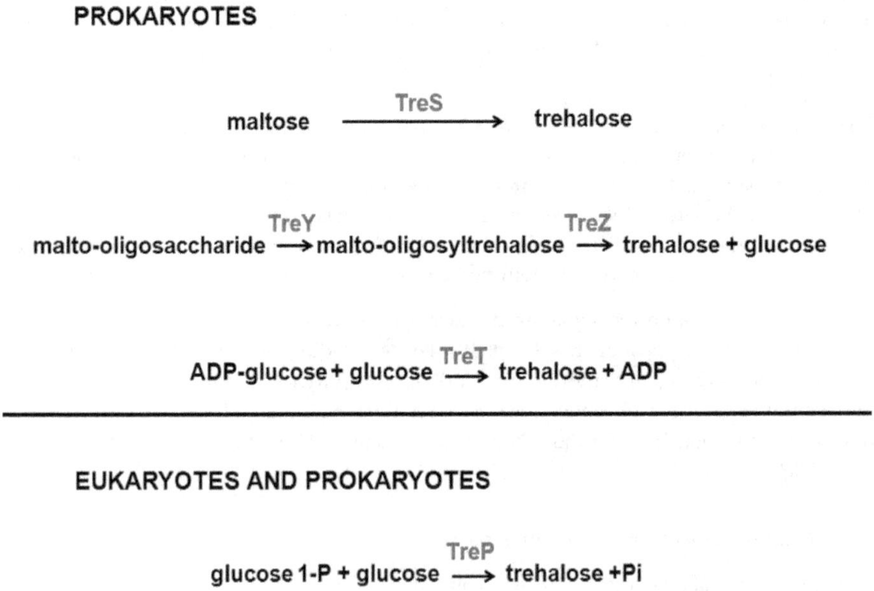

PROKARYOTES

$$\text{maltose} \xrightarrow{\text{TreS}} \text{trehalose}$$

$$\text{malto-oligosaccharide} \xrightarrow{\text{TreY}} \text{malto-oligosyltrehalose} \xrightarrow{\text{TreZ}} \text{trehalose + glucose}$$

$$\text{ADP-glucose + glucose} \xrightarrow{\text{TreT}} \text{trehalose + ADP}$$

EUKARYOTES AND PROKARYOTES

$$\text{glucose 1-P + glucose} \xrightarrow{\text{TreP}} \text{trehalose + Pi}$$

Figure 5. Alternative pathways of trehalose synthesis in prokaryotes and eukaryotes.

3.5. Trehalose in circulation

Once synthesized, trehalose is secreted to the hemolymph. A trehalose transporter cloned from several species, called Tret1, accomplishes this. It has been described in the mosquito *Anopheles gambiae*, the fruit fly *Drosophila melanogaster*, the moth *Bombyx mori* and the bee *Apis mellifera*, besides the sleeping chironomid *Polypedilum vanderplanki*. The *P. vanderplanki* trehalose transporter gets activated by dehydration conditions and that leads to trehalose accumulation such that up to 20% of total body weight can be composed of trehalose [57]. Besides Tardigrades, this last species is the pluricellular eukaryote that can withstand the harshest environmental conditions known, due in no small part to its trehalose accumulation strategy [58, 59].

The Tret1 transporter has twelve predicted membrane-spanning segments, similar to Glut family sugar transporters. Cloned Tret1 transporters from the aforementioned species were expressed in *Xenopus* oocytes. All cloned genes were functional, although kinetic parameters differed [60]. In *Drosophila melanogaster* there are two genes coding for trehalose transporters: Tret1-1 and Tret1-2, with Tret1-1 having several isoforms. Tret 1-2 shows no trehalose uptake and might be a pseudogene [60]. High expression has been noted in the fat body according to FlyAtlas for both genes [61]. Also, expression of both appears to be coincidental in many tissues, besides the fat body [62]. Transport of trehalose appears to be bi-directional for these transporters, such that not only secretion from the fat body cells goes through them, but also trehalose import to other cell types as well. In accordance with this, in *Bombyx mori* expression is also seen in muscle, testis and head, besides the fat body [60].

3.6. Trehalose catabolism

The enzyme responsible for trehalose breakdown is trehalase. This reaction produces two glucose molecules. The *treh* genes encode trehalase. *treh* has been cloned from several insect species, including the fruit fly *Drosophila melanogaster*. There are both membrane attached and soluble forms. In fruit flies it is expressed in many tissues, as expected. In many other insect species expression is seen in hemocytes (hemolymph cells).

There are trehalases in many other organisms, like yeast, thale cress and humans. Trehalase is regulated both at the transcriptional level (depending on nutritional status), and at the post-transcriptional level, by phosphorylation and by the presence of a trehalase inhibitor (this last in the cockroach *P. americana*) [36]. Altogether, trehalase appears to be finely regulated.

3.7. Trehalose and stress

Besides its role as an energy reserve, especially in response to high demands (flight, reproductive efforts like ovogenesis, etc.), trehalose functions as a molecule that aids organisms to cope with stresses from the environment. Organisms are frequently exposed to adverse environmental conditions like lack of water or food, lack of oxygen, oxidative stress and thermal variations, all of which conspire against the organism's survival. In response to these conditions living beings have evolved different strategies, involving the production of anti-freeze, chaperone and heat shock proteins, changes in membrane phospholipid composition, and cryoprotector metabolites such as glycerol, sorbitol, manitol and proline, besides trehalose. In extreme cases, organisms may even enter states of suspended animation, or cryptobiosis, where metabolism is stopped, or all but stopped. Depending on what causes it, cryptobiosis is called differently: anydrobiosis when induced by lack of water, cryobiosis when induced by cold temperatures, chemobiosis when induced by the presence of toxic chemicals, and anoxybiosis when induced by lack of oxygen.

Studies of trehalose have shown it to be a sort of jack-of-all-trades, employed in a variety of stress-provoking situations. This may explain its widespread occurrence and production in nature, despite its metabolic cost. Part of this is, of course, explained by the unusual physicochemical properties of trehalose.

3.7.1. Anhydrobiosis

Some organisms have evolved adaptations to cope with environments of transient very low humidity (mosses, famously the Tadigrades), where vital functions are just maintained or even stopped altogether. The chironomid *Polypedilum vanderplanki* can withstand up to 106°C while dessicated. Experiments with *P. vanderplanki* larvae show that gradual dehydration is accompanied by production and accumulation of trehalose, up to about 18% of total dry weight [59]. In another series of experiments, activity of the synthesis enzymes for trehalose, trehalose phosphate synthase and trehalose phosphate phosphatase was measured. These enzyme's activities increased during dehydration, whereas that of trehalase decreased. This was not only due to altered enzymatic activity, but also to transcriptional control, as the levels of the synthesis enzymes cDNAs increased at the beginning of dehydration, whereas that of trehalase increased only after forty eight hours since inititation of dehydration, perhaps as preparation for subsequent rehydration [63]. Other species also use trehalose as a protectant during anhydrobiosis: the crustacean *Artemia salina* and the nematode *Avelenchus avenae*. It has been suggested that this increase in trehalose is driven by higher concentrations of ions in bodily fluids, rather than stemming from hormonal action.

Fruit flies can also withstand dehydration: 60% of fruit fly larvae subjected to conditions of less than 5% of relative humidity survived after subsequent rehydration. During dehydration and rehydration trehalose levels were measured, and went from 2 to 10 µg per larvae during the eight hours the gradual dehydration procedure lasted in the experiment. This correlated with a decrease in trehalase activity and an increase in trehalose phosphate synthase activity. Conditions reversed upon rehydration [64].

3.7.2. Cryobiosis

Bacteria like *Escherichia coli* use trehalose as a cryoprotectant. Changes in temperature from 37°C to 16°C drive an increase in trehalose levels of about an order of magnitude, and even though this does not affect growth, it does favorably affect survival when cells are subsequently transferred to a medium at 4°C. Bacteria mutant for *otsA* and *otsB*, the homologs of trehalose phosphate synthase and trehalose phosphate phosphatase die faster when transferred to 4°C; this lethality was rescued when transfected with a plasmid carrying these two genes, demonstrating the reliance on trehalose synthesis for survival for survival [65].

The lepidopteran *Cydia pomonella*, whose larvae are the common apple worm pest, also use trehalose as cryoprotectant. In a study that measured trehalose levels throughout the year, and correlated those measurements with temperature changes, they observed higher levels of trehalose (up to 18.4 mg/g of fresh weight) in January, where the temperature dipped to 0.4°C, compared to August (4.8 mg/g), where the temperature reached 23°C [66].

3.7.3. Thermal shock tolerance

Yeast cells (*Saccharomyces cerevisiae*) use trehalose as protection against heat shocks. In a microarray-based study, yeast cells subjected to a 30°C to 40°C heat shock survived better if trehalose accumulated within them. Especially successful were yeast cells with forced trehalose accumulation made by over-expressing *tps1*, the gene that codes for the biosynthetic enzyme for trehalose, together with mutations on the genes *nth1*, *nth2* and *ath1*, that code for trehalose degrading enzymes. The same set of heat shock-responsive genes were induced by heat treatment in both trehalose overexpressing and not overexpressing cells, showing that these heat shock genes are not responsible for increased survival (rather, trehalose is). Hexose transport genes were induced by heat shock, and heat shocked yeast cells transport more glucose into the cell, leading to subsequent increases in trehalose [67].

3.7.4. Oxidative stress

Metabolic activity generates, as a byproduct, reactive oxygen species (ROS) like hydrogen peroxide, superoxide ions, or hydroxyl radicals, among others, that can oxidize several types of biomolecules (like protein oxidation and crosslinking, nucleic acid modification and lipid peroxidation). Due to its prevalence, cells have evolved several different mechanisms to counter these toxic effects: Enzymes like superoxide dismutase or catalase, or molecules like thioredoxin and gluthatione, all counteract ROS toxic effects. Trehalose is also one of these protective mechanisms.

Yeast cells subjected first to a 28°C to 38°C heat shock to induce trehalose accumulation (see paragraph above), and then subjected to oxidative stress with a combination of hydrogen peroxide and ferric chloride incubation, fared much better (survival close to 100%) compared to other yeast cells not subjected to a heat shock first, whose measured trehalose levels were lower. Mutations in the trehalose synthesis genes (*tps1* and *tps2*, coding for trehalose phosphate synthase and trehalose phosphate phosphatase, respectively), did not fare better than control cells without heat shock even though they were subjected to a heat shock before incubation with hydrogen peroxide / ferric chloride. Under these conditions, mutant cells survived better if trehalose was also included in the medium, proving that trehalose accumulation is key for survival under these conditions. In these experiments *tps2* induction required YAP1 (stress response element binding protein), a protein required for catalase and thioredoxin expression, suggesting an integral response to oxidative stress [68].

3.7.5. Anoxybiosis

There is a very large body of literature concerning cellular and organismal responses to oxygen deprivation: cells are generally very sensitive to decreases of oxygen in the medium, especially nervous tissue. Yet some cells and organisms have trehalose-based anoxia tolerance mechanisms. Drosophila kept in a 100% nitrogen atmosphere for five minutes constitutively over-expressing *tps1* (trehalose phosphate synthase) compared to flies not over-expressing *tps1* fared better after the anoxic treatment. *tps1* over-expression lead, of course, to higher trehalose levels. Loss-of-function mutations in *tps1*, conversely, stimulated anoxia sensibility [55].

There are two basic ideas that try to explain in general why trehalose is so helpful in these stressful environments, and they both rely on the lack of reactivity of trehalose and its capacity to form hydrogen bonds with other biomolecules and / or its vitrification at high temperature. Trehalose hydrogen bonding with other biomolecules could substitute for water, preventing or reducing protein aggregation and denaturation, while not otherwise chemically reacting with them. Likewise, trehalose vitrification could lead to formation of trehalose "crystals" that would allow for stabilization of other biomolecules.

3.8. Trehalose in chitin synthesis

In insects, trehalose is also used as a chitin precursor. Chitin is an N-acetyl-β-D-glucosamine polymer. Chitin synthesis is not perfectly understood in detail but is known that it initiates with trehalose being metabolized to glucose in cells, and these glucose molecules, in turn, metabolized to glucosamine sugars, which are then used in chitin synthesis. The biosynthetic pathway is complex, involving at least eight different reactions [69].

4. Conclusion

Why has trehalose become an evolutionary 'winner' molecule? And this, despite the energetic costs associated with its reasonably convoluted metabolism, as a central go-between between the diet and other storage carbohydrates. The reasons seem to stem from the fact that trehalose is not only a high energy storage molecule, with low reactivity and high stability, but also, and precisely because of this stability and lack of reactivity, because it can stabilize and preserve other biomolecules, like proteins and lipids subjected to stressful environments, besides environments; besides, it can be transported it can be transported and accumulated to high concentrations without toxic effects.

Its prevalence in many different organisms of varied evolutionary origins and relatedness strongly suggest that trehalose has been retained in most lineages, rather than introduced, and that it is only in the vertebrate lineage where it is conspicuously missing. Why this is so in vertebrates, is difficult to explain.

Nowadays, trehalose has been subject to a revival in interest, since many of the same characteristics of non reactivity, stability and no toxicity that have propelled it to such a prominent role in many species in the natural world are now being newly appreciated by the pharmaceutical and cosmetic industries. Trehalose is being featured more and more in the excipients of drugs and pills, in creams, and as a sugar substitute in processed food formulas, as it is an energetic, non-toxic, stable meal ingredient, and has a sugary taste. Especially in Japan, where a procedure for industrial synthesis of trehalose was invented and is now being commercially applied, the allure of trehalose is proving to be irresistible, with promises in the near future to likewise take on the world as a whole [38].

Author details

Alejandro Reyes-DelaTorre, María Teresa Peña-Rangel and Juan Rafael Riesgo-Escovar*
Developmental Neurobiology and Neurophysiology Dept., Instituto de Neurobiología, Universidad Nacional Autónoma de México Querétaro, México

Acknowledgement

The corresponding author's laboratory is funded by PAPIIT # IN203110 and CONACYT # 81864. A. R. D. is funded by CONACYT scholarship # 369737.

5. References

[1] Gehring WJ (1998) Master control genes in development and evolution : the homeobox story. New Haven: Yale University Press. xv, 236 p.

[2] Kornberg TB, Krasnow MA (2000) The Drosophila genome sequence: implications for biology and medicine. Science. 287(5461): 2218-2220.

[3] Rubin GM, Yandell MD, Wortman JR, Gabor Miklos GL, Nelson CR, Hariharan IK, et al. (2000) Comparative genomics of the eukaryotes. Science. 287(5461): 2204-2215.

[4] Griffiths AJF (2008) Introduction to genetic analysis. 9th ed. New York: W.H. Freeman and Co. xxiii, 838 p.

[5] St Johnston D (2002) The art and design of genetic screens: Drosophila melanogaster. Nat Rev Genet. 3(3): 176-188.

[6] Ashburner M (1989) Drosophila. Cold Spring Harbor, N.Y.: Cold Spring Harbor Laboratory.

[7] Adams MD, Celniker SE, Holt RA, Evans CA, Gocayne JD, Amanatides PG, et al. (2000) The genome sequence of Drosophila melanogaster. Science. 287(5461): 2185-2195.

[8] Bloomington Drosophila Stock Center at Indiana University [database on the Internet] (2012) Available: http://flystocks.bio.indiana.edu/.

[9] Tickoo S, Russell S (2002) Drosophila melanogaster as a model system for drug discovery and pathway screening. Curr Opin Pharmacol. 2(5): 555-560.

---

* Corresponding Author

[10] Bohni R, Riesgo-Escovar J, Oldham S, Brogiolo W, Stocker H, Andruss BF, et al. (1999) Autonomous control of cell and organ size by CHICO, a Drosophila homolog of vertebrate IRS1-4. Cell. 97(7): 865-875.

[11] Baker KD, Thummel CS (2007) Diabetic larvae and obese flies-emerging studies of metabolism in Drosophila. Cell Metab. 6(4): 257-266.

[12] Musselman LP, Fink JL, Narzinski K, Ramachandran PV, Hathiramani SS, Cagan RL, et al. (2011) A high-sugar diet produces obesity and insulin resistance in wild-type Drosophila. Dis Model Mech. 4(6): 842-849.

[13] Teleman AA (2010) Molecular mechanisms of metabolic regulation by insulin in Drosophila. Biochem J. 425(1): 13-26.

[14] Oldham S, Hafen E (2003) Insulin/IGF and target of rapamycin signaling: a TOR de force in growth control. Trends Cell Biol. 13(2): 79-85.

[15] Becker A, Schloder P, Steele JE, Wegener G (1996) The regulation of trehalose metabolism in insects. Experientia. 52(5): 433-439.

[16] Cheatham B, Vlahos CJ, Cheatham L, Wang L, Blenis J, Kahn CR (1994) Phosphatidylinositol 3-kinase activation is required for insulin stimulation of pp70 S6 kinase, DNA synthesis, and glucose transporter translocation. Mol Cell Biol. 14(7): 4902-4911.

[17] Huang S, Czech MP (2007) The GLUT4 glucose transporter. Cell Metab. 5(4): 237-252.

[18] Wang M, Wang C (1993) Characterization of glucose transport system in Drosophila Kc cells. FEBS Lett. 317(3): 241-244.

[19] Escher SA, Rasmuson-Lestander A (1999) The Drosophila glucose transporter gene: cDNA sequence, phylogenetic comparisons, analysis of functional sites and secondary structures. Hereditas. 130(2): 95-103.

[20] Ceddia RB, Bikopoulos GJ, Hilliker AJ, Sweeney G (2003) Insulin stimulates glucose metabolism via the pentose phosphate pathway in Drosophila Kc cells. FEBS Lett. 555(2): 307-310.

[21] Hall DJ, Grewal SS, de la Cruz AF, Edgar BA (2007) Rheb-TOR signaling promotes protein synthesis, but not glucose or amino acid import, in Drosophila. BMC Biol. 5: 10.

[22] Brogiolo W, Stocker H, Ikeya T, Rintelen F, Fernandez R, Hafen E (2001) An evolutionarily conserved function of the Drosophila insulin receptor and insulin-like peptides in growth control. Curr Biol. 11(4): 213-221.

[23] Rajan A, Perrimon N (2012) An Interleukin-6 like cytokine regulates systemic insulin signaling by conveying the 'fed' state from the fat body to the brain. 53rd Annual Drosophila Conference; Chicago, IL.

[24] Claeys I, Simonet G, Poels J, Van Loy T, Vercammen L, De Loof A, et al. (2002) Insulin-related peptides and their conserved signal transduction pathway. Peptides. 23(4): 807-816.

[25] Garofalo RS (2002) Genetic analysis of insulin signaling in Drosophila. Trends Endocrinol Metab. 13(4): 156-162.

[26] Fernandez R, Tabarini D, Azpiazu N, Frasch M, Schlessinger J (1995) The Drosophila insulin receptor homolog: a gene essential for embryonic development encodes two receptor isoforms with different signaling potential. EMBO J. 14(14): 3373-3384.

[27] Rulifson EJ, Kim SK, Nusse R (2002) Ablation of insulin-producing neurons in flies: growth and diabetic phenotypes. Science. 296(5570): 1118-1120.

[28] Broughton S, Alic N, Slack C, Bass T, Ikeya T, Vinti G, et al. (2008) Reduction of DILP2 in Drosophila triages a metabolic phenotype from lifespan revealing redundancy and compensation among DILPs. PLoS One. 3(11): e3721.

[29] Murillo-Maldonado JM, Sanchez-Chavez G, Salgado LM, Salceda R, Riesgo-Escovar JR (2011) Drosophila insulin pathway mutants affect visual physiology and brain function besides growth, lipid, and carbohydrate metabolism. Diabetes. 60(5): 1632-1636.

[30] Araki E, Lipes MA, Patti ME, Bruning JC, Haag B, 3rd, Johnson RS, et al. (1994) Alternative pathway of insulin signalling in mice with targeted disruption of the IRS-1 gene. Nature. 372(6502): 186-190.

[31] Tamemoto H, Kadowaki T, Tobe K, Yagi T, Sakura H, Hayakawa T, et al. (1994) Insulin resistance and growth retardation in mice lacking insulin receptor substrate-1. Nature. 372(6502): 182-186.

[32] Sone H, Suzuki H, Takahashi A, Yamada N (2001) Disease model: hyperinsulinemia and insulin resistance: Part A–targeted disruption of insulin signaling or glucose transport. Trends in Molecular Medicine. 7(7): 320-322.

[33] Werz C, Kohler K, Hafen E, Stocker H (2009) The Drosophila SH2B family adaptor Lnk acts in parallel to chico in the insulin signaling pathway. PLoS Genet. 5(8): e1000596.

[34] Leevers SJ, Weinkove D, MacDougall LK, Hafen E, Waterfield MD (1996) The Drosophila phosphoinositide 3-kinase Dp110 promotes cell growth. EMBO J. 15(23): 6584-6594.

[35] Huang H, Potter CJ, Tao W, Li DM, Brogiolo W, Hafen E, et al. (1999) PTEN affects cell size, cell proliferation and apoptosis during Drosophila eye development. Development. 126(23): 5365-5372.

[36] Thompson SN (2003) Trehalose – The Insect 'Blood' Sugar. Advances in Insect Physiology: Academic Press. p. 205-285.

[37] Ishihara R, Taketani S, Sasai-Takedatsu M, Kino M, Tokunaga R, Kobayashi Y (1997) Molecular cloning, sequencing and expression of cDNA encoding human trehalase. Gene. 202(1-2): 69-74.

[38] Ohtake S, Wang YJ (2011) Trehalose: current use and future applications. J Pharm Sci. 100(6): 2020-2053.

[39] Wyatt GR, Kale GF (1957) The chemistry of insect hemolymph. II. Trehalose and other carbohydrates. J Gen Physiol. 40(6): 833-847.

[40] Wegener G (1996) Flying insects: model systems in exercise physiology. Experientia. 52(5): 404-412.

[41] Candy DJ, Becker A, Wegener G (1997) Coordination and Integration of Metabolism in Insect Flight*. Comparative Biochemistry and Physiology Part B: Biochemistry and Molecular Biology. 117(4): 497-512.

[42] Chino H, Lum PY, Nagao E, Hiraoka T (1992) The molecular and metabolic essentials for long-distance flight in insects. Journal of Comparative Physiology B: Biochemical, Systemic, and Environmental Physiology. 162(2): 101-106.

[43] Kono Y, Takahashi M, Matsushita K, Nishina M, Kameda Y, Hori E (1994) Inhibition of flight in Periplaneta americana (Linn.) by a trehalase inhibitor, validoxylamine A. Journal of Insect Physiology. 40(6): 455-461.

[44] Arrese EL, Soulages JL (2010) Insect fat body: energy, metabolism, and regulation. Annu Rev Entomol. 55: 207-225.

[45] Jiang Z, Wu XL, Michal JJ, McNamara JP (2005) Pattern profiling and mapping of the fat body transcriptome in Drosophila melanogaster. Obes Res. 13(11): 1898-1904.

[46] Gronke S, Beller M, Fellert S, Ramakrishnan H, Jackle H, Kuhnlein RP (2003) Control of fat storage by a Drosophila PAT domain protein. Curr Biol. 13(7): 603-606.

[47] Becker A, Liewald JF, Stypa H, Wegener G (2001) Antagonistic effects of hypertrehalosemic neuropeptide on the activities of 6-phosphofructo-1-kinase and fructose-1,6-bisphosphatase in cockroach fat body. Insect Biochem Mol Biol. 31(4-5): 381-392.

[48] Lee G, Park JH (2004) Hemolymph sugar homeostasis and starvation-induced hyperactivity affected by genetic manipulations of the adipokinetic hormone-encoding gene in Drosophila melanogaster. Genetics. 167(1): 311-323.

[49] Demerec M (1994) The biology of Drosophila. Facsim. ed. Plainview, N.Y.: Cold Spring Harbor Laboratory Press. x, 632 p.

[50] Staubli F, Jorgensen TJ, Cazzamali G, Williamson M, Lenz C, Sondergaard L, et al. (2002) Molecular identification of the insect adipokinetic hormone receptors. Proc Natl Acad Sci U S A. 99(6): 3446-3451.

[51] Bharucha KN, Tarr P, Zipursky SL (2008) A glucagon-like endocrine pathway in Drosophila modulates both lipid and carbohydrate homeostasis. J Exp Biol. 211(Pt 19): 3103-3110.

[52] Sun D, Garcha K, Steele JE (2002) Stimulation of trehalose efflux from cockroach (Periplaneta americana) fat body by hypertrehalosemic hormone is dependent on protein kinase C and calmodulin. Arch Insect Biochem Physiol. 50(1): 41-51.

[53] Zhiwei W, Hayakawa Y, Downer RGH (1990) Factors influencing cyclic AMP and diacylglycerol levels in fat body of Locusta migratoria. Insect Biochemistry. 20(4): 325-330.

[54] Candy DJ, Kilby BA (1961) The biosynthesis of trehalose in the locust fat body. Biochem J. 78: 531-536.

[55] Chen Q, Ma E, Behar KL, Xu T, Haddad GG (2002) Role of trehalose phosphate synthase in anoxia tolerance and development in Drosophila melanogaster. J Biol Chem. 277(5): 3274-3279.

[56] Paul MJ, Primavesi LF, Jhurreea D, Zhang Y (2008) Trehalose metabolism and signaling. Annu Rev Plant Biol. 59: 417-441.

[57] Kikawada T, Saito A, Kanamori Y, Nakahara Y, Iwata K, Tanaka D, et al. (2007) Trehalose transporter 1, a facilitated and high-capacity trehalose transporter, allows exogenous trehalose uptake into cells. Proc Natl Acad Sci U S A. 104(28): 11585-11590.

[58] Hinton HE (1960) Cryptobiosis in the larva of Polypedilum vanderplanki Hint. (Chironomidae). Journal of Insect Physiology. 5(3–4): 286-300.

[59] Watanabe M, Kikawada T, Minagawa N, Yukuhiro F, Okuda T (2002) Mechanism allowing an insect to survive complete dehydration and extreme temperatures. J Exp Biol. 205(Pt 18): 2799-2802.

[60] Kanamori Y, Saito A, Hagiwara-Komoda Y, Tanaka D, Mitsumasu K, Kikuta S, et al. (2010) The trehalose transporter 1 gene sequence is conserved in insects and encodes proteins with different kinetic properties involved in trehalose import into peripheral tissues. Insect Biochem Mol Biol. 40(1): 30-37.

[61] Chintapalli VR, Wang J, Dow JAT (2007) Using FlyAtlas to identify better Drosophila melanogaster models of human disease. Nature Genetics. 39(6): 715-720.

[62] McQuilton P, St. Pierre SE, Thurmond J, Consortium tF (2011) FlyBase 101 – the basics of navigating FlyBase. Nucleic Acids Research.

[63] Mitsumasu K, Kanamori Y, Fujita M, Iwata K, Tanaka D, Kikuta S, et al. (2010) Enzymatic control of anhydrobiosis-related accumulation of trehalose in the sleeping chironomid, Polypedilum vanderplanki. FEBS J. 277(20): 4215-4228.

[64] Thorat LJ, Gaikwad SM, Nath BB (2012) Trehalose as an indicator of desiccation stress in Drosophila melanogaster larvae: A potential marker of anhydrobiosis. Biochemical and Biophysical Research Communications. 419(4): 638-642.

[65] Kandror O, DeLeon A, Goldberg AL (2002) Trehalose synthesis is induced upon exposure of Escherichia coli to cold and is essential for viability at low temperatures. Proc Natl Acad Sci U S A. 99(15): 9727-9732.

[66] Khani A, Moharramipour S, Barzegar M (2007) Cold tolerance and trehalose accumulation in overwintering larvae of the codling moth, Cydia pomonella (Lepidoptera: Tortricidae). Eur J Entomol. 104(3): 385-392.

[67] Mahmud SA, Hirasawa T, Furusawa C, Yoshikawa K, Shimizu H (2012) Understanding the mechanism of heat stress tolerance caused by high trehalose accumulation in Saccharomyces cerevisiae using DNA microarray. J Biosci Bioeng. 113(4): 526-528.

[68] Benaroudj N, Lee DH, Goldberg AL (2001) Trehalose accumulation during cellular stress protects cells and cellular proteins from damage by oxygen radicals. J Biol Chem. 276(26): 24261-24267.

[69] Merzendorfer H, Zimoch L (2003) Chitin metabolism in insects: structure, function and regulation of chitin synthases and chitinases. J Exp Biol. 206(Pt 24): 4393-4412.

Chapter 4

Adapting the Consumption of Carbohydrates for Diabetic Athletes

Anna Novials and Serafín Murillo

Additional information is available at the end of the chapter

1. Introduction

Nutritional recommendations associated with physical exercise performed by people with type 1 diabetes have changed in recent years, alongside the development of new drug therapies and discoveries in the field of nutrition. Before insulin was discovered in 1921, exercise was considered a dangerous activity, usually discouraged due to the high risk of metabolic disorder involved. Currently, exercise is not only considered safe but is prescribed as a basic treatment of the disease, essential for a healthy lifestyle and recommended for all patients with diabetes. The beneficial effects of regular physical exercise associated with diabetes are numerous, offering improved physical capacity, decreased cardiovascular risk and increased emotional and social wellbeing [1]. However, the specific role of exercise in improving glycemic control is currently under debate. The root of this debate lines in findings that show that the successful treatment of type 1 diabetes consists of creating and maintaining a balance among many different factors involved in glycemic control (kind of exercise, insulin dosage and diet). Therefore, the failure to adapt carbohydrate intake to insulin dosage or the characteristics of the physical activity to be performed may lead to significant glycemic imbalances.

In this sense, exercise-induced hypoglycemia is described as one of the main factors limiting physical exercise, as well as a cause of poor performance in athletes with type 1 diabetes. However, certain strategies, such as supplementing carbohydrate intake, can reduce the severity and duration of these episodes, thereby eliminating the fear of exercise-induced hypoglycemia [2].

The role of carbohydrates in the diet of athletes with diabetes is important as a prerequisite for achieving maximum athletic performance, not only in improving and accelerating the recovery of energy stores in the body, but also in regulating blood glucose levels during training sessions and competitions [3]. In general, the nutritional strategies used by athletes with and without diabetes should not differ. However, certain standards and guidelines must be established for including carbohydrate supplements within the set of specific treatments for diabetes.

2. Carbohydrates and glycemic control

Carbohydrates ingested in food are the main determinant of blood glucose levels following meals. Both the quantity and quality (or kind of carbohydrate) is particularly influential in the variability of blood glucose levels obtained after intake. One must bear in mind that foods rich in carbohydrates represent the basis of nutrition in most cultures, as they are important sources of vitamins, minerals and fiber [4].

Regarding the amount of carbohydrates, the minimum daily requirements are 130 g [5], with no studies available on the safety of nutritional guidelines calling for fewer carbohydrates during periods longer than one year. The quality or kind of carbohydrate ingested can cause significant variations in one's response to food, increasing or decreasing the rate at which the carbohydrates contained can be digested as well as their ability to raise blood sugar levels. Certain food-specific factors can increase the rate at which blood glucose levels rise, such as:

- *Thermal or mechanical processing.* Longer cooking times or certain mechanical processes, such as the milling of grain into flour, increase the food's absorption rate.
- *Degree of starch gelatinization.* Applying heat in the presence of water initiates the process of starch gelatinization, which facilitates the breakdown of food by intestinal enzymes during the digestive process, thereby increasing the absorption rate.
- *Amylose-amylopectin ratio of starches.* Starches are mainly made up of amylose and amylopectin chains. Amylose forms helical structures, unbranched, which are less accessible to digestive enzymes than amylopectin chains.
- *Kind of sugar.* Fructose has a slower absorption rate than glucose or sucrose. Furthermore, once absorbed by the intestine, it must undergo a series of processes by the liver in order to become glucose. Therefore, fructose-rich foods cause blood sugar to increase more slowly than those that contain other kinds of sugar, such as glucose or even starch, in their composition.
- *Other food components.* The presence of high amounts of protein, fiber or fat in food can alter the rate of absorption by slowing down the digestive process. Certain condiments such as vinegar are capable of acidifying food, which slows down the digestive process and therefore the absorption of food.

Furthermore, other factors specific to the individual must be taken into account that may also affect the rate of absorption. These factors include low blood glucose levels or fasting before eating. It should be noted that low carbohydrate intake during the hours or days prior to performing moderate or intense, mid- or long-duration physical activity can cause glycogen reserves in the liver and muscles to decrease, which then lowers the glycemic effect of any food eaten.

One way to know the rate at which carbohydrates contained in a certain food increase blood sugar levels is by looking at its glycemic index. This measurement is made by comparing the blood glucose curve obtained after a quantity of food is ingested that provides 50g of carbohydrates, as compared with the blood glucose curve obtained with a reference food, usually 50g of glucose. Foods are grouped into three different categories, low, moderate and high, according to their glycemic index, with high glycemic index foods producing abrupt spikes in glycemic levels. Recently the concept of glycemic load has also surfaced, which is determined by taking into account the amount of carbohydrates contained in the portion of food consumed.

It is important to remember that measurements are a theoretical concept and that there is a wide range of factors affecting the rate at which carbohydrates are absorbed and raise glycemic levels. These factors include the kind of food, brand or country of origin, cooking methods used, and degree of food processing.

At any rate, in diabetes management it is important to coordinate insulin doses with the carbohydrate content of meals, which can be estimated using a variety of methods including carbohydrate counting, the carbohydrate exchange system, or experience-based serving size approximations.

3. Effect of exercise on blood glucose balance and the risk of hypoglycemia

The effect of exercise on blood glucose levels is determined by the interaction that takes place between metabolic and hormonal effects, as well as changes produced in muscle glucose uptake.

3.1. Energy substrates during rest and exercise

Fat deposits of adipose tissue are the major source of energy in the human body, with a reserve of between 60,000 and 150,000 Kcal. This content is much greater than the energy that carbohydrates can provide, approximately 2,000 Kcal. Most of these, about 1500 kcal, are stored as glycogen in the muscles, and the rest comes from glycogen deposits accumulated in the liver and from glucose found in the blood and extracellular fluids.

During the early stages of exercise [6-8], muscle glycogen is the main energy source for muscle contraction. Subsequently, the depletion of muscle glycogen deposits activates the lipolysis of fatty acids stored in adipose tissue. Thus, the increase of free fatty acids and glycerol in plasma constitutes an additional energy source for muscle contraction. Glucose will then be provided by hepatic glycogenolysis followed by hepatic gluconeogenesis. The substrates used by the liver to synthesize new glucose are lactate, pyruvate and certain amino acids, mainly alanine, together with the glycerol derived from the metabolism of triglycerides. The contribution of different substrates depends on the intensity, duration and kind of exercise performed. The relative proportion of each seeks to maintain three fundamental physiological aspects: 1) to preserve blood glucose balance, 2) to maintain the efficient metabolism and storage of glucose, and 3) to preserve and maintain deposits of muscle glycogen to avoid and/or delay the onset of muscular fatigue.

3.2. Hormonal response to exercise

Metabolic adjustments to exercise are possible thanks to a highly efficient system that integrates nerve impulses and hormonal response. During rest, individuals without diabetes experience what is called a basal insulin secretion, which increases in response to the rise in blood glucose that occurs after ingesting food. It is well known that insulin stimulates glucose uptake by skeletal muscle and by the liver, to later facilitate its storage in the form of glycogen.

Exercise causes adrenergic nerve stimulation, acting on pancreatic beta-cells and leading to inhibited insulin secretion [9-10]. Fortunately, this decrease in insulin levels does not affect muscle glucose uptake, as exercise triggers other mechanisms that can improve muscle glucose uptake.

Stimulation of another pancreatic hormone, glucagon, and, in particular, its close and proper interaction with insulin, are phenomena taking place during the exercise that are essential to maintaining and regulating glucose production. Decreased insulin levels at baseline exercise promote increased glucagon secretion by alpha cells in pancreatic islets. This increase is critical and acts directly on the metabolic pathways of hepatic glucose production (glycogenolysis and gluconeogenesis). Also, during exercise, counter-regulatory hormones increase, such as catecholamines, cortisol and growth hormone, promoting the balance of the aforementioned metabolic pathways and increasing lipolysis in fat cells. High-intensity exercise can activate these counterregulatory hormones in an exaggerated way, resulting in a noticeable increase in hepatic glucose production and a moderate degree of hyperglycemia when exercise is over.

3.3. Metabolic responses to exercise in diabetes

As expected, athletes with type 1 diabetes are different from their non-diabetic counterparts, as they have an insulin secretion deficiency and, consequently, their counterregulatory hormones respond differently. Patients with diabetes treated with insulin should learn to mimic their own natural insulin secretory rhythm, in response to physiological changes induced by exercise.

Any patient with diabetes and, especially, diabetic athletes should prevent problems arising from poor insulin dosage. If an athlete with diabetes starts exercise with a significant insulin deficiency, his/her response to exercise may trigger a hyperglycemic overcompensation, even ketosis, since a lack of insulin induces: 1) increased hepatic glucose production; 2) decreased peripheral glucose utilization; and 3) excessive lipolysis with an increase in free fatty acid production.

If an athlete with diabetes, under any circumstances, experiences an excess of insulin when exercising, his/her response to it can trigger hypoglycaemia, due to: 1) decreased hepatic glucose production; 2) increased peripheral glucose utilization; and 3) reduced lipolysis [11-12].

It is also important to consider the changes that occur after physical activity. Muscle glucose uptake independent of insulin is still stimulated during the hours following activity, and muscles need more quantity of glucose to replenish glycogen stores. Exercise increases insulin sensitivity, which means that the effects of insulin last longer. If an athlete with diabetes not properly increase carbohydrate intake and/or reduce insulin dosage, a significant episode of hypoglycemia may occur hours after exercise has ended [13].

3.4. Effects of insulin and exercise on muscle glucose uptake

Both insulin and muscle contraction help glucose enter the muscles, where it is oxidized and subsequently turned into energy for muscle contraction. However, the mechanisms by which these two stimuli facilitate glucose transport are not entirely understood. There is evidence suggesting that they do not act in a similar manner. For example, exercise, unlike insulin, induces an increase in muscle blood flow and glucose transport, which persists for hours after exercise has ended. Experiments with laboratory animals have demonstrated that the induction of glucose transport produced by exercise is independent of insulin, as tyrosine kinase activity of the insulin receptor is not stimulated. Instead, it appears to act through other pathways, in particular, those involving the AMP kinase (AMPK).

Molecular characterization of glucose transporter GLUT 4, which is specifically expressed in muscle and fat cells, has shed new light on elucidating the mechanisms mentioned above. It appears that the translocation of GLUT 4 [14-15] from the cytosol to the cell membrane is one of the main mechanisms of glucose transport in muscles and can be stimulated both by insulin and by exercise. The existence of another type of transporter in muscle, GLUT 1, has also been demonstrated, although its function is not entirely known. The important role of certain enzymes such as hexokinase and glycogen synthase also highlights the complex system of glucose transport into muscles. AMPK [16] emerges as an important molecule regulating multiple metabolic processes that occur in skeletal muscle, in response to exercise. It appears that muscle contractions induced by exercise increase the activity of the enzyme, which, in turn, stimulates glucose transport.

4. Factors that determine carbohydrate needs during and after physical exercise

A prediction of one's consumption of carbohydrates during and in the hours after physical activity requires a great knowledge and analysis of factors specific to each type of activity. In this sense, the personal observations and experience of each athlete are highly important, assessing the effect of each type of activity on his/her blood sugar levels. It is recommended to monitor blood glucose levels before and after exercise, or, if possible, to use a continuous glucose sensor.

In general, the characteristics of the activity to be performed must be known, as these factors affect blood glucose levels to a high degree:

-   *Type of exercise*: Forms of exercise dominated by the aerobic component, such as walking, running, swimming, skating or cycling, produce greater glucose consumption and, therefore, have a stronger hypoglycaemic effect. In contrast, forms of exercise with an elevated anaerobic component, like sprints, combat sports and working with heavy weights, can produce strong adrenergic stimulation (stimulating liver production of glucose) and thus usually have a lesser hypoglycaemic effect post-exercise. Competitive sports can also be associated with significant emotional stress (high adrenergic stimulation), which causes increased blood sugar after exercise, especially in children and adolescents.
-   *Duration*: During the first 30-60 minutes of moderate- or high-intensity exercise, muscular and hepatic glycogen becomes the primary muscle fuel. Thereafter, the glycogen stores begin to decrease, and muscles increasingly obtain energy from fatty acids and plasmatic glucose. Following this is when the most significant changes in blood glucose are observed.
-   *Intensity*: Glucose is the preferred muscle fuel for exercise performed at a moderate or high intensity, whereas low-intensity exercise uses fatty acids as energy source. Therefore, low-intensity activities such as walking may have a minimal effect on blood sugar, whereas intense activities such as running could cause a stronger and faster blood glucose lowering effect.
-   *Frequency*: Hypoglycaemic effects, especially after exercise, increase after several consecutive days of physical exercise. In this situation, since it is virtually impossible to recover of glycogen stores from one day to another, the body is less able to use the mechanisms of hepatic glycogenolysis to regulate low blood glucose.
-   *Schedule*: The timetable of insulin administration results in the presence of different blood insulin levels throughout the day, which brings about a greater tendency to develop hypoglycemia during physical activity. This effect is more likely when exercise is performed just within 2-3 hours after meals, when rapid-acting insulin acts the strongest.

It is important to keep in mind that an athlete's level of fitness also affects his/her glucose uptake. For this reason, athletes just starting a particular sport usually consume more glucose than those who are more advanced in the sport, as training helps the body to use fat as muscle fuel, preserving most glucose reserves as glycogen.

5. Recommendations for carbohydrate supplementation in diabetic athletes

Nutritional guidelines for athletes with diabetes are similar to those for all athletes, in accordance with the activity performed. Thus, the recommended total amounts of macro and micronutrients should be calculated taking into account the characteristics of each individual as well as the specific requirements of the different stages of training or competition. However, in the case of carbohydrates, in addition to overall quantitative control, the distribution of these carbohydrates throughout the day takes on even more importance, according to the exercise habits and insulin dosage followed by each athlete.

Carbohydrates represent the main nutrient for both ensuring optimal muscle work and for maintaining blood glucose levels as close to normal as possible. Since reserves of carbohydrates in the body are limited, nutritional strategies must aim at maintaining or quickly replenishing these deposits once consumed through physical exercise. In this sense, dietary guidelines that heavily restrict carbohydrates are not recommended, as they would decrease athletic performance and at the same time increase the risk of hypoglycemia.

The kind of carbohydrates recommended depends on the timing of when they are ingested in relation to when exercise is performed. Thus, it is possible to use slow-absorbing carbohydrates (low-glycemic index) in the hours prior to a workout or competition, but also fast-absorbing ones (high-glycemic index) just before or during the activity. Also, continous blood glucose measurements should be used to monitor the rate at which the body needs glucose, bearing in mind the type of food most appropriate for each situation.

On another front, studies in people without diabetes found a better recovery of glycogen stores after exercise when high glycemic index foods were taken [17]. In contrast, in athletes with type 1 diabetes, priority was given to the total amount of carbohydrates over the kind of carbohydrates ingested [18]. Further studies are needed to assess the importance of following nutritional strategies based on the use of high glycemic index foods in the post-exercise recovery of athletes with type 1 diabetes.

5.1. Dietary planning

Carbohydrate intake should be adapted to the needs of each given moment, in accordance with each phase of general conditioning or training for competitions.

5.1.1. Diet during the training phase

In daily training, an athlete's diet must maintain a high percentage of carbohydrates, in order to replenish the muscle and liver glycogen that has been spent. Conserving glycogen stores helps reduce the frequency of hypoglycemia during and in the hours following exercise. It should be noted that hepatic glycogenolysis is one of the main mechanisms involved in restoring blood glucose levels during the process of hypoglycemia and that a lack of liver glycogen stores decisively limits normal functioning.

In individuals without diabetes who train for less than 1 hour a day, the required carbohydrate intake is about 5-6g of carbohydrates / kg of body weight per day, reaching 8-10g/kg weight / day if the training exceeds 2 hours a day. Unfortunately, there are no studies evaluating the effectiveness of these nutritional strategies in athletes with type 1 diabetes, under the assumption that they should be any different for athletes with type 1 diabetes who maintain the disease under optimum control.

5.1.2. Diet during the pre-competition phase

Many athletes have adopted the strategy of carbohydrate loading during the week prior to a running or cycling competition. In general, carbohydrate loading is not recommended, especially when performed after 2-3 days of carbohydrate restriction. This strategy, used by some athletes to increase muscle glycogen stores, puts athletes with diabetes at risk, especially during the period of carbohydrate restriction, leading to even metabolic imbalance and ketosis. On the other hand, ingesting very high amounts of carbohydrates during the loading period can provoke blood sugar fluctuations that could even decrease the athlete's performance during competition. This kind of dietary requires increasing the frequency of capillary blood glucose measurements in order to maintain glycemia within normal range.

At any rate, on the days prior to a competition, a high-carbohydrate diet is recommended, near or, if possible, above the guideline of 8g/kg weight/day.

6. Nutritional composition of foods commonly used in sports

During exercise it is advisable to increase carbohydrate intake, of both low and high glycemic index values. However, glycemic levels must be considered before recommending a certain food or drink or another. Below is a description of the carbohydrate content of some of the foods and drinks most commonly used while performing physical exercise:

-   *Isotonic drinks*: These contain a sugar content of 5 to 8% (between 5 and 8g for every 100cc), the concentration that is most easily digested, allowing for a good tolerance of the drink while practicing sports. Moreover, they provide sodium, potassium or chlorine, electrolytes that help replenish lost minerals, thus, they are especially appropriate for activities lasting 1 hour or longer.
-   *Soft drinks:* This group includes cola and orange soft drinks, tonic water, and carbonated drinks containing sugar. Their sugar content is around 10%, which makes their digestion more difficult when exercise is high-intensity. Furthermore, cola soft drinks contain caffeine, a substance which increases dehydration during exercise.
-   *Energy drinks:* These drinks have a high sugar content (above 10%) and, in addition, contain substances to which anti-fatigue properties are attributed, such as 2-aminoethanesulfonic acid or ginseng. Their high sugar content and stimulating effects do not make them recommendable as a supplement during physical exercise.
-   *Fruit juices:* The distinction between natural and commercial fruit juices must be made. Natural juices have a relatively low carbohydrate content (4-6%), allowing them to be used in a way similar to isotonic drinks. This also applies to commercial juices with "no added sugar", as they contain only the sugar naturally found in fruit (fructose) without adding any other kind of sugar later during processing. In contrast, commercial fruit juices with added sugar during processing have a total sugar content of around 10% (similar to that of soft drinks), in which case consumers are advised to read nutrition labels, as manufacturers add different quantities of sugar.

Another difference between the two is that natural juices raise glycemic levels much more slowly than commercial juices, as they consist primarily of fructose.

- *Glucose tablets:* Pure glucose is the substance that raises glycemic levels most quickly. Occasionally it can produce some digestive discomfort, such as abdominal pain or diarrhea, produced when large quantities of glucose enter the intestine. Tablets are recommended to be taken little by little and always with water, in order to facilitate their absorption.
- *Glucose gels:* These consist of a mixture of glucose (or other forms of sugar) with water and fruit flavorings, forming a textured emulsion similar to that of honey or marmalade, more pleasing than that of glucose tablets or pills. However, they can also produce some digestive discomfort in certain individuals, so they are also recommended to be taken in small servings and always with abundant liquids.
- *Energy bars:* These are made from a grain or flour base, with a certain amount of added sugar or protein. Energy bars serve two different purposes during exercise: in addition to maintaining glycemic levels, they also help ward off hunger during long-duration exercise. They are very practical, as they are easy to transport and store and also easy to digest, rarely creating digestive discomfort.

In terms of their effects on glycemic levels, certain differences are found between energy bars, according to the ingredients used. Thus, grain bars increase glycemia faster than those with a higher protein content. There are also enery bars made with dried fruit as an alternative to traditional grain bars.

Food/drink	Serving	Carbohydrates (g) per serving	Carbohydrates (%)
Isotonic drinks			
Isostar®	200ml	14	7
Gatorade®	200ml	12	6
Aquarius®	200ml	12	6
Powerade®	200ml	13	6,5
Other drinks			
Soft drinks	200ml	20	10
Commercial fruit juices	200ml	24	12
Redbull	200ml	22	11,2
Foods			
Orange	Medium size, 130g	10	8
Apple	Medium size, 130g	12	9
Banana	Small size, 80g	16	20
Graham crackers	3 units, 21g	13	63
Bread	1 large slice, 30g	14	47
Glucose tablets	2 units, 10g	10	99,5
Energy bars	1 unit, 25g	15	60

Table 1. Carbohydrate content of certain foods and drinks used during exercise

7. Carbohydrates in preparing for a competition

7.1. Nutrition during the hours prior to competition

If exercise is performed right after fasting or an initial nutritional deficiency, the reserves of glycogen in the liver are reduced, which also leads to a premature decrease in glycemic levels, implicating a greater risk of hypoglycemia as well as a noticeable decrease in performance. Therefore, diabetic athletes must eat a meal rich in carbohydrates during the 3 hours prior to a training session or competition lasting 45-60 minutes or more. This meal must preferably include foods rich in low-glycemic index carbohydrates.

For short-duration exercise, pre-competition strategies are not as important, as the depletion of glycogen reserves is normally not as crucial to the overall performance of the athlete.

One strategy for controlling glycemic levels consists of measuring the levels of glucose in the blood 15-30 minutes before a competition and, depending on the levels obtained, taking a supplement of 15g of high glycemic index carbohydrates, in order to prevent hypoglycemia during exercise. This supplement is omitted if blood glucose levels are above 150-200mg/dl or higher, in accordance with the kind of exercise performed and the individual needs of each athlete. Including supplements of this nature has been found to reduce the frequency of hypoglycemia also during the hours following physical activity [19].

Before beginning any kind of exercise, the measurement of blood glucose levels is recommended. Thus, the intake of carbohydrates before physical activity begins is conditioned by the glycemic levels detected beforehand. On one hand, physical activity is not advised when levels are above 250mg/dl and ketone bodies are present, or should be performed with caution when glycemic levels are above 300mg/dl. In this case, performing exercise could represent a risk, as physical activity itself increases the production of ketone bodies, which could considerably worsen the metabolic state of the athlete.

On the other hand, glycemic levels before exercise found to be lower than 100mg/dl require the administration of high glycemic index carbohydrates in order to prevent hypoglycemia during the initial minutes of physical activity [21]. This extra supplement should provide 10 to 15 grams of carbohydrates.

7.2. Nutrition during competition

Before beginning mid- or long-duration exercise, reducing insulin doses prior to the activity may not be sufficient but also require an increased intake of carbohydrates during exercise. This is the case of mild- and long-duration exercise (more than 60-90 minutes) and, in particular, unplanned exercise sessions, reducing insulin doses beforehand is not possible.

These carbohydrate supplements should be individualized according to the previously described factors. Table 2 offers approximate target ranges that may prove useful for people with diabetes who are taking up a sport or physical activity.

Exercise intensity (%VO$_{2max}$)	<20 minutes	20-60 minutes	>60 minutes
25 / light	0-10g	10-20g	15-30g/h
50 / moderate	10-20g	20-60g	20-100g/h
75 / intense	0-30g	30-100g	30-100g/h

Table 2. Carbohydrate supplements according to intensity and duration of exercise

These recommendations must be adapted and tailored to the individual needs of each athlete and the doses of carbohydrates confirmed for each of the training sessions or competitions performed. To do so, glycemic levels should be monitored both before and after physical activity, searching for patterns particular to each athlete.

Unplanned physical exercise deserves special attention. In this situation, adjusting insulin doses may not be possible, and hypoglycemia must be prevented exclusively by an increased intake of carbohydrates. The time of day at which the activity is performed must also be considered; when under the maximum effect of a fast-acting insulin analog or hypoglycemic drug, hypoglycemia must be prevented by administering a greater amount of carbohydrates. An initial supplement of 10-30g of carbohydrates should be taken, with continued carbohydrate intake depending on the intensity and duration of the exercise to be performed (see Table 2).

Diabetic athletes, in order to prevent hypoglycemia, often consume an excess of carbohydrates both before and after exercise. In other cases, some tend to overestimate the effects of physical activity on glucose consumption, with the belief that exercise alone will reduce any excesses in their diet. Special emphasis must be placed on adjusting carbohydrate intake to the expected duration and intensity of the exercise, as in some athletes who are very accustomed to a particular kind of exercise the reduced insulin doses and increased carbohydrate intake do not meet the recommendations in clinical guides [22].

7.3. Nutrition after competition

Once the activity is over, the replenishment of glycogen reserves and increased sensitivity to insulin continue for several hours, even until the day after the training session or competition.

Athletes with type 1 diabetes therefore must know that during this period they are more prone to suffer hypoglycemia. The optimal intake of carbohydrates consists of 1,5g per kg of body weight. These carbohydrates should preferably be high glycemic index foods, to be taken during the first 30 minutes immediately following physical activity. Ideally, another dose of 1,5g of carbohydrates per kg of body weight should be taken before 120 minutes have elapsed after the physical activity has ended.

Although this amount of carbohydrates is the recommendation for replenishing glycogen reserves in the muscles and liver in all athletes, the habits of athletes without diabetes make it even more difficult for athletes with type 1 diabetes to achieve proper glycemic control. The recommended quantities may need to be accompanied by insulin in order to prevent hypoglycemia, therefore, nutritional strategies are recommended only for those diabetic athletes with significant experience in handling the disease. At any rate, the amount of carbohydrates in post-exercise supplements should be adjusted to the duration and intensity of the exercise performed, beginning very gradually and in association with a greater number of capillary glycemia measurements.

During the hours following exercise, glucose needs are increased, even 12-16 hours after physical activity has ended. This phenomenon is due in part to an increased sensitivity of muscle cells to glucose and in part to the need for replenishing glycogen reserves depleted during exercise.

As a result, this increase in glucose consumption following exercise raises the risk of suffering episodes of hypoglycemia, which must be prevented by changes in diet and in insulin doses. For this reason, once physical activity has concluded, it is advisable to check glycemic levels and evaluate the need to take food supplements containing carbohydrates, in accordance with the obtained results:

- Glycemia less than 120 mg/dl: take a supplement containing 15-20g of high glycemic index carbohydrates.
- Glycemia between 120 and 200 mg/dl: supplements are usually not required.
- Glycemia greater than 200 mg/dl: carbohydrate supplements are usually not required and the possibility of including a fast-acting insulin analog should be considered[1].

Furthermore, in order to prevent post-exercise hypoglycemia, it may be necessary to compensate for the elevated consumption of glucose by the muscles through reducing insulin doses during the hours following exercise, both in fast-acting insulin as well as in slow-acting insulin for the basal needs of those using an insulin pump. The magnitude of this decrease in insulin depends mainly on the intensity and duration of the exercise performed, keeping in mind that the hypoglycemic effect of exercise is greater during the first 60-90 minutes afterwards [22].

8. Adapting nutrition to different kinds of physical exercise

The following are examples of nutritional strategies recommended for certain kinds of exercise:

- *Short-duration, low-intensity exercise:* Exercise such as walking, riding a bike or light swimming, for a period of time less than 30 minutes, is associated with low glucose consumption, thus, its hypoglycemic effect is small. Normally adjusting insulin treatment and increasing carbohydrate intake are not necessary.

---

[1] Athletes with diabetes must keep in mind that the effects of insulin following exercise are also much greater than normal.

- *Short-duration, high-intensity exercise:* This group includes swimming races or track and field events, such as sprints and hurling (for example, 100 meter dash, shot-putting, etc.), and also combat sports like judo or taekwondo. The brief duration of the exercise means that the complete depletion of glucose is not as significant. Moreover, the elevated intensity of the exercise produces adrenergic activation, which increases the production of glucose by the liver. For these reasons, adjusting insulin doses and supplementing with carbohydrates are usually not necessary.
- *Mid-duration aerobic exercise:* Training for cycling, swimming or track events, as well as spinning, aerobics or step aerobics sessions, lasting approximately one hour, usually produce a more pronounced hypoglycemic effect. Nevertheless, since the sessions are still relatively brief, the decrease in insulin should be in the range of 10-20% prior to the activity, accompanied by an intake of 20-30g of carbohydrates, preferably fast-absorbing.
- *Long-duration exercise:* Sporting events such as cycling competitions, triathlons or marathons involve performing exercise of a moderate or high intensity for prolonged periods of time. The consumption of glucose is quite high, depending on the intensity and duration of the activity. The risk of hypoglycemia is very high, both during the activity and up to 24 hours after the activity has ended.

In this case, adjustments in treatment are much more severe. A decrease in insulin of 25-50% is recommended prior to the onset of activity, which, in the case of long-duration competitions (more than 4 hours), could reach a 75% or greater reduction. The use of fast-acting insulin following the activity should also be reduced, by about 10 to 20%. In a similar fashion, the intake of carbohydrates should increase throughout the activity, according to the intensity of exercise involved.

- *Team sports:* In sports such as soccer, basketball or volleyball, the consumption of glucose can be quite variable, involving factors such as the position of the player on the team, the conditions of the game, and even the scoreboard.

Adjustments in treatment include reducing insulin doses prior to the activity by 10-20%, along with the intake of 10-20g of carbohydrates for every 30 minutes of activity.

9. Conclusions

A proper dietary education proves to be absolutely essential for athletes with type 1 diabetes to achieve optimal results. The objective of nutritional therapy is two-fold: on one hand, to maintain glycemic levels as normal as possible, and, on the other, to achieve the maximum sporting performance possible.

For this purpose, it is necessary to understand the functioning of the energy pathways of the body, as well as its hormonal response to exercise, with the goal of being able to foresee future glycemic response to the activity performed.

Nutritional recommendations regarding the intake of carbohydrates geared toward athletes with type 1 diabetes should not be any different from those geared toward other athletes. Nevertheless, diabetic athletes must alter the way in which they incorporate carbohydrates into their normal diet, in order to produce the smallest changes possible in their glycemic levels. Thus, it is important to properly distribute these carbohydrates among the various meals of the day, keeping in mind the dosage of insulin to be used, while planning different training sessions or competitions.

Finally, dietary changes should take into account the characteristics of the physical activity, such as the kind of exercise, intensity and duration, as well as time of day at which it is performed.

Author details

Anna Novials and Serafín Murillo
Department of Endocrinology and Nutrition, Hospital Clínic de Barcelona, Barcelona, Spain
Institut d'Investigacions Biomèdiques August Pi i Sunyer (IDIBAPS), Barcelona, Spain
Spanish Biomedical Research Centre in Diabetes and Associated Metabolic Disorders (CIBERDEM), Spain

Acknowledgement

We thank K. Katte [from Spanish Biomedical Research Centre in Diabetes and Associated Metabolic Disorders (CIBERDEM)] for language consultancy.

10. References

[1] Schneider SH, Ruderman NB (1990) Exercise and NIDDM (Technical Review). Diabetes Care.13:785–789.

[2] Vrazeau AS, Rabasa-Lhoret R, Strychar&Mircescu H (2008) Barriers to physical activity among patients with type 1 diabetes. Diabetes Care.31:2108-2109.

[3] Nagi D (2006) Exercise and Sport in Diabetes. Wiley & Sons Ltd.Chichester. England.

[4] American Diabetes Association (2008) Position Statement: Nutritional recommendations and principles for individuals with diabetes mellitus. Diabetes Care. 31 (Suppl 1): S61-S78.

[5] Institute of Medicine. (2002) Dietary Reference Intakes: Energy, Carbohydrate, Fiber, Fat, Fatty Acids, Cholesterol, Protein, and Amino Acids. Washington, DC, National Academies Press.

[6] Wasserman DH (1995) Control of glucose fluxes during exercise in the postabsortive state. Annual Review of Physiology. Palo Alto, CA, Annual Reviews, p 191-195.

[7]   Zinker BA, Bracy D, Lacy DB, Jacobs J, Wasserman DH (1993) Regulation of glucose uptake and metabolism during exercise: an in vivo analysis. Diabetes.42: 956-965.

[8]   Horowitz JF, Klein S (2000) Lipid metabolism during endurance exercise. Am J Clin Nutr. 72 (Suppl): 558S-563S.

[9]   Hirsch IB, Marker JC, Smith LJ, Spina R, Parvin CA, Cryer PE (1991) Insulin and glucagon in the prevention of hypoglycemia during exercise in humans. Am J Physiol. 260: E695-704.

[10] Sigal RJ, Fisher SF, Halter JB, Vranic M, Marliss EB (1995) The roles of catecholamines in glucoregulation in intense exercise as defined by the islet cell technique. Diabetes.45: 148-156.

[11] Wasserman DH.,Zinman B (1994)   Exercise in individuals with IDDM (Technical review) Diabetes Care. 17: 924-937.

[12] Davis SN, Galassetti, Wasserman DH, Tate DH (2000) Effects of antecedent hypoglycemia on subsequent counterregulatory responses to exercise. Diabetes. 49: 73-81.

[13] Hernandez JM, Moccia T, Fluckey JD, Ulbrecht JS, Farrell PA (2000) Fluid snacks to avoid late onset post exercise hypoglicemia. Med Sci Sports Exerc. 32: 904-910.

[14] Nakar VA, Downes M, Yu T, Embler E,Wang Y et al. (2008) AMPK and PPARδ agonists are exercise mimetics. Cell.134(3): 405–415.

[15] Goodyear LJ (2008) The Exercise Pill — Too Good to Be True? New Engl J Med.359(17):1842-1844.

[16] Fujii N, Aschenbach WG, Musi N, Hirshman MF, Goodyear LJ (2008) Regulation of glucose transport by the AMP-activated protein kinase. ProcNutr Soc. 63: 205-210.

[17] Burke LM, Collier GR, Hargreaves M (1993) Muscle glycogen storage after prolonged exercise: the effect of glycaemic index of carbohydrate feedings. J Appl Physiol.75:1019-1023.

[18] Diabetes UK (2003) The implementation of nutritional advice for people with diabetes. Diab Med.20:786-807.

[19] Nathan DN, Madnek S, Delahanty L (1985) Programming pre-exercise snacks to prevent post-exercise hypoglycaemia in intensively treated insulin-dependent diabetics. Ann Intern Med. 4:483-86.

[20] American Diabetes Association. (2004) Diabetes and Exercise. Diabetes Care. 27(1):s58-s62.

[21] Grimm JJ, Ybarra J, Berné C, Muchnick S, Golay A (2004) A new table for prevention of hypoglucemia during physical activity in type 1 diabetic patients. Diabetes Metab.30:465-470.

[22] Murillo S, Brugnara L, Novials A (2010) One year follow-up in a group of half-marathon runners with type-1 diabetes treated with insulin analogues. J Sports Med Phys Fitness. 50(4):506-10.

[23] Mc Mahon SK, Ferreira LD, Ratnam N, Davey RJ, Youngs LM, Davis EA, et al. (2007) Glucose requirements to maintain euglycemia after moderate-intensity afternoon exercise in adolescents with type 1 diabetes are increased in a biphasic manner. J Clin Endocrinol Metab.92(3):963-8.

Chapter 5

Starvation Conditions Effects on Carbohydrate Metabolism of Marine Bacteria

Monia El Bour

Additional information is available at the end of the chapter

1. Introduction

In the coastal shorelines terrestrial organic materials transported by river runoff represent an important material source to the ocean and is estimated that 0.4×1015 g yr-1 organic carbon is discharged to ocean by land flows or rivers (Meybeck, 1982, He et al, 2010). This amount of riverine organic carbon is sufficient to support the entire organic carbon turnover in the ocean (Williams and Druffel, 1987) even though, little terrestrial signal by carbon isotopic ratio (δ13C) of the bulk DOC in ocean was shown, suggesting that terrestrial organic carbon undergo rapid removal and decomposition within estuarine mixing (Druffel et al, 1992; Hedges et al, 1997).

Carbohydrates are, among others, the major components of identified organic matter in ocean and account for 3% to 30% of the bulk Dissolved Organic Carbon (DOC) (Gueuen et al, 2006; Hung et al, 2003; Pakulski and Benner, 1994), whereas, in estuarine and marine surface waters they are considered the most labile fractions of bulk organic matter and may play key roles in the geochemical cycles as reported by several studies (Benner et al, 1992; Burdige and Zheng, 1998; Middelboe et al, 1995; Murrell and Hollibaugh, 2000).

In complex macro-aggregates mixtures, the carbohydrates are dominant with proteins and lipids, with evidence that in mucilage samples carbohydrates are relevant components of the organic carbon (17–45% on dry weight basis depending on the age of the aggregates) with heavy colonization by several heterotrophic bacteria and autotrophic organisms which embedded the organic matrix (Simon et al, 2002; Urbani and Sist, 2003). Thus high density of micro organisms enriches the matrix in organic and inorganic nutrients in comparison with the surrounding water and such macro flocks are important hot-spots for bacterial growth and carbon cycling (Kaltenbock and Herndl, 1992; Alldredge, 2000 and Azam and Long, 2001) and then significance of bacteria for the formation and decomposition of aggregates appears to be higher than previously estimated (Simon et al, 2002). .

2. Importance of carbohydrates availability for marine bacteria

Marine micro-organisms have important established roles which may reflect large-scale changes within inter tidal systems and heterotrophic bacteria, for example, are crucial to transformations and re-mineralization of organic carbon, nitrogen and other nutrients throughout oceans (Azam, 1998; Azam et al, 1993).

Bacterial utilization of carbohydrates depends heavily on their chemical composition (Aluwihare and Repeta, 1999; Arnosti, 2000, Zoppini et al, 2010) and slow hydrolysis may provide the pathway for the accumulation in aquatic environments (Cowie and Heges, 1984; Cowie et al, 1995; Benner et al, 1992).

Else, bacteria play an important role in regulation of the rate of organic matter mineralization, nutrient cycling, and energy transfer in aquatic environments and demineralization or the turnover of carbohydrates has been used to evaluate the efficiency of the microbial community and the liability of the organic carbon in coastal waters (Azam and Worden, 2004; Lennon, 2007).

Considering that total number of bacteria on Earth, the largest proportion of bacterial cells presumably reside in oceanic subsurface rather than in terrestrial areas (3.5×10^{30} and $0.25-2.5 \times 10^{30}$ respectively) and therefore bacterial cells are estimated to contain, in total, 350–550 Pg of carbon, up to 60–100% of the total carbon found in plants, as well as large amounts of nitrogen and phosphorous (Whitman *et al* 1998). Thus, despite their modest size as individuals, as a group these organisms not only contribute to the flow of nutrients worldwide, but may also constitute a significant proportion of the nutrients in living biomass (Horner-Devine et al, 2009).

In estuarine areas, biodegradation by heterotrophic bacteria for Organic composition changed rapidly along the estuary, showing a selected removal of carbohydrates and amino acids within the DOC pool in the upper reach and mixing zone, and an autotrophic source of Particular Carbohydrates (PCHO) in the lower estuary, which gave an insight into the DOC estuarine process (He et al, 2010).

For hydrothermal marine compartment, heterotrophic thermophiles in culture far exceeds the number of their autotrophic counterparts, and many are known to metabolize carbohydrates *via* fermentation or respiration nevertheless, the energetic associated with these high-temperature microbial processes have been all but ignored (Amend et al, 2001).

In Coastal marine regions, sediments receive organic matter from a variety of sources, and its reactivity towards anaerobic fermentation and respiration is determined by the relative content of e.g. carbohydrates, lipids and proteins (Kristensen et al, 1995; Kristensen and Holmer, 2001). Proteins and labile carbohydrates derived from algae usually constitute the main organic sources for marine sediments, but coastal areas may also receive detritus rich in structural carbohydrates (e.g. cellulose) from vascular plants (Cowie and Hedges, 1992; Leeuw and Largeau, 1993; Kristensen, 1994). Furthermore, sediments of anthropogenic point sources, such as marine fish farms, may concentrate high loads of protein and lipid rich materials of animal origin (Ackefors and Enell, 1994). Proteins and simple carbohydrates are generally labile towards anaerobic degradation (Arnosti and Repeta, 1994; Holmer and Kristensen, 1994; Arnosti and Holmer, 1999), whereas anaerobic degradation of lipids and structural carbohydrates occur at much lower rates (Boetius and Lochte, 1996; Canuel and Martens, 1996). This difference is most likely caused by differential efficiency of extra-cellular enzymes produced by anaerobic fermenting bacteria (Valdemarsen et al, 2010)

2.1. Bio mineralization of carbohydrates by marine bacteria and production of carbohydrases

Bacteria are the most abundant and most important biological component involved in the transformation and mineralization of organic matter in the biosphere (Cho et al, 1988; , Pomeroy et al, 1991). Heterotrophic bacteria contribute to the cycles of nutrients and carbon in two major ways: by the production of new bacterial biomass (secondary production) and by the re mineralization of organic carbon and nutrients. Understanding this dual character of planktonic bacteria in aquatic ecosystems is a central paradigm of contemporary microbial ecology (Billen et al, 1984; Ducklow et al, 1992 and Del Giorgio and Cole, 1998).

Further, bacteria are capable of out competing, other organisms for organic compounds at low concentrations since they possess high substrate affinities, surface-to volume ratios, and metabolic rates. Natural DOM sources are heterogeneous mixtures of compounds, and bacteria can utilize several molecules simultaneously (Bott et Kaplan, 1985).

Extracellular enzymes, term generally utilised to define enzymes located outside the cytoplasmic membrane, are important catalyst in the decomposition of particulate organic matter (POM) and dissolved organic matter (DOM) including marine snow (Cho and Azam, 1988; Smith et al, 1992) and only small molecules, around 600 Da (Weiss et al, 1991), can be transported across the bacterial membrane and thus bacteria secrete enzymes to hydrolyze high-molecular-weight organic matter. Moreover extra cellular enzymes play an important role in nutrient cycling as they may be produced in order to acquire the limiting nutrient (Hoppe, 1983; Zoppini et al, 2010).

In the aggregates, the activity of the extracellular enzymes transforms non utilizable POM in DOM, thus bringing small nutrient molecules into solution (Smith et al, 1992; Simon et al, 2002). The metabolic activity of colonizing bacteria drives sinking particulate organic matter into non sinking DOM (Azam and Long, 2001; Smith et al, 1992) fueling free-living bacteria (Kiørboe and Jackson, 2001). This bun coupled solubilisation causes rapid POMYDOM transition with important biogeochemical implications for carbon flux in the oceans (Smith et al, 1992; Simon et al, 2002).

Further, almost heterotrophic bacteria have developed several mechanisms which allow the preferred utilization of the most efficiently metabolised carbohydrates when are exposed to a mixture of carbon sources. Interestingly, similar mechanisms are used by some pathogens to control various steps of their infection process. The efficient metabolism of a carbon source might serve as signal for proper fitness. Alternatively, the presence of a specific carbon source might indicate to bacterial cells that they thrive in infection-related organs, tissues or cells and that specific virulence genes should be turned on or switched off. Frequently, virulence gene regulators are affected by changes in carbon source availability. Thus, the activity of PrfA, the major virulence regulator in *Listeria monocytogenes*, seems to be controlled by the phosphorylation state of phosphotransferase system (PTS) components. In *Vibrio cholerae* synthesis of HapR, which regulates the expression of genes required for motility, is controlled via the Crp/cAMP CCR mechanism, whereas synthesis of *Salmonella enterica* HilE, which represses genes in a pathogenicity island, is regulated by the carbohydrate responsive, PTS-controlled Mlc (Poncet et al, 2009).

2.2. Use of carbohydrates mineralization for marine bacteria identification

The metabolic diversity of bacteria is perhaps as remarkable as their taxonomic and evolutionary diversity. Although culture-based studies are limited in their ability to estimate bacterial diversity, and several studies demonstrated varieties of modes of energy conversion of bacteria and wide ranges of substrate uses and metabolic pathways.). The flexibility of bacterial metabolism is similarly best illustrated by their abilities to degrade xenobiotic compounds such as malathion (an insecticide) and 2,4,5-trichlorophenoxyacetic acid (a herbicide), which are toxic to many other organisms (Brock et al, 1987 and Horner et al, 2003).

Metabolic assays mainly based on carbohydrates degradation on either aerobic or anaerobic conditions still effectively used to identify environmental bacteria besides molecular methods. Among these methods, the most widely used commercial identification systems are the API, the Biolog systems and more recently Microgen systems (Awong-Taylor et al, 2007).

The API 20 system included profiles of approximately 40 groups and species of bacteria in addition to the *Enterobacteriaceae* group in the API data set segregated mainly on assimilation of several carbohydrates with varied number depending on the type of API test kit used. (MacDonell et al, 1982; Bertone et al, 1996).

The Biolog identification system (Biolog, Inc., Hayward, CA, USA) is a bacterial identification method that establishes identifications based on the exchange of electrons generated during respiration, leading subsequently to tetrazolium-based color changes. This system tests the ability of micro organisms to oxidize a panel of 95 different carbon sources (Truu et al, 1999).

3. Effects of starvation conditions in carbohydrates metabolism

In their natural environment especially in marine ecosystems, bacteria are frequently exposed to major changes in growth conditions. These changes can include temperature, salt concentrations, essential nutrients, oxygen supply, pressure...etc. In order to sustain these changes, bacteria have developed mechanisms which allow them to adapt to drastic environmental alterations. Mainly, response to carbon source availability seems to be of special importance, because many bacteria dispose of more than one mechanism to adapt Carbon Metabolism and Virulence to changes in carbohydrate composition. These mechanisms include induction of specific carbohydrate transport and utilization system by the presence of the corresponding carbon source and their repression when a more efficiently utilizable carbohydrate is present in addition. The latter phenomenon is called carbon catabolite repression (CCR) and is often mediated by more than one signal transduction pathway, which can include regulation of transcriptional activators or repressors, anti terminators, carbohydrate transporters or metabolic enzymes (Poncet et al, 2009).

3.1. *In vivo* starvation effect on catabolic carbohydrates profiles of *E.coli*

Escherichia coli, a fecal coliform, was found to survive for longer periods of time in non sterile natural seawater when sediment material was present than in seawater without sediments , and growth was observed to occur by one occasion (Gerba et al, 1976). This enteric bacterium was found to increase rapidly in number in autoclaved natural seawater and autoclaved sediment taken from areas receiving domestic wastes, even when the seawater had salinities as high as 34 g/kg. Else, longer survival of *E. coli* in sediment is revealed and attributed to greater content of organic matter present in sediment than the seawater.

Experimental, mainly *in vitro*, studies have been performed on the survival of different enteric bacterial species in seawater and results showed solar radiation is the most adverse factor for enteric bacteria. Its effect was found to be wavelength-dependent (most inactivation caused by UVB solar spectrum with a range of dose from 0.98 to 4 kJ/m22) and thus restricted to shallow (45 to 90 cm) or clear waters (Rozen and Belkin, 2001; Sinton et al, 1994; Villarino et al, 2003; Whitman et al, 2004). Moreover, in cases of its release from wastewater into marine environment, enteric bacteria are subject to osmotic stress with effects depending on water salinity (a high loss of viability for *E. coli* occurred for salinities between 15 and 30 g/L) (Anderson et al, 1979; Pereira and Alcantara, 1993). Most coastal waters provide low concentrations of nutrients essential to the growth and survival of enteric bacteria (Barcina et al, 1997). These stresses induce different resistance mechanisms for at least part of the cellular population (Martin et al 1998; Rozen et al 2002). As a consequence, different cellular states occur in the stressed population, including the viable but nonculturable state (VBNC). The ability of such cells to recover their growth capacity is still under debate (Arana et al, 2007).

Thus, we investigated survival and virulence of *Escherichia coli* strains (*E.coli* O126:B16 and *E.coli* O55:B5) exposed to natural conditions in brackish water by incubation in water microcosms in the Bizerte lagoon in Northern Tunisia and exposed for 12 days to natural sunlight in June (231 to 386 W/m2, 26 6 1 uC, 30 g/L) and in April (227 to 330 W/m2, 17 6 1 uC, 27 g/L) or maintained in darkness for 21 days (17 6 1 uC, 27 g/L). The results revealed sunlight as the most significant inactivating factor (decrease of 3 Ulog within 48 hours for the two strains) compared to salinity and temperature (in darkness). Survival time of the strains was prolonged as they were maintained in darkness. Local strain (*E. coli* O55:B5) showed better survival capacity (T90 5 52 hours) than *E. coli* O126:B16 (T90 5 11 h).

For both, modifications were noted only for some metabolic activities of carbohydrates hydrolysis (Table 1). Thus, *E. coli* O126:B16 culturable cells lost the ability to assimilate amygdalin, and *E. coli* O55:B5 lost ability to assimilate melibiose, amygdalin, rhamnose, and saccharose.

Although, in previous data revealed exposure increase to sunlight, particularly UV radiation, alleviates some of the negative effects of altered organic matter quality through selective photolytic degradation. Both ultraviolet radiation (UV-B, 280–320 nm and UV-A, 320–400 nm) and visible light (photo synthetically active radiation, PAR, 400–700 nm) and induce major photolytic changes to complex organic molecules and generate large quantities of readily utilizable substrates for bacterial metabolism (Lindell et al 1995, Espeland & Wetzel 2001, Tietjen & Wetzel 2003). Thus, rates of carbon and nutrient cycling are often accelerated, stimulating ecosystem productivity. Interestingly, this enhanced microbial respiration may lead to increased CO_2 production and evasion to the atmosphere, creating a potential positive feedback to atmospheric CO_2 concentrations (Wetzel 2001; Lingo et al, 2007)

Strains	ONPG	ADH	LDC	ODC	CIT	H2S	URE	TDA	IND	VP	GEL	GLU	MAN	INO	SOR	RHA	SAC	MEL	AMY	ARA	OX	CAT
Sunlight incubation (April and June) (n=2)																						
E.coli O126B16[a]	+	+	+	+	-	-	-	+	+	-	-	+	+	-	+	-	-	-	+	+	-	+
after 3 days	+	+	+	+	-	-	-	+	+	-	-	+	+	-	+	-	-	-	+	+	-	+
E.coli O55B5[a]	+	+	+	+	-	-	-	+	+	-	-	+	+	-	+	+	+	+	+	+	-	+
after 12 days	+	+	+	+	-	-	-	+	+	-	-	+	+	-	+	+	+	+	+	+	-	+
Darkness incubation (n=2)																						
E.coli O126B16[a]	+	+	+	+	-	-	-	+	+	-	-	+	+	-	+	-	-	-	+	+	-	+
after 2 days	+	+	+	+	-	-	-	+	+	-	-	+	+	-	+	-	-	-	(-)	+	-	+
after 21 days	+	+	+	+	-	-	-	+	+	-	-	+	+	-	+	-	-	-	(-)	+	-	+
E.coli O55B5[a]	+	+	+	+	-	-	-	+	+	-	-	+	+	-	+	+	+	+	+	+	-	+
after 2 days	+	+	+	+	-	-	-	+	+	-	-	+	+	-	+	+	+	(-)	(-)	+	-	+
after 16 days	+	+	+	+	-	-	-	+	+	-	-	+	+	-	+	(-)	(-)	(-)	(-)	+	-	+
after 21 days	+	+	+	+	-	-	-	+	+	-	-	+	+	-	+	(-)	(-)	(-)	(-)	+	-	+

(+) positive reaction; (-) : negative reaction; *n*: number of colony tested. ONPG: β-galactosidase, ADH: arginine dihydrolase, ODC: ornithine decarboxylase, CIT: use of citrate, URE: urease, TDA: tryptophane desaminase, IND: indole production, VP: acetoin production, GEL: gelatinase, GLU: glucose assimilation, MAN: mannitol, SOR: sorbitol, RHA: rhamnose, SAC: saccharose, MEL: melibiose, ARA: arabinose

Table 1. Characteristics of *E.coli* O55B5 and *E.coli* O126B16 before and during seawater incubation

3.2. *In vitro* starvation effect on catabolic carbohydrates profiles of *E.coli*

Depending on the nature of aquatic environmental receptor (seawater, wastewater or brackishwater), different physiological, metabolic and pathogenic modifications related to stress conditions have been reported by much studies (Gauthier *et al*, 1993; Dupray and Derrien, 1995; Baleux *et al*, 1998; Troussellier *et al*, 1998; Monfort and Baleux, 1999). And different studies showed that adaptation forms of enterobacteria to drastic conditions occurring gradually during discharging in wastewater (Dupray and Derrien, 1995), or during the transfer from a previous culture to a saline medium or to mixed wastewater/seawater medium (Munro *et al*, 1987), allowing better survival in seawater conditions. In this report, *E. coli* was chosen as a representative strain of Thermotolerant Coliform (TTC), used as fecal contamination indicator. Most of *E. coli* strains are normal intestinal flora components; however, certain pathogenic strains are important disease factors. Those pathogenic strains are classified into at least six distinct groups: enteropathogenic *E.coli* (EPEC), enterotoxigenic *E.coli* (ETEC), enteroinvasive *E.coli* (EIEC), diffuse-adhering *E.coli* (DAEC), enteroaggregative *E.coli* (EAEC), enterohemorrhagic *E.coli* (EHEC) (Schroeder *et al*, 2004). Thus, we studied the behavior of these 6 different types of *E.coli* when discharged in seawater with or without a previous incubation in wastewater microcosms. Else, we examined their metabolic, antibiotic and virulence resistance patterns after exposure to salinity gradient.

Therefore, the studies were carried out using membrane chambers with 120ml capacity, constituted by assembly of glass tubes allowing dissolved matter to penetrate, through 0.2 μm- pore- size, 25 mm-diameter filters, fixed between two Teflon joints. For each *E.coli* strains two types of microcosms were prepared: one type was inoculated by seawater and the second with wastewater.

For microcosm inoculation, the seawater used was brought from Bizerte lagoon (northern part in Tunisia of Tunis). Wastewater was brought from sewage treatment station (near Tunis town).

The results obtained (Table 2) showed most important modifications noted for the strains previously inoculated in wastewater and O126B16 *E.coli* strain lost ability of fermenting glucose and mannitol, producing indole and to use citrate, but it acquired urease and gelatinase activities. All the other pathogenic *E.coli* strains maintained their major metabolic characteristics except for the ability to use citrate and to produce indole.

In further study, we compared the behavior of *E.coli* (ATCC 14948), *Vibrio paraheamolyticus and Salmonella Typhymurium* (ATCC 17802) when discharged in seawater with or without previous incubation in wastewater microcosms and we examined mainly their metabolic, patterns after exposure to salinity gradient. The results obtained showed metabolic profiles changes both for *E.coli* and *Salmonella Typhymurium* and loss for production of several carbolases mainly glucose, mannose, rhamnase, sorbitol, sucrose, melibiose and arabinose hydrolases whereas for *Vibrio paraheamolyticus* no changes were observed.

Similar results were found previously by Ben Kahla –Nakbi et al (2007) for *V. alginolyticus* strains incubated in seawater microcosms for long period of starvation and for what the result of metabolic carbohydrates profiles tested by results of API system tests, showed no modification. These findings were explained by preparation of cells to enter into VBNC state characterized by low metabolic rate, reduction in size, morphological changes or synthesis of specific proteins. This adaptation strategy developed by bacteria in aquatic environments allows them to survive during long period of time. Modifications of cellular morphology should be related to modifications with carbohydrates metabolites contained in pariatal membranes for these species.

| | API 20E* | | | | | | | | | | | | | | | |
Strains	ONPG	ADH	ODC	CIT	URE	TDA	IND	VP	GEL	GLU	MAN	SOR	RHA	SAC	MEL	ARA
O126B16	+	+	-	+	-	+	+	-	-	+	+	+	+	+	+	+
S.W.	-	-	-	-	-	+	+	-	-	+	+	+	-	-	+	+
W.W.	-	-	-	+	+	+	-	+	+	-	-	-	-	+	-	-
ECEAgg	+	-	+	+	-	+	+	-	-	+	+	+	+	+	+	+
S.W.	+	-	+	-	-	+	-	-	-	+	+	+	+	+	+	+
W.W.	+	+	+	-	-	+	-	-	-	+	+	+	+	+	+	+
ECEI	+	-	+	+	-	+	+	-	-	+	+	+	+	-	+	+
S.W.	+	-	+	-	-	+	+	-	-	+	+	+	+	-	+	+
W.W.	+	+	+	+	-	-	-	-	-	+	+	+	-	-	-	-
ECEH	+	-	+	+	-	+	+	-	-	+	+	+	+	-	+	+
S.W.	+	-	-	-	-	+	+	+	-	+	+	+	+	-	-	+
W.W.	+	+	-	+	-	+	-	-	-	+	+	+	+	-	-	+
ECEP	+	-	+	+	-	+	+	-	-	+	+	+	+	-	+	+
S.W.	+	-	-	-	-	+	+	+	-	+	+	+	+	-	-	+
W.W.	+	-	-	+	-	+	-	-	-	+	+	+	+	-	-	+
ECET	+	-	+	+	-	+	+	-	-	+	+	+	+	+	+	+
S.W.	+	-	+	-	-	+	+	+	-	+	+	+	+	+	+	+
W.W.	+	+	+	+	-	+	-	-	-	+	+	+	-	+	-	+

*ONPG: β-galactosidase, ADH: arginine dihydrolase, ODC: ornithine decarboxylase, CIT: use of citrate, URE: urease, TDA: tryptophane desaminase, IND: indole production, VP: acetoin production GEL: gelatinase, GLU: glucose, MAN: mannitol, SOR: sorbitol, RHA: rhamnose, SAC: saccharose, MEL: melibiose, ARA: arabinose.

Table 2. Biochemical patterns. Modifications of some biochemical characters on APi 20E strips of different *E.coli* strains after survival in seawater with previous incubation in seawater (SW) or in wastewater (WW). O126B16 (reference *E.coli* strain), EaggEC (Enteroaggregative *E.coli* strain), EIEC (Enteroinvasive *E.coli* strain), EHEC (Enterohemorrhagic *E.coli* strain), EPEC (Enteropathogenic *E.coli* strain), ETEC (Entero toxigenic *E.coli* strain).

Keymer et al (2007) described similar results for *V.cholerae* and pointed that environmental parameters, measured *in situ* during sample collection, should correlated to the presence of specific dispensable genes and metabolic capabilities, including utilization of mannose, sialic acid, citrate, and chitosan oligosaccharides. Thus, gene content identified and metabolic pathways that are likely selected for in certain coastal environments and may influence *V. cholerae* population structure in aquatic environments.

Our previous data described *A.hydrophila* cultured for 30 days in marine water microcosms which mainained the ability to metabolize all the carbohydrates and continued to produce several enzymes with differences of its homologous cultured in waste water previous to incubation in marine water microcosm which failed to produce numerous enzymes and metabolize carbohydrates (El Mejri et al, 2008). Thus, for marine heterotrophic bacterial populations as *Vibrio sp* and *Aeromonas sp* halotolerant conserve major potential of their carbohydrate metabolism.

For high salinity or halotolerant bacteria, many strains that grew fermentatively on carbohydrates in the presence of air (determined by the acidification of the medium) grew especially well on glucose, totally by substrate- level phosphorylation in the absence of air. Tomlinson and Hochstein (1976) demonstrated that O2 consumption by *Halobacterium saccharovorum* in presence of glucose was 18% of the theoretical amount required for its complete oxidation it was for 83% in presence of galactose or fructose which like galactose, requires more oxygen for its oxidation and therefore was generally a poorer substrate for fermentation (Javor et al, 1984). The measurement of O2 consumption, growth rates, and fermentation products in the presence of glycerol, pyruvate, and acetate by strains that demonstrated relatively good anaerobic growth on these substrates would provide further evidence of the importance of fermentative metabolism in extreme halophiles.

In the water column, biofillm microenvironments in suspended flocks may form a stabilizing refugee that enhances the survival and propagation of pathogenic (i.e., disease-causing) bacteria entering coastal waters from terrestrial and freshwater sources. The EPS matrix offers microbial cells a tremendous potential for resiliency during periods of stress, and may enhance the overall physiological activities of bacteria as it's emphasized by Giller et al(1994) that influences small-scale microbial biofilms must be addressed in understanding larger-scale processes within inter tidal systems. Intertidal systems are a key interface of the ocean, atmosphere, and terrestrial environments, and as such, are characterized by frequent fluctuations in temperature, ion concentration, desiccation, UV-irradiation, and wave action. The relative frequency of these fluctuations poses both physical and biochemical challenges to micro organisms which inhabit this environments such estuarine areas. The characteristics and intensities of such stresses may vary substantially (Decho et al, 2000).

Therefore, in estuarine region with with salinity of 35 ppt it is been reported that the growth rate of lingo-cellulose-degrading populations (50%) is inhibited by NaCl (Liu and Boone, 1991). According to Park et al (2006) the high salinity seems to significantly reduce ecto enzyme activities. Thus, it should be possible that type of bacterial population found at particular station can be influenced by salinity changes. Also, it has been reported in most of the aquatic environments there is a significant relationship between extracellular enzyme activities, their corresponding substrates (polymers) and their hydrolysis products (monomers) (Münster et al, 1992). Thus, presence of cellulolytic and hemi-cellulolytic bacteria in one particular region and their absence in other region should indicated presence of wide-ranging organic matter in each region (Khanderparker et al, 2011). Previous research has also put forth a widely accepted concept that hydrolytic enzymes are induced by presence of polymeric substrates (Chrost, 1991; Vetter and Deming, 1999). It was observed that the genus *Bacillus* and *Vibrio* were the dominant hemicellulase and cellulase producers in both the estuaries. According to Ruger (1989) the genus *Bacillus* comprised phylogenetically and phenotypically diverse species, which are ubiquitous in terrestrial and fresh water habitats and are also widely distributed in seawater ecosystems. While the familyof *Vibrionaceae* represent the most important bacterial autochtonous groups in marine environments. Members of this family often predominate in the bacterial flora of seawater, plankton, and fish. In the West Pacific Ocean, , Vibrios accounted for nearly 80% of the bacterial population in surface seawater (Simidu et al, 1980) and Vibrio phylogenetic diversity of their culturable forms in Bulgarian hot springs as described along with their abilities to metabolize carbohydrates (Derekova et al,2008)..

For Moari et al, (2011), marine models tested showed, prokaryotic abundance in superficial marine sediments controlled by organic trophic resources, while in sub-surface sediments, prokaryotic activities and abundance were driven by environmental factors and predatory pressure, suggesting that the shift in prokaryotic community structure could be coupled to a change in life-style of microbial assemblages.

Previously, Bouvier et al, (2002), discussed the relationship between the compositional succession and changes in single-cell metabolic activitiesd in the Choptank River, and suggested that profound phylogenetic shifts were linked to cell stress, loss of activity, and death (Del Giorgio and Bouvier, 2002). Similarly both types of succession, i.e., activation/ inactivation and replacement, occured simultaneously, and understanding the relative importance of such processes in determining bacterial succession and community composition remain to be clarified considering environmental conditions that trigger bacterial succession observed within the mixing areas. .

4. Conclusion

Carbohydrates metabolism still basic to marine microbiota mainly for autochtonous communities as *Vibrio, Aeromonas* and *Bacillus* genus with high potential of degradation and considerable supply of big amounts of different carbolases and pathways which should take more attention in order to be clarified.

Increase in significant amounts of organic materials are being exchanged between the land and marine ecosystems with higher amounts of dissolved carbohydrates provided especially the polymeric substrates which represent the most abundant molecules used by marine bacteria as structural and /or storage compounds under extreme variations in environmental parametres and biological marine habitats.

The intrinseque ability of bacterial cells to enter in to viable but non culturable state (VBNC) characterized by low metabolic rate and morphological modifications (reduction in cell size and membrane alterations) should be related with carbohydrates metabolism and change observed experimentally. Therefore, investigations in carbohydrates metabolites which continue going on will certainly elucidate more molecular adaptations forms of bacteria (for mainly enteric species) in variable marine ecosystems .

Author details

Monia El Bour
National Institute of Sea Sciences and Technologies, Salammbô, Tunisia

5. References

Ackefors H. & Enell M. (1994) The release of nutrients and organic matter from aquaculture systems in Nordic countries. *Journal of Applied Ichthyology*. Vol.10, pp 225–241.Ackefors and Enell, 1994

Alldredge AL.(2000). Interstitial dissolved organic carbon (DOC) concentrations within sinking marine aggregates and their potential contribution to carbon flux. *Limnologie and Oceanography*. Vol45, pp1245–1253

Aluwihare, L.I., Repeta, D.J. & Chen, R.F., (1997). A major biopolymeric component to dissolved organic carbon in surface sea water. *Nature*. Vol387, pp166–169.

Amend J.P. & Plyasunov A.V. (2001). Carbohydrates in thermophile metabolism: Calculation of the standard molal thermodynamic properties of aqueous pentoses and hexoses at elevated temperatures and pressures. *Cosmochimica Acta*. Vol 65, pp. 3901-3917.

Anderson, I.C., Rhodes, M.W., Kator, H.I., (1979). Sublethal stress in *Escherichia coli*: a function of salinity. *Applied and Environmental Microbiology* Vol.38, pp1147-1152. Pereira and Alcantara, 1993

Arnosti C. and Repeta D. J. (1994). Oligosaccharide degradation by anaerobic marine bacteria : characterization of an experimental system to study polymer degradation in sediments. *Limnology and Oceanography*. Vol 39, pp1865–1877.

Arnosti C. and Holmer M. (1999). Carbohydrate dynamics and contributions to the carbon budget of an organic-rich coastal sediment. *Geochimistry and. Cosmochimestry*. Acta. Vol. 63, pp 393–403.

Arnosti C.(2000). Substrate specificity in polysaccharide hydrolysis: contrast between bottom water and sediments. *Limnologie and Oceanography*. Vol.45, pp1112– 9.

Azam F, & Long R. (2001). Sea snow microcosm. Nature. Vol. 8, pp414:495..

Azam, F. & Worden, A.Z., (2004). Microbes, molecules, and marine ecosystems. *Science*. Vol. 303, pp 1622-1624.

Baleux, B., Caro, A., Lesne, J., Got P., Binard, S., Delpeuch, B., 1998. Survie et maintien de la virulence de *Salmonella Typhimurium* VNC exposée simultanément à trois facteurs stressants expérimentaux. *Océanologica Acta* Vol.21, pp939-950.

Barcina I., LeBaron P., & Vives-Rego J. (1997). Survival of allochthonous bacteria in aquatic systems: A biological approach. *FEMS Microbiology and Ecology*. Vol.23, pp 1–9.

Ben Kahla-Nakbi A, Besbes A, Chaieb K, Rouabhia M, & Bakhrouf A. (2007). *Marine Environmental Research*, Vol. 64,pp 469

Benner, R., Pakulski, J.D., McCarty, M., Hedges, J.I. & Hatcher, P.G. (1992). Bulk chemical characteristics of dissolved organic matter in the ocean. *Science* Vol. 255, pp 1561-1564.

Billen G. (1984). Heterotrophic utilization and regeneration of nitrogen. In *Heterotrophic Activity in the Sea*, ed. JE Hobbie, PJleB Williams, pp. 313–55.New York: Plenum

Boetius, A., Ravenschlag, K., Schubert, C.J., Rickert, D., Widdel, F., Gieseke, A., Amann, R., Jørgensen, B.B., Witte, U. & Pfannkuche, O. (2000). A marine microbial consortium apparently mediating anaerobic oxidation of methane. *Nature*. Vol. 407, pp 623-626. .

Bott T.L.& Kaplan L.(1985). Bacterial biomass, metabolic state and activity in stream sediments: relation to environmental variables and multiple assay comparisons. *Applied and Environmental Microbiology*. Vol. 50, pp 508-522.

Bouvier T.C. & Del Giogio P.A.(2002). Compositional changes in free-living bacterial communities along a salinity gradient in two temperate estuaries. *Limnologie and Oceanography*. Vol. 2002, pp. 453-470.

Brock, T. D. (1987). The study of microorganisms *in situ*: progress and problems. *Symposium Of Society Of. General Microbiologyl* Vol.. 41, pp1–17.

Burdige, D.J. & Zheng, S.L.(1998). The biogeochemical cycling of dissolved organic nitrogen in estuarine sediments. *Limnology and Oceanography.* Vol.43, pp1796–1813.

Canuel E. A. and Martens C. S. (1996). Reactivity of recently deposited organic matter: degradation of lipid compounds near the sediment–water interface. *Geochimestry and Cosmochimestry* Acta 60, pp 1793–1806.

Cho BC & Azam F.(1988) Major role of bacteria in biogeochemical fluxes in the ocean's interior. *Nature.* Vol.332, pp441–3.

Chrost, R.J., 1991. Environmental control of the synthesis and activity of aquatic microbial ectoenzymes. In: *Microbial Enzymes in Aquatic Environments.* Springer, pp. 29-59.

Cowie G. L. and Hedges J. I. (1992) Sources and reactivities of amino-acids in a coastal marine environment. *Limnologie Oceanography.* Vol.37, pp703–724.

Cowie GL, Heges JI, Prahl FG & De Lange GJ.(1995) Elemental and major biochemical changes across an oxidation front in a relict turbidite: an oxygen effect. *Geochimestry Cosmochim Acta.* Vol.59. pp33– 46..

Del Giorgio P.A. & Cole J.J.(1998). Bacterial growth efficiency in natural aquatic systems. *Annual Review of Ecology.* Vol. 29, pp 503-541.

Decho A.W.,(2000). Microbial biofilms in intertidal systems: an overview. *Continental Shelf Research.* Vol.20, pp. 1257-1273.

Derekova, A., Mandeva, R., Kambourova, M., (2008). Phylogenetic diversity of thermophilic carbohydrate degrading Bacilli from Bulgarian hot springs. *World Journal of Microbiology and Biotechnology.* Vol. 24, pp 1697-1702..

Druffel, E.R.M., Williams, P.M., Bauer, J.E. & Ertel, J.R. (1992). Cycling of dissolved and particulate organic matter in the open ocean. *Journal of Geophysical Research.* Vol.97, pp15639–15659.

Ducklow HW & Carlson CA. (1992). Oceanic bacterial production *Advances in Microbiol Ecologie.* Vol.12, pp113–81

Dupray E., Derrien A. (1995). Influence du passage de *Salmonella* spp. Et *Escherichia coli* en eaux usées sur leur survie ulterieure en eau de mer. *Water Research.* Vol. 29, pp1005-1011.

Espeland, E. M. & Wetzel, R. G., (2001). Complexation, stabilization and UV photolysis of extracellular and surface-bound glucosidase and alkaline phosphatase: implications for biofilm microbiota. – *Microbiology and Ecology.* Vol. 42, pp 572–585.

Gauthier M.J., Munro P.M., Flatau G.N., Clément R.L. & Breittmayer V.A. (1993). Nouvelles perspectives sur l'adaptation des entérobactéries dans le milieu marin. *Marine Life,* Vol.3, pp1-18.

Gerba C.P. & McLeod J.S. (1976).Effect of Sediments on the Survival of *Escherichia coli* in Marine Waters. *Applied and Environmental Microbiology.* Vol.32, pp 114-120.

Giller, P.S., Hildrew, A.G., Ra!aelli, D.G. (Eds). (1994). Aquatic Ecology: Scale, Pattern and Processes. Blackwell Publishers, Oxford, 649 pp..

Gueuen, C., Guo, L., Wang, D., Tanaka, N., Hung, C.C. (2006). Chemical characteristics and origin of dissolved organic matter in the Yukon River. *Biochemistry* Vol. 77, pp139–155.

Holmer M. & Kristensen E. (1994) Organic-matter mineralization in an organic-rich sediment – experimental stimulation of sulfate reduction by fish food pellets. *FEMS Microbiology and Ecology.* Vol.14, pp 33–44.

Hedges, J.I., Keil, R.G., Benner, R., 1997. What happens to terrestrial organic matter in the ocean? *Organic Geochemistry.* Vol. 27, pp195–212.

Hoppe HG. (1983). Significance of exoenzymatic activities in the ecology of brackish water: measurements by means of methylumbelliferylsubstrates. Marine Ecological Progressive Series Vol.11, pp 299–308.

He B., Dai M., Zhai W., Wng L., Wang L., Chen J., Lin J., Han A. & Xu Y. (2010). Distribution, degradation and dynamics of dissolved organic carbon and its major compound classes in the Pearl River estuary, China. *Marine Chemistry.* Vol. 119, pp 52–64.

Horner-Devine M.C., Carney K.M. &. Bohannan .M.J. (2004). An ecological perspective on bacterial biodiversity. *Proceedings of the Royal Society of London.* Vol.271, 113-122

Hung, C.-C., Guo, L., Santschi, P.H., Alvarado-Quiroz, N. & Haye, J.M. (2003). Distributions of carbohydrate species in the Gulf of Mexico. Marine Chemistry. Vol. 81, pp119–135.

Javor B.J.,(1984).Growth Potential of Halophilic Bacteria Isolated from Solar Salt Environments: Carbon Sources and Salt Requirements. *Applied and Environmental Microbiology.* Vol.48, pp. 352-360.

Kaltenbock E & Herndl GJ.(1992). Ecology of amorphous aggregates (marine snow) in the northern Adriatic Sea IV Dissolved nutrients and autotrophic community associated with marine snow. *Marine Ecological Progressive Series* Vol.87, pp147–59.

Keymer D.P., Miller M.C., Schoolnik G.K, Boehm A.B. (2007). Genomic and phenotypic diversity of costal Vibrio cholerae strains is linked to environmental factors. *Applied and Environmental Microbiology.* Vol 73, pp 3705-3714.

Khanderparker R.,Verna P., Meena R.M. & Deobagkar D.D. (2011). Phylogenetic diversity of carbohydrate degrading culturable bacteria from Mandovi and Zuari estuaries, Goa, west coast of India. *Estuarine, Coastal and Shelf Science.* Vol. 95, pp 359-366.

Kiørboe T. & Jackson GA.(2001). Marine snow, organic solute plumes, and optimal chemosensory behaviour of bacteria. *Limnology Oceanography.* Vol.46, pp1309– 18.

Kristensen E. (1994). Decomposition of macroalgae, vascular plants and sediment detritus in seawater: use of step wise thermogravimetry. *Biogeochemistry* Vol.26, pp1–24.

Kristensen E., Ahmed S. I. and Devol A. H. (1995) Aerobic and anaerobic decomposition of organic matter in marine sediment: which is fastest? *Limnology and Oceanography.* Vol 40, pp1430–1437.

Kristensen E. and Holmer M. (2001) Decomposition of plant materials in marine sediment exposed to different electron acceptors (O2, NO3, and SO42), with emphasis on substrate origin, degradation kinetics, and the role of sulfate reduction. *Geochimestry and Cosmochimestry.* Acta Vol.65, pp419–433.

Lango Z., Leech D.M., Wetzel R.G., (2007). Indirect effects of elevated atmospheric CO2 and solar radiation on the growth of culturable bacteria. *Fundamental and Applied Limnology Archiv Für Hydrobiologie,* Vol. 168, pp. 327-333..

Lennon, J.T., 2007. Diversity and metabolism of marine bacteria cultivated on dissolved DNA. *Applied and Environmental Microbiology* Vol. 73, pp 2799-2805.

Leeuw J. W. D. and Largeau C. (1993). A review of macromolecular organic compounds that comprise living organisms and their role in kerogen, coal and petroleum formation. In Organic (eds. M. H. Engel and S. A. Macko). Plenum Press, New York.

Lindell, M. J., Granéli, H. W. & Tranvik, L. J., (1995). Enhanced bacterial growth in response to photochemical transformation of dissolved organic matter. *Limnologie and Oceanography*, Vol. 40, pp 195–199.

Liu, Y., Boone, D.R., 1991. Effects of salinity on methanogenic decomposition. *Bioresource Technology*. Vol. 35, pp 271-274.

Martin, Y., Troussellier, M., Bonnefont, J-L., 1998. Adaptative responses of *Escherichia coli* to marine environmental stresses: a modelling approach based on viability and dormancy concepts. *Oceanologica Acta* 21, 951-964.

Meybeck, M., (1982). Carbon, nitrogen, and phosphorus transport by world rivers. *American Journal of Science*. Vol.282, pp401–450.

Middelboe, M., Borch, N.H., Kirchman, D.L., 1995. Bacterial utilization of dissolved free amino acids, dissolved combined amino acids and ammonium in the Delaware Bay estuary: effects of carbon and nitrogen limitation. *Marine Ecology Progress Series*. Vol.28, pp109–120.

Molari M., Giovannelli D., D'Errico G. & Manini E. (2011). Factors influencing prokaryotic community structure composition in sub-surface coastal sediments. *Estuarine, Coastal and Shelf Science*. Vol. xxx, pp1-8.

Monfort P., Baleux B. (1999). Bactéries viables non cultivables: réalité et conséquences. *Bulletin de la Société Française de Microbiologie*, Vol.14, pp 201-207.

Murrell, M.C. & Hollibaugh, J.T. (2000). Distribution and composition of dissolved and particulate organic carbon in northern San Francisco Bay during low flow conditions. *Estuarine, Coastal and Shelf Science*. Vol. 51, pp75–90.

Munro P., Gauthier J., Laumond F. (1987). Changes in *Escherichia coli* cells starved in seawater or grown in seawater-wastewater mixtures. *Applied Environmental Microbiologie*, Vol.53, pp1476-1481.

Münster, U., Einiö, P., Nurminen, J. & Overbeck, J. (1992). Extracellular enzymes in a polyhumic lake: important regulators in detritus processing. *Hydrobiologia*. Vol. 229, pp. 225-238.

Pakulski, J.D. & Benner, R. (1994). Abundance and distribution of carbohydrates in the ocean. *Limnology and Oceanography*. Vol. 39, pp930–940..

Park, J.S., Choi, D.H., Hwang, G.J., & Park, B.C. (2006). Seasonal study on ectoenzyme activities, carbohydrate concentrations, prokaryotic abundance and production in a solar saltern in Korea. *Aquatic Microbial Ecology*. Vol. 43, pp 153-163.

Pereira, M.G., Alcantara, F., 1993. Culturability of *Escherichia coli* and *Streptococcus faecalis* in batch culture and "*in situ*" in estuarine water (Portugal). *Water Research* Vol.27, pp.1351-1360

Pomeroy LR, Wiebe WJ, Deibel D, Thompson RJ & Rowe GT,(1991). Bacterial responses to temperature and substrate concentration during the Newfoundland spring bloom. *Marine Ecology Progressive Series*.Vol 75, pp143–59

Poncet S., Milohanic E., Mazé A., Nait Abdallah J., Aké F., Larribe M., Deghmane A.E., Taha M.K., Dozot M., De Bolle X., Letesson. J.J & Deutscher J.(2009). Correlations between Carbon Metabolism and Virulence in Bacteria. In Contributions to Microbiology (Mattias Collins & Raymond Scush (Editors). Vol. 16, pp.88-102.

Rozen, Y. & Belkin, S., 2001. Survival of enteric bacteria in seawater. Microbiology Review 25, 513-529.

Rozen, Y., LaRossa, R.A., Templeton, L.J., Smulski, D.R. & Belkin, S. (2002). Gene expression analysis of the response by *Escherichia coli* to seawater. *Antonie van Leeuwenhoek*. Vol.81, pp15-25.

Ruger, H.J., 1989. Benthic studies of the northwest African upwelling region psychrophilic bacterial communities from areas with different upwelling intensities. *Marine Ecological Progress Series*. Vol. 57, pp. 45-52.

Simidu, U., Taga, N., Cohvell, R.R., Schwartz, J.R., (1980). Heterotrophic bacterial flora of the seawater from the Nansei Shoto (Ryukyu Retto) area. *Bulletine Japanese Society of Science and Fisheries*. Vol. 46, pp.505-510.

Simon M, Grossart HP, Schweitzer B & Ploug H.(2002).Microbial ecology of organic aggregates in aquatic ecosystems. Aquat Microb Ecol. Vol.28, pp175– 211.

Smith, D.C., Simon, M., Alldredge, A.L. & Azam, F., (1992). Intense hydrolytic enzyme activity on marine aggregates and implications for rapid particle dissolution. *Nature* Vol. 359, pp. 139-142.

Sinton, L.W., Davies-Colley, R.J., Bell, R., 1994. Inactivation of enterococci and fecal coliforms from sewage and meatworks effluents in seawater chambers, *Applied and Environmental Microbiology*. Vol60, pp2040-2048.

Tietjen, T. & Wetzel, R. G., (2003) Extracellular enzyme-clay mineral complexes: enzyme adsorption, alteration of enzym activity, and protection from photodegradation. *Aquatic Ecology*, Vol. 37, pp 331–339.

Tomlinson G. A., & Hochstein L.I. (1976). *Halobacterium saccharovorum sp*. nov., a carbohydrate-metabolizing, extremely halophilic bacterium. *Canadian Journal of Microbiologie*. Vol. 22, pp.587-591.

Troussellier, M., Got, P., Bouvy, M., M'Boup, M., Arfi, R., Lebihan, F., Monfort, P., Corbin, D. & Bernard, C., (2004). Water quality and health status of the Senegal River estuary. *Marine Pollution Bulletin*. Vol. 48, pp 852-862.

Truu J., Talpsep E., Heimrn E., Stottmeister U., Wand H., Hermarn A. (1999). Comparison system of Api 20NE and Biolog GN identification systems assessed by techniques of multivariante analyses. *Journal of Microbiological Methods*. Vol.36, pp 193-201.

Urbani, R., Sist, P., 2003. Studio di caratterizzazione chimica delle componenti polisaccaridiche e del loro ruolo nel meccanismo di formazione degli aggregati gelatinosi. In: Istituto per la Ricerca Scientifica e Tecnologica Applicata al Mare, Ministero dell'Ambiente e della Tutela del Territorio, editors. Programma di monitoraggio e studio sui processi di formazione delle mucillagini nell'Adriatico e nel Tirreno — MAT. Final Report, June 2003. pp.135–70.

Valdemarsen T., & Kristensen E.(2010). Degradation of dissolved organic monomers and short-chain fatty acids in sandy marine sediment by fermentation and sulfate reduction. *Geochimica et Cosmochimica Acta,* Vol. 74, pp1593–1605

Vetter, Y.A. & Deming, J.W. (1999). Growth rates of marine bacterial isolates on particulate organic substrates solubilized by freely released extracellular enzymes. *Microbial Ecology,* Vol. 37, pp 86-94..

Weiss MS, Abele J, Weckesser W, Welte W, Schultz E, Schulz GE. (1991) Molecular architecture and electrostatic properties of a bacterial porin. *Science.* Vol.254, pp1627– 30.

Wetzel, R. G. & Tuchman, N. C. (2005) Effects of atmospheric CO2 enrichment and sunlight on degradation of plant particulate and dissolved organic matter and microbial utilization – *Archives of. Hydrobiology,* Vol. 162, pp 287–308.

Williams PJleB. (1981). Microbial contribution to overall marine plankton metabolism: direct measurements of respiration. *Oceanologica. Acta* Vol.4, pp359–64

Williams, P.M. & Druffel, E.R.M. (1987). Radiocarbon in dissolved organic matter in the Central North Pacific Ocean. *Nature.* Vol. 330, pp246–248.

Whitman, W. B., Coleman, D. C. & Wiebe, W. J. (1998). Prokaryotes: the unseen majority. *Proceedings of the. Natural Academic of Sciences USA,* Vol. 95, pp 6578–6583.

Zoppini A., Puddu A., Fagi S., Rosati M. & Sist P. (2005). Extracellular enzyme activity and dynamics of bacterial community in mucilaginous aggregates of the northern Adriatic Sea. *Science of the Total Environment.* Vol.353, pp 270– 286

Animal and Plant

Chapter 6

Plant Responses to Sugar Starvation

Iwona Morkunas, Sławomir Borek, Magda Formela and Lech Ratajczak

Additional information is available at the end of the chapter

1. Introduction

The production of carbohydrates via photosynthesis is the most fundamental activity in plant life. Carbohydrate synthesis, transport, utilization, and storage are dynamic processes, strongly dependent on cell physiology, plant organ, environmental conditions, and developmental stage of the plant. The plant's ability to monitor and respond to the level of carbohydrates may act as a controlling mechanism, integrating the influence of environmental conditions (e.g. light, nutrients, biotic and abiotic stress factors) with internal developmental programs, controlled directly by hormones (Koch 2004, Rolland et al. 2006, Hammond and White 2008, Loreti et al. 2008, Ramon et al. 2008, Agulló-Antón et al. 2011). Studies conducted in recent years have provided an extensive body of information on the participation of carbohydrates in metabolic reactions of plant cells. In plants, sugars are essential as respiratory substrates for the generation of energy and metabolic intermediates that are then used for the synthesis of macromolecules. Binding to sugar is required for proper functioning of many proteins and lipids. Moreover, carbohydrates have important hormone-like functions as physiological signals, which cause activation or repression of many plant genes, and this in turn leads to specific metabolic effects. Sugar-signaling networks have the ability to regulate directly the expression of genes and to interact with other signaling pathways. The progress in research on molecular mechanisms of sugar sensing and signaling in plants shows that signal molecules include glucose, fructose, sucrose, and trehalose (Jang and Sheen 1994, Koch 1996, Müller et al. 1999, Rolland et al. 2002, Koch 2004, Gibson 2005, Gonzali et al. 2006, Rolland et al. 2006, Ramon et al. 2008, Rosa et al. 2009, Cho and Yoo 2011). Plants have developed effective mechanisms of perception and transduction of sugar signals. Perception of the sugar signal may take place already in the apoplast during transport across membranes or within the cell, in the cytosol. Sugar signal perception and transduction may involve cell wall invertases (CW-INV), sucrose and glucose transporters (and specific sugar receptors), and hexokinase (HXK) (Sheen et al. 1999, Smeekens 2000, Loreti et al. 2001, Rolland et al. 2002, Harrington and Bush 2003, Moore et al. 2003, Sherson et al. 2003, Koch 2004, Rolland and Sheen 2005,

Rolland et al. 2006, Ramon et al. 2008, Cho et al. 2009, Hanson and Smeekens 2009, Smeekens et al. 2010). Additionally, biochemical studies provide evidence for the involvement of a variety of protein kinases, i.e. Snf1-related kinases (SnRKs) (Rolland et al. 2006, Smeekens et al. 2010), calcium-dependent protein kinases (CDPKs), mitogen-activated protein kinases (MAPKs), and protein phosphatases, 14-3-3 proteins, Ca^{2+} ions as a second messenger, and G-proteins (Rolland et al. 2006) in sugar signal transduction (Rolland et al. 2002, Sinha et al. 2002). However, it must be stressed that HXK plays a significant role as a component of the sugar sensing machinery. Genetic analyses have revealed a central role for HXK as a conserved glucose sensor (Moore et al. 2003, Harrington and Bush 2003, Rolland and Sheen 2005, Rolland et al. 2006, Ramon et al. 2008, Cho et al. 2009, Hanson and Smeekens 2009, Smeekens et al. 2010). HXK sensing and signaling functions are probably dependent on HXK's subcellular localization, translocation, and/or interactions with downstream effectors. The HXK sugar sensor, as a cytosolic protein or associated with mitochondria or other organelles, then could activate a signaling cascade through HXK-interacting proteins (HIPs) or affect transcription directly after nuclear translocation (Rolland et al. 2002, Rolland et al. 2006, Hanson and Smeekens 2009). Various sugar signals activate many HXK-dependent and HXK-independent pathways and use different molecular mechanisms to control transcription, translation, protein stability, and enzymatic activity. It has been shown that in 7-day-old *Arabidopsis* seedlings, glucose (3%) during 6 h increased the expression of 983 genes and decreased the expression of 771 genes (Li et al. 2006). Sucrose (3%) in *Arabidopsis* cell suspension during 48 h modulated the expression of 243 genes in light and 193 genes in darkness (Nicolaï et al. 2006). Glucose-regulated transcription factors (TFs) account for 8.3% (82 factors among 978) of all glucose-regulated genes, where most of TFs showed repression (Price et al. 2004).

Genes encoding the enzymes of the phenylpropanoid biosynthesis pathway (Hara et al. 2003, Solfanelli et al. 2006, Morkunas et al. 2011) as well as sink-specific enzymes, such as sucrose synthase (Salanoubat and Belliard 1989), granule-bound starch synthase (Visser et al. 1991), and extracellular invertase (Roitsch et al. 1995), are induced by sucrose or glucose. Sugar-induced gene expression has also been detected for enzymes involved in pathogen and stress response, such as proteinase inhibitor II of potato (Johnson and Ryan 1990), chalcone synthase (Tsukaya et al. 1991), and flavonoid biosynthetic enzymes (Morkunas et al. 2011). In contrast, sugar repression of photosynthetic genes was observed. For example, genes encoding photosynthetic proteins, e.g. the small subunit of the Calvin cycle enzyme ribulose bisphosphate carboxylase/oxygenase (Rubisco) and the chlorophyll *a* binding protein, are repressed by carbohydrates (Sheen 1990, Krapp et al. 1993).

Due to their regulatory and signal function, sugars affect all phases of the plant life cycle by controlling the number of essential metabolic processes. In animals, the level of sugar in cells is strictly controlled, but in plants the level and composition of carbohydrates varies widely, depending on tissue type and environmental conditions. An excess or loss of carbohydrates or their derivatives triggers various reactions in plants and significantly affects the metabolism, growth, and development. Moreover, all abiotic and biotic stress responses are regulated, at least in part, by sugars (Koch 1996, Rolland et al. 2002). Plants are generally considered to be autotrophs but sometimes they can be heterotrophs, e.g. at some stages of development (such as seed germination, as long as the seedling grows in darkness, before it emerges above the soil surface) and in non-photosynthetic organs, such as most roots, stems, and flowers. In germinating seeds, in periods of unfavorable environmental conditions or after too deep sowing, mobilization of storage materials in cotyledons may be delayed or there may be some disturbances in distribution of carbohydrates. This may lead to a decrease in available carbohydrates in embryo axes (i.e. sugar starvation) and in a lower seed germination rate. Additionally, high carbohydrate losses can be observed in most plant species under the influence of environmental conditions, such as water or temperature, attack of pathogens or herbivores. These factors may lead to a remarkable decrease in photosynthetic rate in donor tissues (i.e. leaves that synthesize and export carbohydrates), and this reduces the supply of carbohydrates to acceptor tissues (i.e. non-photosynthetic tissues importing carbohydrates for respiration, growth, and development). Besides, in some conditions, e.g. during dormancy or leaf shedding, photosynthesis is switched off or slowed down. In such conditions, the stored carbohydrates must be used, so their reserves may be greatly diminished in non-photosynthetic tissues then. Knowledge of the response to sugar starvation and of adaptive mechanisms in plants is both fundamental and agronomically important.

2. Materials and methods of research on sugar starvation

Carbohydrate starvation has been studied in many of plant species, e.g. in common wheat (Wittenbach 1977, Wittenbach et al. 1982), maize (Saglio and Pradet 1980, Pace et al. 1990), barley (Farrar 1981), pearl millet (Baysdorfer et al. 1988), pea (Webster and van't Hof 1973, Sahulka and Lisa 1978, Webster and Henry 1987, Morkunas et al. 1999, 2000), soybean (Kerr et al. 1985, Walsh et al. 1987), sycamore (Journet et al. 1986, Dorne et al. 1987, Roby et al. 1987, Genix et al. 1990), tobacco (Moriyasu and Ohsumi 1996), lupine (Morkunas et al. 1999, Borek and Ratajczak 2002, Morkunas et. al. 2003, Borek et al. 2006, Borek and Nuc 2011, Borek et al. 2011, 2012a, 2012b), *Arabidopsis thaliana* (Rose et al. 2006). Metabolic changes caused by sugar starvation are usually analysed in cell cultures, callus tissue cultures, and in isolated plant organs cultured *in vitro*. In all types of the cultures it is easy to control nutrition of the plant material by manipulating the medium composition. Cell cultures clearly react by metabolic changes when sugar is omitted from the nutrient medium, e.g. in sycamore (*Acer pseudoplatanus*) cells (Journet et al. 1986, Gout et al. 2011), but similar sensitivity is observed also in excised root tips (Brouquisse et al. 1991, 1992). *In vitro* cultured embryo axes (isolated from seeds), cut off from the natural source of nutrients, i.e. cotyledons, are a good model for research on the effect of sugars on seedling development and metabolism (Borek et al. 2001, Borek and Ratajczak 2002, Borek et al. 2003, 2006, Borek and Nuc 2011, Borek et al. 2012a, 2012b), or even for research on the role of sugar in plant resistance to pathogens (Morkunas et al. 2005, Morkunas and Gmerek 2007, Morkunas et al. 2007, Morkunas and Bednarski 2008, Morkunas et al. 2008, 2011). The influence of sugar on leaf senescence was assessed on excised leaves and on leaves of intact plants. The effect of sugar starvation was achieved by keeping the plant material in darkness or by its spraying with photosynthetic inhibitors (Mohapatra et al. 2010). To examine the effects of sugar on senescence of various flower parts, cut flowers were treated with various sucrose concentrations in the vase solution (Azad et al. 2008, Arrom and Munné-Bosch 2012). Brouquisse et al. (1998) show that in roots (the main sink organs of maize plants) the consequences of carbon depletion induced by extended darkness are identical to those observed in the excised root tip, but some aspects of sugar starvation (e.g. effect on translocation of assimilates, links with the metabolism of nitrogen and phosphorus, interactions with plant hormones) were investigated in intact plants (Ciereszko et al. 2005). By spraying of sugar-starved plants with various sugars, it can be determined which of them are signal molecules and which of them are sugar sensors (Lothier et al. 2010).

Particularly valuable information is obtained as a result of research on various types of mutants, which sometimes supplies surprising information about the role of sugars as signal molecules (Gibson 2005, Ramon et al. 2008, Usadel et al. 2008, Hanson and Smeekens 2009).

Plant reactions to sugar deficits have been studied with the use of a wide range of research methods: electron microscopy (Borek and Ratajczak 2002, Borek et al. 2006, 2011, 2012a), confocal microscopy (Morkunas and Bednarski 2008, Morkunas et al. 2011, 2012), enzymatic activity assays and metabolite assays, e.g. by using HPLC (Morkunas et al. 2010), nuclear 1HNMR spectroscopy (Brouquisse et al. 2007, Kim et al. 2007), and ^{13}C and ^{31}P-NMR spectroscopy (Vauclare et al. 2010, Gout et al. 2011). Electron paramagnetic resonance (EPR) spectroscopy is applied to measure free radicals (Morkunas et al. 2004, Morkunas and Bednarski 2008, Morkunas et al. 2008, 2012). Novelties include various methods of molecular biology: assays of gene expression (Buchanan-Wollaston et al. 2005) and application of reporter genes (Lee et al. 2007). Moreover, sugar-starvation-induced promoters have been used to obtain useful recombinant proteins (Xu et al. 2011a, 2011b).

3. Morphological, anatomical, and ultrastructural changes under sugar starvation

Sugar starvation causes changes in growth, as shown in organ cultures *in vitro* and in cell suspensions. For example, significant differences are observed in morphological structure of embryo axes of narrowleaf lupine (*Lupinus angustifolius* L.), white lupine (*Lupinus albus* L.), yellow lupine (*Lupinus luteus* L.), Andean lupine (*Lupinus mutabilis* Sweet) and garden pea (*Pisum sativum* L.) cultured *in vitro* under sugar starvation and with 60 mM sucrose. Radicles of *Pisum* and *Lupinus* spp. cultured under sugar starvation were shorter and thicker than sugar-fed radicles. The length and fresh weight of starved isolated embryo axes was about 2-fold lower than those of sugar-fed axes (Morkunas 1997, Borek et al. 2012a). In contrast to lupine isolated embryo axes, lupine seedlings were much bigger when they were grown on medium without sucrose (Borek et al. 2012a).

Light micrographs of radicle cross sections revealed that starved radicles were thicker than sugar-fed radicles. Cortical cells of sugar-fed embryo axes were rounded, with large intercellular spaces. Many of the observed cells were dividing or have just divided. The cells lacked a central vacuole; instead, they contained very numerous small vacuoles. The major feature distinguishing cortical cells of starved lupine (as compared to sugar-fed lupine) was the presence of a large central vacuole with marginal cytoplasm.

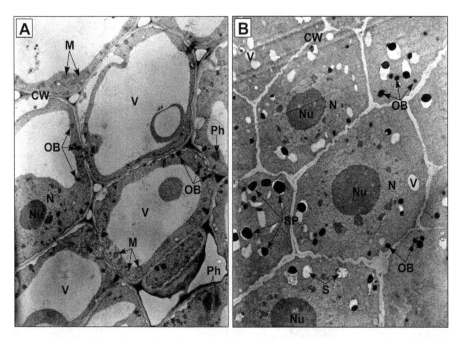

Figure 1. Ultrastructure of root meristematic zone cells of yellow lupine embryo axes grown for 96 h *in vitro* on medium without sucrose (A; sugar starvation) and on medium with 60 mM sucrose (B). CW cell wall, M mitochondrion, N nucleus, Nu nucleolus, OB oil body, Ph phytoferritin, S starch, SP storage protein, V vacuole.

Ultrastructural studies of the root meristematic zone of sucrose-starved lupine embryo axes showed that (in contrast to cells of embryo axes fed with sucrose) their cytoplasm, along with endoplasmic reticulum, mitochondria, plastids, ribosomes, and nuclei, were forced to the periphery by growing vacuoles (Fig. 1A, B). In spite of gradual autolysis of cytoplasmic proteins, in starved cells mitochondria were protected as organelles maintaining cellular respiration. Electron micrographs reveal that the inner mitochondrial membrane in starved pea cells is well developed, forming numerous cristae (Fig. 2A), in contrast to sucrose-fed cells, with less developed cristae (Fig. 2B). Oxygraphic studies of mitochondria isolated from starved embryos of *Pisum sativum* and *Lupinus angustifolius* have shown that they are active and exercise respiratory control (Morkunas et al. 2000, 2003). The phenomenon of mitochondria protection against autolysis under sugar starvation was observed also in other species and plant tissues (Baysdorfer et al. 1988, Brouquisse et al. 1991). However, Couée et al. (1992) reported that under conditions of carbohydrate starvation, mitochondria could be degraded. Those authors revealed heterogeneity of mitochondria isolated from root tips of maize (*Zea mays*) and fractionated on Percoll density gradients, i.e. higher- and lower-density mitochondria. The higher-density mitochondria from glucose-starved maize root tips retained the ultrastructure and most of the respiratory properties of non-starved mitochondria. By contrast, lower- and intermediate-density mitochondria were absent in the mitochondrial fractions from glucose-starved maize root tips and were not detected *in situ*. Interestingly, in plastids of starved pea cells, large amounts of phytoferritin were accumulated (Fig. 3A), while only starch grains and tubular structures were found in plastids of cells with a high level of sucrose (Fig. 3B) (Morkunas 1997). Lowering carbohydrate level in storage tissues of germinating seeds enhances the breakdown of reserves. In cotyledons of 4-day-old lupine seedlings and in excised lupine cotyledons grown on medium without sucrose, the deposits of storage protein were clearly smaller than in organs fed with sucrose. Starch granules and oil bodies were also smaller and less numerous (Fig. 4A, B). Similarly, the cell walls, containing hemicelluloses (important storage compounds in lupine seeds) were thinner (Borek et al. 2006, 2011, 2012a).

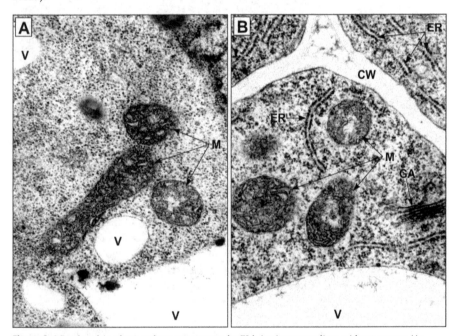

Figure 2. Mitochondria of pea embryo axes grown for 72 h *in vitro* on medium without sucrose (A; sugar starvation) and on medium with 60 mM sucrose (B). CW cell wall, ER endoplasmic reticulum, GA Golgi apparatus, M mitochondrion, V vacuole.

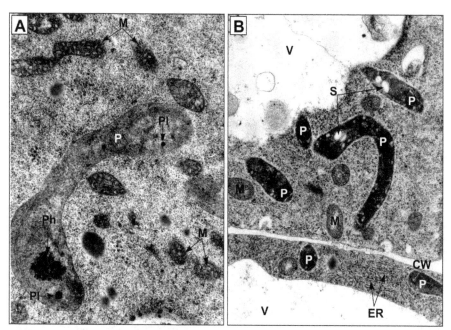

Figure 3. Plastids of pea embryo axes grown for 72 h *in vitro* on medium without sucrose (A; sugar starvation) and on medium with 60 mM sucrose (B). CW cell wall, ER endoplasmic reticulum, M mitochondrion, P plastid, Ph phytoferritin, Pl plastoglobule, S starch, V vacuole.

Research conducted on *Arabidopsis* cell suspension revealed that sugar starvation caused an immediate arrest of cell growth, together with a rapid degradation of cellular proteins. Cell divisions, as indicated by cell number and accumulation of biomass, were stopped immediately after the initiation of sucrose starvation. Observations of the morphology of starved cells using Nomarski interference microscopy and confocal laser-scanning microscopy revealed that in sucrose-free medium, the width of the cytoplasm was markedly decreased, while the vacuole was enlarged (Rose et al. 2006). Carbon starvation induced autophagy in plant cells, as confirmed by experiments on model systems, e.g. on cell suspensions of sycamore (Journet et al. 1986), rice (Chen et al. 1994), *Arabidopsis* (Rose et al. 2006) and in pea and lupine embryo axes grown *in vitro* (Morkunas 1997, Morkunas et al. 2003, Borek et al. 2011, 2012a, 2012b). Vacuolar autophagy starts as early as several hours after sugar starvation. Sequestration of portions of the cytoplasm with organelles (except for the nucleus) by endomembranes results in formation of autophagic vacuoles, followed by their fusion with other vacuoles. During periods of nutrient starvation, the autophagic process can be reinitiated in plant cells that are already vacuolated (Chen et al. 1994, Aubert et al. 1996, Moriyasu and Ohsumi 1996, Moriyasu and Klionsky 2003, Moriyasu et al. 2003). Carbohydrate starvation-induced autophagy is associated with an increase in intracellular proteolysis (James et al. 1996, Moriyasu and Ohsumi 1996, Thompson and Vierstra 2005) and a marked degradation of membrane polar lipids (Aubert et al. 1996, Inoue and

Moriyasu 2006). In addition, Rose et al. (2006) revealed that concomitantly, the number of transvacuolar strands is decreased dramatically. Starvation-induced acidic compartments were most frequent in the dense perinuclear cytoplasm, close to the large central vacuole. In vacuoles of starved cells, quinacrine was strongly accumulated. The role of autophagy in adaptation to sugar starvation has been investigated primarily because of its impact on basal metabolism and biomass production. The intracellular degradation of cytoplasmic components by means of autophagy produces amino acids, phospholipids, and other elements that are necessary for basic metabolism and essential biosynthetic pathways. In starved cells the initiation of autophagy is needed to obtain respiratory substrates (Journet et al. 1986, Chen et al. 1994, Aubert et al. 1996).

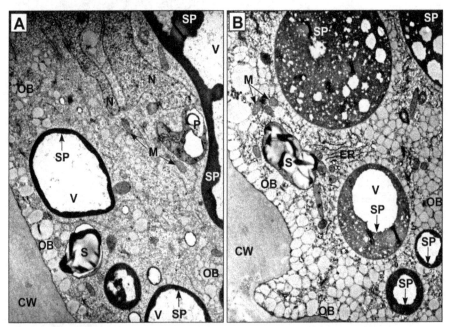

Figure 4. Ultrastructure of parenchyma cells of white lupine excised cotyledons grown for 96 h *in vitro* on medium without sucrose (A) and on medium with 60 mM sucrose (B). CW cell wall, ER endoplasmic reticulum, M mitochondrion, N nucleus, Nu nucleolus, OB oil body, Ph phytoferritin, S starch, SP storage protein, V vacuole.

4. Plant metabolism under sugar starvation

The studies have shown that in most cases, sugar starvation triggers a specific sequence of events in plant cells. The cells subjected to sugar starvation at first adapt to the lack of carbohydrates through gradual replacement of carbohydrate metabolism by protein and lipid metabolism. Such metabolic reorganization may result in autophagy (Saglio and Pradet 1980, Journet et al. 1986, Brouquisse et al. 1991, Rose et al. 2006). Brouquisse et al.

(1992) report that during sugar starvation of maize root meristems, 3 phases can be distinguished:

1. acclimation (from 0 to 30-35 h), when cellular carbohydrate levels and respiration rate decrease, while nitrogen is released from storage proteins through their degradation;
2. survival phase (from 30-35 to 90-100 h), involving intensive breakdown of proteins and lipids, release of P_i phosphorylcholine, and free amino acids; this phase can be reversed by sugar feeding;
3. cell disorganization (more than 100 h), when the level of all metabolites and enzymatic activity are significantly decreased and the changes are irreversible, leading to death.

During sugar starvation at the acclimation and survival phases, the total protein content decreases. This is associated with a temporary increase in free amino acids and an increase in proteolytic activity (Tassi et al. 1992, James et al. 1993, Moriyasu and Ohsumi 1996, Borek and Ratajczak 2002). In vitro studies of embryo axes of lupines and garden pea have revealed a dramatic decrease in protein concentration (especially the soluble fraction) in starved cells (Morkunas 1997, Borek et al. 2012a). The decrease in cytosolic proteins in starved cells was correlated with an increase in the activity of proteolytic enzymes, i.e. endo- and exopeptidases (Morkunas et al. 1999, Borek and Ratajczak 2002). The increased activity of proteases could result from activation of enzymatic proteins that were already present in cells but in a bound or inactive form. Their increased activity could also result from de novo synthesis of the enzymes, caused by release of their genes from catabolite repression. The increased activity of proteases does not confirm unambiguously that the lack of sugar causes an enhanced expression of genes of these enzymes, because enzymatic activity can be controlled in other ways, e.g. by allosteric regulation or by phosphorylation and dephosphorylation, which is the most common type of reversible covalent modification (Berg et al. 2002). However, a study of starved maize root meristems shows that the increased activity of proteolytic enzymes under sugar starvation is an effect of enhanced transcription and translation (James et al. 1993). It also has been evidenced that proteases are closely associated with some cell compartments. Optimal acidic pH for endo- and carboxypeptidase, which increased under starvation, suggested that proteases were located in vacuoles or in lysosomes (Feller 1986, Mikola and Mikola 1986, Huffaker 1990). Accumulation of autolysosomes was observed in tobacco suspension cells cultured under sucrose starvation conditions in the presence of a cysteine protease inhibitor E-64. Experiments with fluorescent dyes, green fluorescent protein (GFP), and endocytosis markers (FM4-64 and Lucifer Yellow CH) suggested that there is a membrane flow from the plasma membrane to autolysosomes. Using fluorescent dyes and markers of the central vacuole (GFP-AtVam3p, sporamin-GFP and gamma-VM23-GFP), the transport of components of the central vacuole to autolysosomes was displayed (Yano et al. 2004). When cells of tobacco were treated with E-64c cysteine protease inhibitor, then both protein degradation and protease activation were inhibited. Simultaneously, many spherical bodies accumulated in the cytosol, beside the nucleus. The bodies, with acidic pH inside, contained acid phosphatase, which is a marker enzyme of autolysosomes. The bodies accumulated because E-64c inhibitor blocked the degradation of the proteins contained in them.

Consequently, the normally short life span of these bodies was significantly prolonged, and their observation was possible. In all probability, the observed spherical bodies were autolysosomes participating in protein degradation during autophagy (Moriyasu and Ohsumi 1996). During advanced sugar starvation, autophagy involves cytosolic proteins and even whole organelles can be degraded. Plastids, ribosomes, and endoplasmic reticulum are consumed relatively early, whereas the plasma membrane, mitochondria, and peroxisomes persist longer (Baker and Graham 2002). The increased activity of proteolytic enzymes and autophagic processes observed under sugar starvation are certainly adaptive processes enabling the cells to maintain continuous energy supply under sugar starvation. Amino acids released from degraded cytosolic proteins, storage proteins, and organelles are used as respiratory substrates (Brouquisse et al. 1991, Ratajczak et al. 1996, Yu 1999, Borek et al. 2001, Gonzali et al. 2006).

A perfect example of enzymes induced under sugar starvation is glutamate dehydrogenase (GDH). This is a mitochondrial enzyme that catalyzes the synthesis and degradation of glutamate, which is one of the central amino acids of nitrogen metabolism in plants (Lehmann and Ratajczak 2008, Borek et al. 2011). The reversibility of the reaction catalyzed by GDH *in vivo* is considered as alternative and depends on carbon and ammonium status (Lehmann et al. 2011). In sucrose-starved pea and lupine embryo axes, a remarkable increase in GDH activity was observed (Morkunas et al. 2000, Lehmann et al. 2003, 2010, Borek and Nuc 2011, Borek et al. 2012a). Sugar starvation caused also a significant increase in a number of GDH isoenzymes (Morkunas et al. 2000, Lehmann et al. 2003, 2010). The increased GDH activity caused by sugar deficit was also recorded in callus of *Nicotiana plumbaginifolia* (Maestri et al. 1991) and in carrot cell suspension (Robinson et al. 1992). Under sugar starvation, plant mitochondria may adapt to the stress conditions by increasing their ability to use amino acids as respiratory substrates. Thus the increase in GDH activity under sugar starvation may be the reason for the higher oxidation of glutamate by mitochondria. GDH catalyzing the oxidative deamination of glutamate, in cooperation with the corresponding aminotransferases, is the main pathway of amino acid catabolism in plants (Lea 1993, Lehmann and Ratajczak 2008). During this process, a large quantity of ammonia is produced. Nitrogen in this form is toxic, so its utilization is necessary. One of the possibilities of ammonia detoxification is the synthesis of asparagine, whose content increases during sugar starvation. In sugar-starved suspension cells of sycamore, asparagine content increased steadily, whereas the cell protein content declined progressively (Genix et al. 1990).

The higher activity of the enzymes of protein and amino acid catabolism under sugar starvation probably results from the release of the genes encoding them from catabolite repression (caused by sugar). Catabolite repression is of fundamental importance since it allows cells to adapt their metabolism to carbon sources other than sugar when they are in a habitat where sugar is unavailable. If sugar is present in the medium, then the synthesis of enzymes of the catabolism of carbon sources other than carbohydrates is inhibited. This applies to proteolytic enzymes, amino acid dehydrogenases (mostly GDH), alcohol dehydrogenase, and isocitrate dehydrogenase. Other studies of the influence of sugar deficit on plant cell metabolism show that already in the initial phase of starvation (after 1.5 h), specific proteins appear, known as carbohydrate-responsive proteins (Baysdorfer et al. 1988). They may be the proteins associated with the phase of tissue acclimation to sugar starvation. Prolonged starvation results in synthesis of many specific proteins known as starvation-related proteins (STP) (Tassi et al. 1992). The genes encoding them (at least some of them) are normally under catabolite repression caused by carbohydrates. Their expression is initiated when the level of sugar in the cell falls below a critical level.

During sugar starvation, respiration rate declines rapidly in embryos of garden pea (Morkunas et al. 2000) and yellow lupine (Borek et al. 2011). The lower respiration rate is caused by the deficit of respiratory substrates rather than by mitochondrial degradation. In mitochondria isolated from starved pea embryos, respiratory activity (with glutamate as a substrate) is 60% higher than in those isolated from embryos fed with 60 mM sucrose (Morkunas et al. 2000). A similar relationship was observed in mitochondrial fractions from starved and control (sugar-fed) embryos of narrowleaf lupine (Morkunas et al. 2003). The concentration of mitochondrial proteins in starved cells was slightly lower than in control cells, and the activity of a mitochondrial marker enzyme, NAD^+-dependent isocitrate dehydrogenase, was reduced in starved embryos only slightly. Besides, mitochondria isolated from lupine embryo axes cultivated without sucrose, exhibited respiration coupled with oxidative phosphorylation. They exhibited a respiration control ratio of 2 for succinate as a respiratory substrate. Mitochondria isolated from starved embryo axes oxidized glutamate and malate more intensively than mitochondria from embryo axes fed with sucrose. Ultrastructural analysis of cells of starved pea embryo axes shows that the inner mitochondrial membrane is highly convoluted, forming many folds (cristae, Fig. 2A). A similar finding was reported by Journet et al. (1986) who studied sycamore cell suspension. Mitochondria were protected in sycamore cells even as late as after 60 h of starvation, and then an addition of sugar to the medium resulted in intensive respiration. In starved cultures, sugar reserves were used up very quickly. Respiration rate was maintained at a high level for up to 30 h of starvation. In that period, respiratory substrates were provided by degradation of starch, phosphate esters, and proteins. Morkunas et al. (2000) also showed that the transfer of pea embryo axes from sugar starvation conditions onto the medium with sucrose after 48 h recovered their respiration activity. Metabolic adaptations of starved embryos of lupine (Morkunas et al. 2003) and pea (Morkunas et al. 2000) were quite effective physiologically, enabling their survival, and their transfer to a medium with sucrose allowed restoration of normal growth. Similarly, Couée et al. (1992), who studied maize root tips cultured *in vitro*, found that after 48 h of sugar starvation they still had a significant pool of functional mitochondria. Carbohydrate stress caused degeneration of lower-density mitochondria, but not of higher-density mitochondria, which was proved by measurements of respiratory activity of mitochondria and of fumarase activity (mitochondrial marker enzyme).

Under sugar starvation, the increased utilization of cytosolic proteins and storage proteins, is accompanied by a more intensive catabolism of lipids, aimed to supply respiratory substrates. In turn, sugar feeding significantly retards the breakdown of storage lipid. In the research on the regulatory function of sugars in lipid metabolism, sucrose and glucose are predominantly taken under consideration. Sucrose is particularly important in seed tissues because storage lipid is synthesized from this sugar in developing seeds and sucrose is one of the main end products of storage lipid breakdown during seed germination. In seedlings of *Arabidopsis* (a model oil-seed plant), 1% sucrose significantly retards the breakdown of storage lipid (Eastmond et al. 2000), while 0.3 M glucose nearly eliminates the mobilization of this storage compound (To et al. 2002). Seedlings grown for 22 days on glucose solution contained about 80% of their seed storage lipid. In contrast, 22-day-old seedlings grown in equi-molar sorbitol retained only 4-5% of their seed storage lipid. This result additionally proves that the effect of glucose is not due to the osmotic potential of the media (To et al. 2002).

Transcriptome analysis of sucrose-starved rice suspension cells has revealed a decrease in the expression of transcripts involved in fatty acid synthesis but induced those involved in fatty acid degradation, such as 3-ketoacyl-CoA thiolase, fatty acid multifunctional proteins, and acyl-CoA oxidase (Wang et al. 2007). An increased expression of genes encoding lipase and enzymes linked to fatty acid β-oxidation was noted in sucrose-starved *Arabidopsis* suspension cells (Contento et al. 2004). Similarly, in dark-induced senescent *Arabidopsis* leaves an enhanced gene expression of lipase and enzymes involved in fatty acid β-oxidation was observed (Buchanan-Wollaston et al. 2005). Glucose starvation remarkably enhances the intensity of lipid degradation in isolated maize root tips. After 24 h of glucose starvation, the rate of oxidation of palmitic acid to CO_2 was increased 2.5-fold. The overall β-oxidation of fatty acids (measured as acetyl-CoA formation) was increased up to 5-fold in glucose-starved roots. An increase in catalase activity was observed as well (Dieuaide et al. 1992). Catalase is not directly involved in fatty acid β-oxidation but is involved in detoxification of H_2O_2 produced in glyoxysomes by the activity of acyl-CoA oxidase when acyl-CoA is converted into trans-2-enoyl-CoA. Sucrose starvation in *Arabidopsis* suspension cells caused a 5.7-fold increase in the expression of acyl-CoA oxidase-4, which is the first committed step of short-chain fatty acid β-oxidation. A significant increase in acyl-CoA oxidase-4 activity in suspension cells was observed after 12 h of sucrose starvation, whereas a similar effect in one-week-old seedlings was noted after 2 days of sucrose starvation. The significant increase in gene expression of catalase-3 (glyoxysomal isoform) was observed already after 6 h of the removal of sucrose from the medium of *Arabidopsis* suspension cells. An increase in catalase activity was observed 6 h later, i.e. after 12 h of sucrose starvation. In seedlings the increase in catalase activity was observed after 2 days of sucrose starvation. About half of the catalase activity observed during sucrose starvation was due to glyoxysomal isoform 3 (Contento and Bassham 2010). In cucumber suspension cells, the expression of genes encoding 2 marker enzymes of the glyoxylate cycle (isocitrate lyase and malate synthase) was induced during deficits of sucrose, mannose and fructose in the medium (Graham et al. 1994). The above-mentioned results suggest that the enhanced gene expression patterns and enzymatic activities involved in lipolysis, fatty acid β-oxidation, and glyoxylate cycle, facilitate the production of acetyl-CoA from the lipid, which may be directly utilized by respiration to sustain energy production and growth when carbohydrates are exhausted.

Another set of data concerns the regulation of lipid metabolism by sucrose in developing and germinating protein lupine seeds (storage protein content up to 45% of seed dry matter). Restriction in sucrose feeding of developing embryos of yellow lupine (lipid content about 6% of seed dry matter), white lupine (lipid content 7-14%), and Andean lupine (lipid content about 20%) caused a decrease in storage lipid accumulation during seed development (Borek et al. 2009). Ultrastructural investigations of cotyledons of 4-day-old yellow, white, and Andean lupine seedlings showed that oil bodies were smaller and less numerous when sugar level was decreased in tissues (Borek et al. 2006, 2011, 2012a). A peculiar and puzzling feature was observed in 4-day-old sucrose-starved isolated embryo axes of yellow, white, and Andean lupine grown *in vitro*. Their lipid level was by 43, 44, and 70% higher, respectively, than in sucrose-fed organs (Borek et al. 2012b). Lipase and catalase activity increased when sugar reserves were depleted in yellow lupine germinating seed organs (Borek et al. 2006). However, the enzymes involved in further steps of lipid breakdown (cytosolic aconitase, isocitrate lyase, NADP$^+$-dependent cytosolic isocitrate dehydrogenase) were more active in sucrose-fed tissues (Borek and Nuc 2011). Changes in enzymatic activity caused by sucrose starvation were accompanied by modification of gene expression. Lipase mRNA level was higher, while mRNA levels for cytosolic aconitase and NADP$^+$-dependent isocitrate dehydrogenase were lower in yellow lupine cotyledons and embryo axes whose sugar content was decreased (Borek and Nuc 2011). During yellow lupine seed germination, lipid-derived carbon skeletons are used for sugar synthesis but some of them are also used for amino acid formation (mainly asparagine, glutamine, and glutamate). In sugar-deficient conditions in tissues, this carbon flow from storage lipid to amino acids is significantly inhibited (Borek et al. 2003, Borek and Ratajczak 2010).

5. The role of sugar starvation in plant growth and development

As mentioned above, sugar in plants is not only an energetic and structural substrate but it also regulates the expression of many genes. Sugar may play a role of hormonal-like signal in a variety of eukaryotic cell types. The best known are the signaling pathways in yeasts (*Saccharomyces cerevisiae*), but many similar pathways have also been discovered in plants (Rolland et al. 2006). The time-course monitoring of gene expression profiles in suspension cells was investigated by Wang et al. (2007). After 12, 24, and 48 h of sucrose-starvation in rice cells, the expression of 867 genes was increased, while the expression of 855 genes was inhibited. Most of these genes encoded enzymes associated with metabolism. It is characteristic that genes associated with catabolism were up-regulated, while genes involved in biosynthetic pathways were down-regulated. Importantly, Wang et al. (2007) succeeded to identify the genes responsible for transcription and translation, which were the second and third major categories of genes in each group significantly affected by sucrose starvation. Response-to-stress genes were the fourth largest group of regulated genes, which suggests that most of the stress-associated genes were also significantly up-regulated by sucrose starvation. Admittedly, the model of cell suspension used by Wang and his research team is a rather "artificial" system, but identification of the genes affected by sugar starvation proved to be very useful for interpretation of results of research on the role of sugar starvation in the regulation of various processes of plant growth and development and in plant responses to biotic and abiotic stresses.

Most of the data on the regulatory role of sugar starvation in plant development was obtained as a result of research on senescence of plant organs, particularly of leaves and flowers. Controversies between groups of researchers on the role of sugar in leaf senescence may partly result from selection of various experimental models. Buchanan-Wollaston et al. (2005) identified over 800 genes of *Arabidopsis*, whose expression was up- or down-regulated during senescence. Those authors came to a conclusion that gene expression patterns during natural leaf senescence, compared with those identified when senescence is artificially induced in leaves by darkness or sucrose starvation in cell suspension cultures, showed not only similarities but also considerable differences. In this chapter, the role of sugar starvation in senescence is discussed in section 6.

There is also a rich literature on participation of the sugar signal in mobilization of storage materials during seed germination. Interaction between sugar and gibberellin in germinating barley grain is well documented (Thomas and Rodriquez 1994). Depending on the seedling's demand for carbohydrates, the rate of starch mobilization in barley endosperm is regulated by a combination of the sugar signal with gibberellin. This regulation is a textbook example of the role of feedback in metabolic adaptation to developmental needs of the plant (Bewley and Black 1994). Results of research on the role of sugar starvation in the expression of amylase genes have some practical applications. Promoter of the gene of rice α-amylase 3D (RAmy3D), which is induced by sugar starvation, has proved to be very useful in transgenic plant cell cultures producing recombinant proteins, including plant-made pharmaceutical and plant-made industrial proteins (Park et al. 2010, Huang and McDonald 2012). For example, this promoter has been used for achieving high-level expression in rice cell suspension cultures of many therapeutic proteins, including human growth hormone, human α1-antitrypsin, human granulocyte-macrophage colony stimulating factor, bryodin-1, lysozyme and human serum albumin (Xu et al. 2011a, 2011b). Virtually every recombinant protein expressed in this rice suspension cell system has produced significantly higher secreted protein levels than attained with any other plant cell expression system tested (Xu et al. 2011a).

The role of sugar starvation in the control of storage lipid mobilization in germinating seeds was mentioned above. Storage protein mobilization in germinating seeds is also affected by sugar starvation. A good model for research on this phenomenon is yellow lupine seed. Its main storage compounds are proteins, which account for up to 45% of seed dry weight (Duranti et al. 2008, Borek et al. 2012a, 2012b). In ripe lupine seeds, starch is absent (Duranti et al. 2008, Borek et al. 2006, 2011). Sugar starvation releases from catabolite repression not only the genes of proteolytic enzymes but also the genes of amino acid catabolism. This allows the use of amino acid carbon skeletons both as sources of energy and as structural components. Interestingly, transitional starch appears during germination of yellow lupine seeds both in cotyledons and in seedling embryo axes (Borek et al. 2006, 2011, 2012a). This phenomenon can be explained by analogy with the appearance of transitional starch in leaves during photosynthesis. Sugar signal inhibits the expression of genes of some photosynthetic proteins (Sheen 1990, Krapp et al. 1993), so carbohydrates must be quickly converted into the neutral starch. In cotyledons of germinating lupine, carbohydrates are synthesized through gluconeogenesis (Borek et al. 2003, Borek and Ratajczak 2010, Borek et al. 2011), so transitional starch synthesis prevents the inhibition by sugar of the continued mobilization of storage lipids and proteins. Transitional starch synthesis in seedling embryo axes may help to maintain the gradient necessary for sugar transport between the source (cotyledons) and the sink (embryo axes). Hypocotyl elongation of etiolated seedlings was suppressed by sugar. However, it is possible that maintenance of an appropriate level of carbohydrates in the embryo axis prevents the inhibition of hypocotyl elongation in etiolated seedlings (Jang et al. 1997, Rolland et al. 2002, Yang et al. 2004, Borek et al. 2012a). The major task of the hypocotyl during epigeal germination (e.g. in lupine) is to bring the cotyledons above the ground. Thus excessively deep sowing may lead to a loss of the ability to transform cotyledons from storage materials into photosynthetic organs (Elamrani et al. 1994).

Sugar signal in cooperation with hormone signaling pathways, may control a wide range of processes related to plant growth and development (Rolland et al. 2006). Sugar and nitrogen are the main factors that modify plant morphology, as they determine the growth rate and transition to the next developmental stages. Transcription analyses show that sugar and inorganic nitrogen act as both metabolites and signaling molecules (Price et al. 2004). There are many recent reports about the role of sugar starvation in processes of shoot growth reduction and root proliferation enhancement during phosphorus deficits (Ciereszko et al. 2005, Karthikeyan et al. 2007, Polit and Ciereszko 2009). The importance of sugar in plant responses to biotic and abiotic stresses is discussed in section 7.

6. Sugar starvation versus senescence

Shaded leaves age more quickly. Their sugar content declines as a result of limited photosynthesis, so their senescence may be induced by sugar starvation. Chung et al. (1997) discovered in *Arabidopsis* plants a gene called *sen 1*, which was expressed during leaf senescence. A particularly strong expression of the gene was observed in excised leaves kept in darkness, which simultaneously showed symptoms of accelerated senescence. When the excised leaves kept in darkness were treated with sugar, then the *sen 1* gene promoter was repressed. On the basis of the results, the cited authors postulated that sugar starvation induces leaf senescence. Further research on plant tissue senescence, e.g. with the use of a hexokinase mutant (as mentioned above, hexokinase is a sugar sensor), confirmed that a low level of sugar may induce leaf senescence (Moore et al. 2003). However, some other authors reported that leaf senescence is induced by a high level of sugar (Parrott et al. 2005, Pourtau et al. 2006). The dispute on which of these contradictory hypotheses is true was one of the most interesting scientific debates in the field of plant biology in the last decade. A detailed outline of the debate was presented by van Doorn (2008). He critically analyzed numerous works whose results support one or the other hypothesis. He did not question the results, but warned against their simplistic interpretation and unjustified generalization. Van Doorn did not support unambiguously any of the hypotheses, but concluded that "it is quite possible that neither of the two hypotheses is correct". On the basis of gene expression analyses during natural and induced leaf senescence, Buchanan-Wollaston et al. (2005) proposed a model of cooperation of the sugar signaling pathway with abscisic acid, jasmonate, ethylene, salicylic acid, and cytokinin signaling pathways.

As in the case of leaf senescence, there is some controversy about the role of sugar starvation in flower senescence. Sugar starvation may be involved in the process of flower senescence because application of sugars to cut flowers generally delays its visible symptoms (van Doorn 2004). However, petals showing symptoms of senescence have a relatively high sugar content. Arrom and Munné-Bosch (2012) assayed a large number of sugars and hormones in various organs of uncut and cut lily flowers. Addition of sucrose to the vase solution elevated the level of sugar in various organs of cut flowers, but primarily it altered the hormonal balance of floral tissues. Those results confirm the earlier findings that sucrose added to the vase solution accelerates the senescence of cut flowers with the participation of hormones. Hoeberichts et al. (2007) used cDNA microarrays to characterize senescence-associated gene expression in petals of cut carnation (*Dianthus caryophyllus*) flowers. The plants were treated with ethylene, silver thiosulphate (which blocks the ethylene receptor), or sucrose. Ethylene accelerated flower senescence and enhanced senescence-associated gene expression. Silver thiosulphate and sucrose had an opposite effect. The cited authors conclude that sucrose may slow down flower senescence by a negative regulation of the ethylene signal pathway. Cut flowers proved to be a good model for research on senescence regulation in plants, participation of hormones in this process (Tripathi and Tuteja 2007, and references cited there) and the role of sugar as a regulator of gene expression (van Doorn and Woltering 2008). Those studies enriched our knowledge but also provided some useful tips for horticultural practice.

At the cellular level, sugar starvation in senescent organs and in organs that have not initiated the senescence program causes similar or identical symptoms: degradation of cellular structures and degradation of many cellular components, such as proteins and lipids. Nevertheless, the goals of these transformations differ. In the case of senescence, valuable metabolites are removed from the cells destined to die, to feed other cells in other plant organs, e.g. in developing leaves or maturing seeds. During sugar starvation of non-senescent organs, protein and lipid degradation is aimed to supply substrates for respiratory processes, which provide the energy necessary for survival. Senescence is a stage of programmed death, during which a low level of sugar may be necessary for activation of some senescence-associated genes, whose expression is repressed by sugar. However, sugar starvation can be used in plant development in many other processes, not only in programmed cell death. As mentioned above, sugar signaling may, in cooperation with plant hormones, play a major role in plant response to environmental factors and in shaping of plant morphology. However, even at the level of cells, sugar starvation, if not associated with programmed cell death, may show other symptoms than during senescence. For example, generation of free radicals (which play an important role in programmed cell death) may be neutralized by initiation of scavenging free radicals whenever cells need to be protected against the negative effects of sugar starvation (Morkunas et al. 1999, 2003). The role of sugars in plant defense response against free radicals is discussed in section 7.

7. Effect of sugar starvation on plant response to biotic and abiotic stress factors

Depending on the duration of sugar starvation, considerable ultrastructural and metabolic changes were observed in plant cells (see above). Moreover, it was interesting whether sugar starvation (as a nutritional stress) is accompanied by changes in the redox status of the cell, like in the case of plant response to other stress factors (biotic and abiotic). In plant response to abiotic stress factors, the increased generation of free radicals, including reactive oxygen species (ROS), attests to a greater influence of the stress factor, whereas during the plant-pathogen interaction, particularly at an early stage of the infection, an increased generation of free radicals in plant cells is favorable (Morkunas and Bednarski 2008). Free radicals may be toxic to the pathogenic microorganism, hamper the penetration of host tissues (through initiation of lignification of host cell walls), activate the expression of host defense genes, and participate in signal transduction over short and long distances.

Under sugar starvation, synthesis of free radicals is intensified, particularly after 72 and 96 h of starvation of lupine embryo axes *in vitro* (Morkunas et al. 2003). Electron Paramagnetic Resonance (EPR) analyses of free radicals detected the presence of free radicals with a signal at $g = 2.0060$. Sugar starvation caused a 2-fold increase in the concentration of free radicals in embryo axes cultured *in vitro* for 72 h. Spectroscopic splitting factors (g-values) indicated that these radicals might be quinones involved in electron transport pathways. Thus, intensified synthesis of free radicals during nutritional stress and other stresses on plant cells may result in a transient oxidative stress due to enhanced ROS generation (Polle et al. 1990, Schraudner 1992, del Rio 1996, Polle 1996, Purvis 1997, Becana et al. 1998, Chamnongpol 1998, Prasad et al. 1999). Concentrations of free radicals increased also in aging plants (Leshem 1981, Thompson et al. 1987, Pastori and del Rio 1997). However, generation of free radicals during sugar starvation may be neutralized by a number of adaptive mechanisms protecting the cells from oxidative damage (Morkunas et al. 1999, 2003). An example of such a mechanism is the accumulation of large amounts of phytoferritin in plastids of starved lupine and pea cells. This fact was interpreted as a defense mechanism aimed at inactivation of iron ions (Fe^{2+}), as it prevents ROS generation. Results of malondialdehyde (MDA) assays supported this interpretation (Morkunas et al. 1999). Ferritin, which binds iron ions (Fe^{2+}) prevented ROS generation in Fenton reaction (Becana et al. 1998). Besides, in starved cells an antioxidant system was activated, i.e. superoxide dismutase (SOD), catalase (CAT) and peroxidase (POX) activity was higher than in sugar-fed tissues (Borek et al. 2006, Morkunas and Gmerek 2007, Morkunas and Bednarski 2008). Additionally, an increase was observed in the ability of starved embryo axes to take up Mn^{2+} from the growth medium, which seems to be an interesting adaptation, probably associated with increased SOD activity (Morkunas et al. 2003). All stresses generate ROS, potentially leading to oxidative damage. However, it must be emphasized that during nutritional stress, such as sugar starvation, specific defense mechanisms are induced, alleviating the results of oxidative stress.

Apart from the well-known classical antioxidant mechanisms, sugars and sugar-metabolizing enzymes are important players in the defense against oxidative stress (Bolouri-Moghaddam et al. 2010). Those authors hypothesized that the synergistic interaction of sugars (or sugar-like compounds) and phenolic compounds is a part of an integrated redox system. It quenches ROS and contributes to stress tolerance, especially in tissues or organelles with high soluble sugar concentrations.

Moreover, sugar-signaling pathways interact with stress pathways in a complex network, which modulates metabolic plant responses (Ho et al. 2001, Tran et al. 2007). Soluble sugars may either act directly as negative signals or as modulators of plant sensitivity, so they can also play important roles in cell responses to stress-induced remote signals (Rosa et al. 2009). Descriptive ecological and agronomic studies have revealed a strong correlation between soluble sugar concentrations and stress tolerance. However, since energy and resources are required for plants to cope with abiotic and biotic stress conditions, the source-sink partitioning between different organs is a crucial component of the mechanisms of stress tolerance (Ho 1988, Krapp and Stitt 1995). Recent studies for increasing tolerance to environmental stress, through metabolic engineering of compatible solutes have shown that increased concentrations of soluble sugars and/or other osmolytes increase plant tolerance to abiotic stresses, such as drought, salinity, and cold (Rathinasabapathi 2000).

Energy and resources are required for plants to cope with abiotic and biotic stress conditions, so in sugar starvation conditions, when metabolic processes are slowed down (e.g. respiration), the plants are more sensitive to stress factors. For example, the link between low sugar content and plant sensitivity to fungal diseases known as "low-sugar diseases", was first reported over 50 years ago by Horsfall and Dimond (1957). It must be emphasized that many environmental factors, such as insufficient light, high humidity, excessive nitrogen fertilization, and excessively deep sowing, may cause a decrease in the level of carbohydrates in host plant tissues, contributing to an increase in plant sensitivity to fungal infection (Vidhyasekaran 1974a, 1974b, Morkunas et al. 2004, Huber and Thompson 2007, Yoshida et al. 2008, Morkunas et al. 2010, Morkunas et al. 2012). By contrast, low temperature can induce plant resistance to specific pathogens, because it results in accumulation of soluble carbohydrates, affecting the cell water potential and other osmotically-active molecules, as well as accumulation of pathogenesis-related (PR) proteins (Tronsmo 1993, Thomashow 1998, Hiilovaara-Teijo et al. 1999, Płażek and Żur 2003, Yuanyuan et al. 2009). Sugar starvation of host plant cells facilitates the development of fungal pathogens because water potential is much higher there than when the concentration of soluble carbohydrates in cells is high. A lack of carbohydrates in the growth medium caused a decrease in endogenous levels of soluble carbohydrates in lupine embryo axes (Morkunas et al. 2005, Borek et al. 2006), and facilitated the development of *Fusarium* infection. This was reflected in an increased area of necrotic spots, more intensive post-infection changes in cell ultrastructure, and in limited elongation and fresh weight increments of embryo axes, as compared to inoculated axes with a high endogenous level of carbohydrates (Morkunas et al. 2005, Morkunas and Bednarski 2008). Besides, Morkunas and Bednarski (2008) observed on transmission electron micrographs the occurrence of hyphae of the pathogenic fungus *Fusarium oxysporum* in intercellular spaces of inoculated embryo axes cultured without sucrose as early as 72 h after inoculation, while no fungi were found in intercellular spaces of inoculated embryo axes cultured with sucrose. Cook and Papendick (1978) report that a high level of carbohydrates is a strong osmoticum in host plant cells, decreasing the water potential and inhibiting fungal growth. Moreover, in the absence of sugar, genes encoding cell wall-degrading extracellular enzymes secreted by fungal pathogens are released from catabolite repression (Akimitsu et al. 2004). These enzymes may contribute to pathogenesis by degrading wax, cuticle, and cell walls, which aids in tissue invasion and pathogen dissemination (Carapito et al. 2009, Kikot et al. 2009). Moreover, the induction of defense in host plant cells is cost-intensive (Swarbrick et al. 2006). Thus, under sugar starvation, where respiration rate is lowered, acquisition of the energy necessary to initiate defense responses, such as synthesis of PR proteins, phenylpropanoids, and papilla formation, is limited (Morkunas et al. 2005, Morkunas and Gmerek 2007, Morkunas et al. 2007, Morkunas and Bednarski 2008, Morkunas et al. 2011). It needs to be emphasized that a very important defense strategy against *Fusarium oxysporum* in inoculated lupine embryo axes cultured with sucrose involved a high accumulation of flavonoids, including isoflavonoids (Morkunas et al. 2005, 2007), fast stimulation of tissue lignification (Morkunas and Gmerek 2007), and intensive generation of superoxide anion radical (Morkunas and Bednarski 2008). This mechanism was not observed in inoculated embryo axes cultured without sucrose.

8. Conclusions

At some stages of plant development, sugar content may decrease considerably (so-called sugar starvation), usually under the influence of adverse environmental factors (biotic or abiotic). In this chapter, we have shown that sugar starvation in plant cells, depending on its duration, causes both ultrastructural and physiological-biochemical changes of varied severity. In non-senescent cells, the changes are aimed to maintain respiration and other basic metabolic processes at a specified level. Thus under conditions of sugar starvation, a priority is the acquisition of energy necessary for cell survival, even at the expense of organelles except the nucleus and the system supplying energy, i.e. mitochondria. The changes taking place during sugar starvation in the acclimation phase or even in the survival phase can be reversed, as long as the respiratory system functions properly. Carbon, nitrogen, and phosphorus are the major elements whose availability affects plant growth and development, so the sugar deficit signal in cooperation with nitrogen and phosphorus signaling pathways may influence plant morphology. Moreover, it is interesting that during sugar starvation, the observed processes resemble senescence. Many symptoms are identical, e.g. autophagy, decrease in protein content, and strong activation of proteolytic enzymes. However, senescence is controlled by programmed cell death. By contrast, if sugar starvation concerns non-senescent cells, then the initiated reactions cause cell acclimation to such conditions. At first these are reactions that protect cells against irreversible degradation of cell structures, especially those responsible for conservation and realization of genetic information as well as energy supply. At the second phase of sugar starvation, defense strategies are initiated, e.g. activation of the enzymatic antioxidant system, manganese ion uptake, phytoferritin accumulation, protecting cells against the destructive influence of an excess of free radicals (e.g. membrane damage). However, the mechanisms underlying the processes used by plant cells to survive sugar starvation require further research. Currently we should focus on explanation of the controlling mechanisms, which may involve signal molecules. Their synthesis may, to some extent, alleviate the influence of stress factors, and thus play an important role in plant acclimation to stress conditions. In plant cells, in response to environmental factors (either abiotic or biotic), many signaling pathways are initiated. Knowledge of the signaling pathways initiated under conditions of sugar starvation, and their cross-talk in the network, would certainly greatly help to explain the molecular mechanisms underlying plant reactions to sugar starvation.

Author details

Iwona Morkunas* and Magda Formela
Department of Plant Physiology, Faculty of Horticulture and Landscape Architecture, Poznań University of Life Sciences, Poznań, Poland

Sławomir Borek and Lech Ratajczak
Department of Plant Physiology, Faculty of Biology, Adam Mickiewicz University, Poznań, Poland

Acknowledgement

This study was supported by the Polish Committee for Scientific Research (KBN, grant no. N N303 414437).

9. References

Agulló-Antón MÁ, Sánchez-Bravo J, Acosta M, Druege U (2011) Auxins or sugars: what makes the difference in the adventitious rooting of stored carnation cuttings? J Plant Growth Reg 30: 100-113.

Akimitsu K, Isshikia A, Ohtanib K, Yamamotoa H, Eshelc D, Prusky D (2004) Sugars and pH: A clue to the regulation of fungal cell wall-degrading enzymes in plants. Physiol Mol Plant Pathol 65: 271-275.

Arrom L, Munné-Bosch S (2012) Sucrose accelerates flower opening and delays senescence through a hormonal effect in cut lily flowers. Plant Sci 188-189: 41-47.

Aubert S, Gout E, Bligny R, Marty-Mazars D, Barrieu F, Alabouvette J, Marty F, Douce R (1996) Ultrastructural and biochemical characterization of autophagy in higher plant cells subjected to carbon deprivation: control by the supply of mitochondria with respiratory substrates. J Cell Biol 133: 1251-1263.

* Corresponding Author

Azad AK, Ishikawa T, Ishikawa T, Sawa Y, Shibata H (2008) Intracellular energy depletion triggers programmed cell death during petal senescence in tulip. J Exp Bot 59: 2085-2095.

Baker A, Graham IA (2002) Plant peroxisomes: Biochemistry, cell biology and biotechnological applications. Kluwer Academic Publishers, Dordrecht, The Netherlands.

Baysdorfer C, Warmbrodt RD, Van Der Woude WJ (1988) Mechanisms of starvation tolerance in pearl millet. Plant Physiol 88: 1381-1387.

Becana M, Moran JF, Iturbe-Ormaetxe I (1998) Iron-dependent oxygen free radical generation in plants subjected to environmental stress: toxicity and antioxidant protection. Plant Soil 201: 137-147.

Berg JM, Tymoczko JL, Stryer L (2002) Biochemistry. Fifth edition, WH Freeman, New York.

Bewley JD, Black M (1994) Seeds. Physiology of development and germination. Second edition, Plenum Press, New York, London.

Bolouri-Moghaddam MR, Le Roy K, Xiang L, Rolland F, Van den Ende W (2010) Sugar signalling and antioxidant network connections in plant cells. FEBS J 277: 2022-2037.

Borek S, Kubala S, Kubala S (2012a) Regulation by sucrose of storage compounds breakdown in germinating seeds of yellow lupine (*Lupinus luteus* L.), white lupine (*Lupinus albus* L.) and Andean lupine (*Lupinus mutabilis* Sweet). I. Mobilization of storage protein. Acta Physiol Plant 34: 701-711.

Borek S, Kubala S, Kubala S, Ratajczak L (2011) Comparative study of storage compound breakdown in germinating seeds of three lupine species. Acta Physiol Plant 33: 1953-1968.

Borek S, Morkunas I, Ratajczak W, Ratajczak L (2001) Metabolism of amino acids in germinating yellow lupin seeds. III Breakdown of arginine in sugar-starved organs cultivated in vitro. Acta Physiol Plant 23: 141-148.

Borek S, Nuc K (2011) Sucrose controls storage lipid breakdown on gene expression level in germinating yellow lupine (*Lupinus luteus* L.) seeds. J Plant Physiol 168: 1795-1803.

Borek S, Pukacka S, Michalski K (2012b) Regulation by sucrose of storage compounds breakdown in germinating seeds of yellow lupine (*Lupinus luteus* L.), white lupine (*Lupinus albus* L.) and Andean lupine (*Lupinus mutabilis* Sweet). II. Mobilization of storage lipid. Acta Physiol Plant 34: 1199-1206.

Borek S, Pukacka S, Michalski K, Ratajczak L (2009) Lipid and protein accumulation in developing seeds of three lupine species: *Lupinus luteus* L., *Lupinus albus* L., and *Lupinus mutabilis* Sweet. J Exp Bot 60: 3453-3466.

Borek S, Ratajczak L (2010) Storage lipids as a source of carbon skeletons for asparagine synthesis in germinating seeds of yellow lupine (*Lupinus luteus* L.). J Plant Physiol 167: 717-724.

Borek S, Ratajczak W (2002) Sugars as a metabolic regulator of storage protein mobilization in germinating seeds of yellow lupine (*Lupinus luteus* L.). Acta Physiol Plant 24: 425-434.

Borek S, Ratajczak W, Ratajczak L (2003) A transfer of carbon atoms from fatty acids to sugars and amino acids in yellow lupine (*Lupinus luteus* L.) seedlings. J Plant Physiol 160: 539-545.

Borek S, Ratajczak W, Ratajczak L (2006) Ultrastructural and enzymatic research on the role of sucrose in mobilization of storage lipids in germinating yellow lupine seeds. Plant Sci 170: 441-452.

Brouquisse R, Gaudillère JP, Raymond P (1998) Induction of a carbon-starvation-related proteolysis in whole maize plants submitted to light/dark cycles and to extended darkness. Plant Physiol 117: 1281-1291.

Brouquisse R, James F, Pradet A, Raymond P (1992) Asparagine metabolism and nitrogen distribution during protein degradation in sugar-starved maize root tips. Planta 188: 384-395.

Brouquisse R, Rolin D, Cortès S, Gaudillère M, Evrard A, Roby C (2007) A metabolic study of the regulation of proteolysis by sugars in maize root tips: effects of glycerol and dihydroxyacetone. Planta 225: 693-709.

Brouquisse R. James F, Rajmond P, Pradet A (1991) Study of glucose starvation in excised maize root tips. Plant Physiol 96: 619-626.

Buchanan-Wollaston V, Page T, Harrison E, Breeze E, Lim PO, Nam HG, Lin JF, Wu SH, Swidzinski J, Ishizaki K, Leaveret CJ (2005) Comparative transcriptome analysis reveals significant differences in gene expression and signalling pathways between developmental and dark/starvation-induced senescence in Arabidopsis. Plant J 42: 567-585.

Carapito RI, Imberty A, Jeltsch JM, Byrns SC, Tam PH, Lowary TL, Varrot A, Phalip V (2009) Molecular basis of arabinobio-hydrolase activity in phytopathogenic fungi. Crystal structure and catalytic mechanism of *Fusarium graminearum* GH93 exo-α-L-arabinanase. J Biol Chem 284: 12285-12296.

Chamnongpol S (1998) Defense activation and enhanced pathogen tolerance induced by H_2O_2 in transgenic tobacco. Proc Natl Acad Sci USA 95: 5818-5823.

Chen MH, Liu LF, Chen YR, Wu HK, Yu SM (1994) Expression of α-amylases, carbohydrate metabolism, and autophagy in cultured rice cells is coordinately regulated by sugar nutrient. Plant J 6: 625-636.

Cho JI, Ryoo N, Eom JS, Lee DW, Kim HB, Jeong SW, Lee YH, Kwon YK, Cho MH, Bhoo SH, Hahn TR, Park YI, Hwang I, Sheen J, Jeon JS (2009) Role of the rice hexokinases OsHXK5 and OsHXK6 as glucose sensors. Plant Physiol 149: 745-759.

Cho YH, Yoo SD (2011) Signaling role of fructose mediated by FINS1/FBP in *Arabidopsis thaliana*. PLoS Genet 7: e1001263. doi:10.1371/journal.pgen.1001263.

Chung BC, Lee SY, Oh SA, Rhew TH, Nam HG, Lee CH (1997) The promoter activity of *sen1*, a senescence-associated gene of *Arabidopsis*, is repressed by sugars. J Plant Physiol 151: 339-345.

Ciereszko I, Johansson H, Kleczkowski LA (2005) Interactive effects of phosphate deficiency, sugar and light/dark conditions on gene expression of UDP-glucose pyrophosphorylase in Arabidopsis. J Plant Physiol 162: 343-353.

Contento AL, Bassham DC (2010) Increase in catalase-3 activity as a response to use of alternative catabolic substrates during sucrose starvation. Plant Physiol Biochem 48: 232-238.

Contento AL, Kim SJ, Bassham DC (2004) Transcriptome profiling of the response of Arabidopsis suspension culture cells to Suc starvation. Plant Physiol 135: 2330-2347.

Cook RJ, Papendick RI (1978) Role of water potential in microbial growth and development of plant disease with special reference to postharvest pathology. Hortsciences 13: 559-564.

Couée I, Murielle J, Carde JP, Brouquisse R, Raymond P, Pradet A (1992) Effects of glucose starvation on mitochondrial subpopulations in the meristematic and submeristematic regions of maize root. Plant Physiol 100: 1891-1900.

del Rio LA (1996) Peroxisomes as a source of superoxide and hydrogen peroxide in stressed plants. Biochem Soc Trans 24: 434-438.

Dieuaide M, Brouquisse R, Pradet A, Raymond P (1992) Increased fatty acid β-oxidation after glucose starvation in maize root tips. Plant Physiol 99: 595-600.

Dorne AJ, Bligny R, Rebeille F, Roby C, Douce R (1987) Fatty acid disappearance and phosphorylcholine accumulation in higher plant cells after a long period of sucrose deprivation. Plant Physiol Biochem 25: 589-595.

Duranti M, Consonni A, Magni C, Sessa F, Scarafoni A (2008) The major proteins of lupin seed: characterisation and molecular properties for use as functional and nutraceutical ingredients. Trends Food Sci Tech 19: 624-633.

Eastmond PJ, Germain V, Lange PR, Bryce JH, Smith SM, Graham IA (2000) Postgerminative growth and lipid catabolism in oilseeds lacking the glyoxylate cycle. Proc Natl Acad Sci USA 97: 5669-5674.

Elamrani A, Gaudillère JP, Raymond P (1994) Carbohydrate starvation is a major determinant of the loss of greening capacity in cotyledons of dark-grown sugar beet seedlings. Physiol Plant 91:56-64.

Farrar JF (1981) Respiration rate of barley roots: its relation to growth, substrate supply and the illumination of the shoot. Ann Bot 48: 53-63.

Feller U (1986) Proteolytic enzymes in relation to leaf senescence. In: Dalling MJ (ed.) Plant proteolytic enzymes. 2. CRC Press, Boca Raton, pp. 49-68.

Genix P, Bligny R, Martin JB, Douce R (1990) Transient accumulation of asparagine in sycamore cells after a long period of sucrose starvation. Plant Physiol 94: 717-722.

Gibson SI (2005) Control of plant development and gene expression by sugar signaling. Curr Opin Plant Biol 8: 93-102.

Gonzali S, Loreti E, Solfanelli C, Novi G, Alpi A, Perata P (2006) Identification of sugar-modulated genes and evidence for in vivo sugar sensing in Arabidopsis. J Plant Res 119: 115-123.

Gout E, Bligny R, Douce R, Boisson AM, Rivasseau C (2011) Early response of plant cell to carbon deprivation: in vivo ^{31}P-NMR spectroscopy shows a quasi-instantaneous disruption on cytosolic sugars, phosphorylated intermediates of energy metabolism, phosphate partitioning, and intracellular pHs. New Phytologist 189: 135-147.

Graham IA, Derby KJ, Leaver CJ (1994) Carbon catabolite repression regulates glyoxylate cycle gene expression in cucumber. Plant Cell 6: 761-772.

Hammond JP, White PJ (2008) Sucrose transport in the phloem: integrating root responses to phosphorus starvation. J Exp Bot 59: 93-109.

Hanson J, Smeekens S (2009) Sugar perception and signaling - an update. Curr Opin Plant Biol 12: 562-567.

Hara M, Oki K, Hoshino K, Kuboi T (2003) Enhancement of anthocyanin biosynthesis by sugar in radish (Raphanus sativus) hypocotyls. Plant Sci 164: 259-265.

Harrington GN, Bush DR (2003) The bifunctional role of hexokinase in metabolism and glucose signaling. Plant Cell 15: 2493-2496.

Hiilovaara-Teijo M, Hannukkala A, Griffith M, Yu XM, Pihakaski-Maunsbach K (1999) Snow-mold-induced apoplastic proteins in winter rye leaves lack antifreeze activity. Plant Physiol 121: 665-673.

Ho LC (1988) Metabolism and compartmentation of imported sugars in sink organs in relation to sink strength. Annu Rev Plant Physiol Plant Mol Biol 39: 355-378.

Ho SL, Chao YC, Tong WF, Yu SM (2001) Sugar coordinately and differentially regulates growth- and stress-regulated gene expression via a complex signal transduction network and multiple control mechanisms. Plant Physiol 125: 877-890.

Hoeberichts FA, van Doorn WG, Vorst O, Hall RD, van Wordragen MF (2007) Sucrose prevents up-regulation of senescence-associated genes in carnation petals. J Exp Bot 58: 2873-2885.

Horsfall JG, Dimond AE (1957) Interactions of tissue sugar, growth substances, and disease susceptibility. Z Pflanzenkr Pflanzenschutz 64: 415-421.

Huang TK, McDonald KA (2012) Bioreactor systems for *in vitro* production of foreign proteins using plant cell cultures. Biotechnol Adv 30: 398-409.

Huber DM, Thompson IA (2007) Nitrogen and plant disease. In: Datnoff LE, Elmer WH, Huber DM (eds.) Mineral nutrition and plant disease. APS, St Paul, MN, pp. 31-44.

Huffaker RC (1990) Proteolytic activity during senescence of plants. New Phytol 116: 199-231.

Inoue Y, Moriyasu Y (2006) Autophagy is not a main contributor to the degradation of phospholipids in tobacco cells cultured under sucrose starvation conditions. Plant Cell Physiol 47: 471-480.

James F, Brouquisse R, Pradet A, Raymond P (1993) Changes in proteolytic activities in glucose-starved maize root tips. Regulation by sugars. Plant Physiol Biochem 31: 845-856.

James F, Brouquisse R, Suire C, Pradet A, Raymond P (1996) Purification and biochemical characterization of a vacuolar serine endopeptidase induced by glucose starvation in maize roots. Biochem J 320: 283-292.

Jang JC, León P, Zhou L, Sheen J (1997) Hexokinase as a sugar sensor in higher plants. Plant Cell 9: 5-19.

Jang JC, Sheen J (1994) Sugar sensing in higher plants. Plant Cell 6: 1665-1679.

Johnson R, Ryan CA (1990) Wound-inducible potato inhibitor II genes: enhancement of expression by sucrose. Plant Mol Biol 14: 527-536.

Journet EP, Blingy R, Douce R (1986) Biochemical changes during sucrose deprivation in higher plant cells. J Biol Chem 261: 3193-3199.

Karthikeyan AS, Varadarajan DK, Jain A, Held MA, Carpita NC, Raghothama KG (2007) Phosphate starvation responses are mediated by sugar signaling in *Arabidopsis*. Planta 225: 907-918.

Kerr PS, Rufty TW Jr, Huber SC (1985) Changes in nonstructural carbohydrates in different parts of soybean (*Glycine max* [L.] Merr.) plants during a light/dark cycle and in extended darkness. Plant Physiol 78: 576-581.

Kikot GE, Hours RA, Alconada TM (2009) Contribution of cell wall degrading enzymes to pathogenesis of *Fusarium graminearum*: a review. J Basic Microbiol 49: 231-241.

Kim SW, Koo BC, Kim J, Liu JR (2007) Metabolic discrimination of sucrose starvation from *Arabidopsis* cell suspension by ^{1}H NMR based metabolomics. Biotech Bioprocess Eng 12: 653-661.

Koch K (2004) Sucrose metabolism: regulatory mechanisms and pivotal roles in sugar sensing and plant development. Curr Opin Plant Biol 7: 235-246.

Koch KE (1996) Carbohydrate-modulated gene expression in plants. Annu Rev Plant Physiol Plant Mol Biol 47: 509-540.

Krapp A, Hofmann B, Schäfer C, Stitt M (1993) Regulation of the expression of rbcSandother photosynthetic genes by carbohydrates: a mechanism for the "sink regulation" of photosynthesis? Plant J 3: 817-828.

Krapp A, Stitt M (1995) An evaluation of direct and indirect mechanisms for the "sink-regulation" of photosynthesis in spinach: changes in gas exchange, carbohydrates, metabolites, enzyme activities and steady state transcript levels after cold-girdling source leaves. Planta 195: 313-323.

Lea PJ (1993) Nitrogen metabolism. In: Plant Biochemistry and Molecular Biology (ed.) Lea PJ, Leegood RC, John Wiley and Sons, Chester, 5: 155-180.

Lee EJ, Matsumura Y, Soga K, Hoson T, Koizumi N (2007) Glycosyl hydrolases of cell wall are induced by sugar starvation in Arabidopsis. Plant Cell Physiol 48: 405-413.

Lehmann T, Dabert M, Nowak W (2011) Organ-specific expression of glutamate dehydrogenase (GDH) subunits in yellow lupine. J Plant Physiol 168: 1060-1066.

Lehmann T, Ratajczak L (2008) The pivotal role of glutamate dehydrogenase (GDH) in the mobilization of N and C from storage material to asparagine in germinating seeds of yellow lupine. J Plant Physiol 165: 149-158.

Lehmann T, Ratajczak L, Deckert J, Przybylska M (2003) The modifying effect of sucrose on glutamate dehydrogenase (GDH) activity in lupine embryos treated with inhibitors of RNA and protein synthesis. Acta Physiol Plant 25: 325-335.

Lehmann T, Skrok A, Dabert M (2010) Stress-induced changes in glutamate dehydrogenase activity imply its role in adaptation to C and N metabolism in lupine embryos. Physiol Plant 138: 35-47.

Leshem YY (1981) Oxy free radicals and plant senescence. In: What's new in Plant Physiol 12: 1-4.

Li Y, Lee KK, Walsh S, Smith C, Handingham S, Sorefan K, Cawley G, Bevan MW (2006) Estabilishing glucose- and ABA-regulated transcription networks in Arabidopsis by microarray analysis and promoter classification using a Relevance Vector Machine. Genome Res 16: 414-427.

Loreti E, De Bellis L, Alpi A, Perata P (2001) Why and how do plant cells sense sugars? Annals Bot 88: 803-812.

Loreti E, Povero G, Novi G, Solfanelli C, Alpi A, Perata P (2008) Gibberellins, jasmonate and abscisic acid modulate the sucrose-induced expression of anthocyanin biosynthetic genes in Arabidopsis. New Phytologist 179: 1004-1016.

Lothier J, Lasseur B, Prudhomme MP, Morvan-Bertrand A (2010) Hexokinase-dependent sugar signaling represses fructan exohydrolase activity in Lolium perenne. Func Plant Biol 37: 1151-1160.

Maestri E, Restivo FM, Gulli M, Tassi I (1991) Glutamate dehydrogenase regulation in callus cultures of Nicotiana plumbaginifolia: effect of glucose feeding and carbon source starvation of the isoezymatic pattern. Plant Cell Environm 14: 613-618.

Mikola L, Mikola J (1986) Occurrence and properties of different types of peptidases in higher plants. In: Dalling MJ (ed.) Plant proteolytic enzymes. 1. Boca Raton, CRC Press, pp. 97-117.

Mohapatra PK, Patro L, Raval MK, Ramaswamy NK, Biswal UC, Biswal B (2010) Senescence-induced loss in photosynthesis enhances cell wall β-glucosidase activity. Physiol Plant 138: 346-355.

Moore B, Zhou L, Rolland F, Hall Q, Cheng WH, Liu YX, Hwang I, Jones T, Sheen J (2003) Role of the Arabidopsis glucose sensor HXK1 in nutrient, light, and hormonal signaling. Science 300: 332-336.

Moriyasu Y, Hattori M, Jauh GY, Rogers JC (2003) Alpha tonoplast intrinsic protein is specifically associated with vacuole membrane involved in an autophagic process. Plant Cell Physiol 44: 795-802.

Moriyasu Y, Klionsky DJ (2003) Autophagy in plants. In: Klionsky DJ (ed.) Autophagy. Landes Bioscience, Georgetown, TX, pp: 208-215.

Moriyasu Y, Ohsumi Y (1996) Autophagy in tobacco suspension-cultured cells in response to sucrose starvation. Plant Physiol 111: 1233-1241.

Morkunas I (1997) Regulation by carbohydrates of germinating legume seeds metabolism. PhD thesis, Department of Plant Physiology, Adam Mickiewicz University, Poznań, Poland.

Morkunas I, Bednarski W (2008) Fusarium oxysporum induced oxidative stress and antioxidative defenses of yellow lupine embryo axes with different level of sugars. J Plant Physiol 165: 262-277.

Morkunas I, Bednarski W, Kopyra M (2008) Defense strategies of pea embryo axes with different levels of sucrose to Fusarium oxysporum and Ascochyta pisi. Physiol Mol Plant Pathol 72: 167-178.

Morkunas I, Bednarski W, Kozłowska M (2004) Response of embryo axes of germinating seeds of yellow lupine to Fusarium oxysporum. Plant Physiol Biochem 42: 493-499.

Morkunas I, Borek S, Ratajczak W, Ratajczak L, Antonowicz J (1999) Accumulation of phytoferritin caused by carbohydrate stress in plastids of embryos in lupin and pea cultivated in vitro. In: Zesz Probl Post Nauk Rol, Wyd. ZFR PAN, Warszawa, 469: 247-256.

Morkunas I, Formela M, Marczak Ł, Stobiecki M, Bednarski W (2012) The mobilization of defense mechanisms in the early stages of pea seed germination against Ascochyta pisi. Protoplasma doi: 10.1007/s00709-012-0374-x.

Morkunas I, Garnczarska M, Bednarski W, Ratajczak W, Waplak S (2003) Metabolic and ultrastructural responses of lupine embryo axes to sugar starvation. J Plant Physiol 160: 311-319.

Morkunas I, Gmerek J (2007) The possible involvement of peroxidase in defense of yellow lupine embryo axes against Fusarium oxysporum. J Plant Physiol 164: 185-194.

Morkunas I, Kozłowska M, Ratajczak L, Marczak Ł (2007) Role of sucrose in the development of Fusarium wilt in lupine embryo axes. Physiol Mol Plant Pathol 70: 25-37.

Morkunas I, Lehmann T, Ratajczak W, Ratajczak L, Tomaszewska B (2000) The involvement of glutamate dehydrogenase in the adaptation of mitochondria to oxidize glutamate in sucrose starved pea embryos. Acta Physiol Plant 22: 389-394.

Morkunas I, Marczak J, Stachowiak M, Stobiecki (2005) Sucrose-induced lupine defense against Fusarium oxysporum: Sucrose-stimulated accumulation of isoflavonoids as a defense response of lupine to Fusarium oxysporum. Plant Physiol Biochem 43: 363-373.

Morkunas I, Narożna D, Nowak W, Samardakiewicz S, Remlein-Starosta D (2011) Cross-talk interactions of sucrose and *Fusarium oxysporum* in the phenylpropanoid pathway and the accumulation and localization of flavonoids in embryo axes of yellow lupine. J Plant Physiol 168: 424-433.

Morkunas I, Stobiecki M, Marczak Ł , Stachowiak J, Narożna D, Remlein-Starosta D (2010) Changes in carbohydrate and isoflavonoid metabolism in yellow lupine in response to infection by *Fusarium oxysporum* during the stages of seed germination and early seedling growth. Physiol Mol Plant Pathol 75: 46-55.

Müller J, Wiemken A, Aeschbacher RA (1999) Trehalose metabolism in sugar sensing and plant development. Plant Sci 147: 37-47.

Nicolaï M, Roncato MA, Canoy AS, Rouquié D, Sarda X, Freyssinet G, Robaglia C (2006) Large-scale analysis of mRNA translation states during sucrose starvation in Arabidopsis cells identifies cell proliferation and chromatin structure as targets of translational control. Plant Physiol 141: 663-673.

Pace GGM, Volk RJ, Jackson WA (1990) Nitrate reduction in response to CO_2-limited photosynthesis. Plant Physiol 92: 286-292.

Park CI, Lee SJ, Kang SH, Jung HS, Kim DI, Lim SM (2010) Fed-batch cultivation of transgenic rice cells for the production of hCTLA4Ig using concentrated amino acids. Process Biochem 45: 67-74.

Parrott D, Yang L, Shama L, Fischer AM (2005) Senescence is accelerated, and several proteases are induced by carbon "feast" conditions in barley (*Hordeum vulgare* L.) leaves. Planta 222: 989-1000.

Pastori GM, del Rio LA (1997) Natural senescence of pea leaves. Plant Physiol 113: 411-418.

Płażek A, Żur I (2003) Cold-induced plant resistance to necrotrophic pathogens and antioxidant enzyme activities and cell membrane permeability. Plant Sci 164: 1019-1028.

Polit JT, Ciereszko I (2009) In situ activities of hexokinase and fructokinase in relation to phosphorylation status of root meristem cells of *Vicia faba* during reactivation from sugar starvation. Physiol Plant 135: 342-350.

Polle A (1996) Developmental changes of antioxidative systems in tobacco leaves as the affected by limited sucrose export in transgenic plants expressing yeast invertase in apoplastic space. Planta 198: 253-262.

Polle A, Chakrabarti K, Schürmann W, Rennenberg H (1990) Composition and properties of hydrogen peroxide decomposing systems in extracellular and total extracts from needles of Norway spruce (*Picea abies* L., Karst). Plant Physiol 94: 312-319.

Pourtau N, Jennings R, Pelzer E, Pallas J, Wingler A (2006) Effect of sugar-induced senescence on gene expression and implications for the regulation of senescence in *Arabidopsis*. Planta 224: 556-568.

Prasad KVSK, Saradhi PP, Sharmila P (1999) Concerted action of antioxidant enzymes and curtailed growth under zinc toxicity in *Brassica juncea*. Environ Exp Bot 42: 1-10.

Price J, Laxmi A, Martin SKST, Jang JC (2004) Global transcription profiling reveals multiple sugar signal transduction mechanisms in *Arabidopsis*. Plant Cell 2004 16: 2128-2150.

Purvis AC (1997) Role of the alternative oxidase in limiting superoxide production by plant mitochondria. Physiol Plant 100: 165-170.

Ramon M, Rolland F, Sheen J (2008) Sugar sensing and signaling. In: The Arabidopsis book. Rockville, MD: American Society of Plant Biologists; doi:10.1199/tab.0117.

Ratajczak W, Lehmann T, Polcyn W, Ratajczak L (1996) Metabolism of amino acids in germinating yellow lupin seeds. I. The decomposition of ^{14}C-aspartate and ^{14}C-glutamate during the imbibition. Acta Physiol Plant 18: 13-18.

Rathinasabapathi B (2000) Metabolic engineering for stress tolerance: Installing osmoprotectant synthesis pathways. Annals Bot 86: 709-716.

Robinson SA, Stewart GR, Phillips R (1992) Regulation of glutamate dehydrogenase activity in relation to carbon limitation and protein catabolism in carrot cell suspension cultures. Plant Physiol 98: 1190-1195.

Roby C, Martin JB, Bligny R, Douce R (1987) Biochemical changes during sucrose deprivation in higher plant cells. J Biol Chem 262: 5000-5007.

Roitsch T, Bittner M, Godt DE (1995) Induction of apoplastic invertase of *Chenopodium rubrum* by D-glucose and a glucose analog and tissue specific expression suggest a role in sink source regulation. Plant Physiol 108: 285-294.

Rolland F, Baena-Gonzalez E, Sheen J (2006) Sugar sensing and signaling in plants: conserved and novel mechanisms. Annu Rev Plant Biol 57: 675-709.

Rolland F, Moore B, Sheen J (2002) Sugar sensing and signaling in plants. Plant Cell 14 (suppl.): S185-S205.

Rolland F, Sheen J (2005) Sugar sensing and signalling networks in plants. Biochem Soc Trans 33: 269-270.

Rosa M, Prado C, Podazza G, Interdonato R, González JA, Hilal M, Prado FE (2009) Soluble sugars - metabolism, sensing and abiotic stress. Plant Signaling Behavior 4: 388-393.

Rose TL, Bonneau L, Der C, Marty-Mazars D, Marty F (2006) Starvation-induced expression of autophagy-related genes in Arabidopsis. Biol Cell 98: 53-67.

Saglio PH, Pradet A (1980) Soluble sugars, respiration, and energy charge during aging of excised maize root tips. Plant Physiol 66: 516-519.

Sahulka J, Lisa L (1978) The influence of exogenously supplied sucrose on glutamine synthetase and glutamate dehydrogenase levels in excised *Pisum sativum* roots. Biol Plant 20: 446-452.

Salanoubat M, Belliard G (1989) The steady-state level of potato sucrose synthase mRNA is dependent on wounding, anaerobiosis and sucrose concentration. Gene 84: 181-185.

Schraudner M (1992) Biochemical plant responses to ozone. III. Activation of the defense related proteins β-1,3-glucanase and chitinase in tabacco leaves. Plant Physiol 99: 1321-1328.

Sheen J (1990) Metabolic repression of transcription in higher plants. Plant Cell 2: 1027-1038.

Sheen J, Zhou L, Jang JC (1999) Sugars as signaling molecules. Curr Opin Plant Biol 2: 410-418.

Sherson SM, Alford HL, Forbes SM, Wallace G, Smith SM (2003) Roles of cell-wall invertases and monosaccharide transporters in the growth and development of *Arabidopsis*. J Exp Bot 54: 525-531.

Sinha AK, Romer U, Kockenberger W, Hofmann M, Elling L, Roitsch T (2002) Metabolism and non-metabolizable sugars activate different signal transduction pathways in tomato. Plant Physiol 128: 1480-1489.

Smeekens S (2000) Sugar induced signal transduction in plants. Annu Rev Plant Physiol Plant Mol Biol 51: 49-81.

Smeekens S, Ma J, Hanson J, Rolland F (2010) Sugar signals and molecular networks controlling plant growth. Curr Opin Plant Biol 13: 274-279.

Solfanelli C, Poggi A, Loreti E, Alpi A, Perata P (2006) Sucrose-specific induction of the anthocyanin biosynthetic pathway in Arabidopsis. Plant Physiol 140: 637-246.

Swarbrick PJ, Schulze-Lefert P, Scholes JD (2006) Metabolic consequences of susceptibility and resistance in barley leaves challenged with powdery mildew. Plant Cell Environ 29: 1061-1076.

Tassi F, Maestri E, Restivo FM, Marminoli N (1992) The effects of carbon starvation on cellular metabolism and protein and RNA synthesis in Gerbera callus cultures. Plant Sci 83: 127-136.

Thomas BR, Rodriquez R (1994) Metabolite signal regulate gene expression and source/sink relations in cereal seedlings. Plant Physiol 106: 1235-1239.

Thomashow MF (1998) Role of cold-responsive genes in plant freezing tolerance. Plant Physiol 118: 1-7.

Thompson AR, Vierstra RD (2005) Autophagic recycling: lessons from yeast help define the process in plants. Curr Opin Plant Biol 8: 165-173.

Thompson JE, Legge RL, Barber RF (1987). The role of free radicals in senescence and wounding. New Phytol 105: 317-344.

To JPC, Reiter WD, Gibson SI (2002) Mobilization of seed storage lipid by Arabidopsis seedlings is retarded in the presence of exogenous sugars. BMC Plant Biol 2: 4.

Tran LS, Nakashima K, Shinozaki K, Yamaguchi Shinozaki K (2007) Plant gene networks in osmotic stress response: from genes to regulatory networks. Methods Enzymol 428: 109-128.

Tripathi SK, Tuteja N (2007) Integrated signaling in flower senescence. Plant Signal Beh 2: 437-445.

Tronsmo AM (1993) Resistance to winter stress factors in half-sib families of Dactylis glomerata, tested in a controlled environment. Acta Agric Scand 43: 89-96.

Tsukaya H, Ohshima T, Naito S, Chino M, Komeda Y (1991) Sugar dependent expression of the CHS-A gene for chalcone synthase from petunia in transgenic Arabidopsis. Plant Physiol 97: 1414-1421.

Usadel B, Blasing OE, Gibon Y, Retzlaff K, Hohne M, Gunther M, Stitt M (2008) Global transcript levels respond to small changes of the carbon status during progressive exhaustion of carbohydrates in Arabidopsis rosettes. Plant Physiol 146: 1834-1861.

van Doorn WG (2004) Is petal senescence due to sugar starvation? Plant Physiol 134: 35-42.

van Doorn WG (2008) Is the onset of senescence in leaf cells of intact plants due to low or high sugar levels? J Exp Bot 59: 1963-1972.

van Doorn WG, Woltering EJ (2008) Physiology and molecular biology of petal senescence. J Exp Bot 59: 453-480.

Vauclare P, Bligny R, Gout E, De Meuron V, Widmer F (2010) Metabolic and structural rearrangement during dark-induced autophagy in soybean (Glycine max L.) nodules: an electron microscopy and ^{31}P and ^{13}C nuclear magnetic resonance study. Planta 231: 1495-1504.

Vidhyasekaran P (1974a) Possible role of sugars in restriction of lesion development in finger millet leaves infected with *Helminthosporium tetramera*. Physiol Plant Pathol 4: 457-467.

Vidhyasekaran P (1974b) Finger millet helminthosporiose, a low sugar disease. Z Pflanzenkr Pflanzenschutz 81: 28-38.

Visser RG, Stolte A, Jacobsen E (1991) Expression of a chimaeric granule-bound starch synthase-GUS gene in transgenic potato plants. Plant Mol Biol 4: 691-699.

Walsh KB, Vessey JK, Layzell DB (1987) Carbohydrate supply and N_2 fixation in soybean. Plant Physiol 85: 137-144.

Wang HJ, Wan AR, Hsu CM, Lee KW, Yu SM, Jauh GY (2007) Transcriptomic adaptations in rice suspension cells under sucrose starvation. Plant Mol Biol 63: 441-463.

Webster PL, Henry M (1987) Sucrose regulation of protein synthesis in pea root meristem cells. Environ Exp Bot 27: 253-262.

Webster PL, van't Hof J (1973) Polyribosomes in proliferating and non-proliferating root meristem cells. Am J Bot 60: 117-121.

Wittenbach VA (1977) Induced senescence of intact wheat seedlings and its reversibility. Plant Physiol 59: 1039-1042.

Wittenbach VA, Lin W, Hebert RR (1982) Vacuolar localization of proteases and degradation of chloroplasts in mesophyll protoplasts from senescing primary wheat leaves. Plant Physiol 69: 98-102.

Xu J, Dolan MC, Medrano G, Cramer CL, Weathers PJ (2011a) Green factory: Plants as bioproduction platforms for recombinant proteins. Biotechnol Adv doi:10.1016/ j.biotechadv.2011.08.020 (in press).

Xu J, Ge X, Dolan MC (2011b) Towards high-yield production of pharmaceutical proteins with plant cell suspension cultures. Biotechnol Adv 29: 278-299.

Yang Z, Zhang L, Diao F, Huang M Wu N (2004) Sucrose regulates elongation of carrot somatic embryo radicles as a signal molecule. Plant Mol Biol 54: 441-459.

Yano K, Matsui S, Tsuchiya T, Maeshima M, Kutsuna N, Hasezawa S, Moriyasu Y (2004) Contribution of the plasma membrane and central vacuole in the formation of autolysosomes in cultured tobacco cells. Plant Cell Physiol 45: 951-957.

Yoshida M, Nakajima T, Tonooka T (2008) Effect of nitrogen application at anthesis on *Fusarium* head blight and mycotoxin accumulation in breadmaking wheat in the western part of Japan. J Gen Plant Pathol 74: 355-363.

Yu SM (1999) Cellular and genetic responses of plants to sugar starvation. Plant Physiol 121: 687-693.

Yuanyuan M, Yali Z, Jiang L, Hongbo S (2009) Roles of plant soluble sugars and their responses to plant cold stress. African J Biotech 8: 2004-2010.

Chapter 7

Alterations in Root Morphology of Rootstock Peach Trees Caused by Mycorrhizal Fungi

José Luis da Silva Nunes, Paulo Vitor Dutra de Souza,
Gilmar Arduino Bettio Marodin, José Carlos Fachinello
and Jorge Ernesto de Araújo Mariath

Additional information is available at the end of the chapter

1. Introduction

The morphology of plant roots have gained prominence in various branches of knowledge, especially in the Biological and Agricultural Sciences, according to the same being one of the main features of the plant body related to the supply and support of plant (Marschner, 1995). Agricultural practices of soil management require special attention in the relations of the roots of different plants with different managements employees, because the health of plants is dynamically linked to these delicate relatio (Silva et al., 2005). This is because the management practices linked monocultures allow the reproduction of micro-organisms that cause crop damage, and the common use of pesticides to alleviate this problem (Bressan & Vasconcelos, 2002).

In this sense, studies are being conducted with the objective of evaluating the possibility to reduce the use of these chemicals in the control of harmful micro-organisms, ranging from research on structural strength of the plant, past the front of the dynamic plant managements, to the use of microorganisms considered beneficial plants (Bressan & Vasconcelos, 2002). On this last point, the arbuscular mycorrhizal fungi (AMF) colonize the root system of most plants, and one of the most reported benefits has been a greater phosphorus absorption by the mycorrhized plants (Nunes et al, 2006), forming a mutualistic symbiosis type biotrophic (Dodd, 2000). This symbiosis is widely distributed in the plant kingdom, occurring in 83% of dicotyledonous plants, in 79% of monocots and in all Gymnosperms, without altering the external appearance of the root (Wilcox, 2002). Moreover, the occurrence of symbiosis is widespread in most habitats, both natural ecosystems and in ecosystems altered by human activities (Sylvia et al., 2001).

In this respect, mutualism is manifested in the bidirectional exchange of nutrients, where the plant comes from carbohydrates to the fungus, while it provides you with water and nutrients, especially for the case of phosphorus (Smith et al., 2003). Although the result of symbiosis be beneficial for the phytobionts, the effectiveness varies in function of the combination the vegetal species and fungus involved in the association (Smith et al., 2003).

By mechanisms promoted by the AMF, the external hypha and mycelia increase the root capacity to exploit the soil results in greater nutrient absorption (Siqueira et al., 2002). However, this absorption has also been related to alterations in the morphological properties of the root of the host plant (Moreira & Siqueira, 2002).

The root system morphology is determined genetically, and can vary among species and individuals in function of environmental factors, such as water availability, nutrients and temperature (Tokeshi, 2000) and the plasticity of the root system can also be influenced by AMF (Berta et al., 1995). The root morphology influences the fast development of the root system and is critical for the successful establishment of most horticultural and fruit plants (Bressan & Vasconcellos, 2002).

This fact is fundamental to the understanding of the effects of the AMF on root development, especially in the case of rootstock plants (Berta et al., 1995). However, the relationships involved in the formation of this symbiosis, since the signaling between the phytobionts, the early stages of the colonization process, as well as possible alterations in the morphological structure of the roots (Berta et al., 1995), in order to be considered a complete understanding relations between the symbionts

There is little information about such relationships, as well as morphological changes produced by mycorrhizal infection in plant tissues (Souza et al., 2000). Some authors report that the AMF does not cause major morphological changes in roots (Cooper, 1984), but studies showed that the AMF induces changes in the architecture (Berta et al., 1995; Norman et al., 1996), especially in the increase of the root ramification, in the morphology (Berta et al., 1995; Bressan & Vasconcelos, 2002; Kothari et al., 1990; Norman et al., 1996,) and the anatomy (Berta et al., 1995) the roots of different plant species.

Most infections of the root system of plants by soil microorganisms imply relations between the actors involved, these relations are based on compatibility between symbionts or the ability of the microorganism to overcome the defense mechanisms of plants (Paszkowski, 2006). The study of morphological relationships between the symbionts highlight the determinants of compatibility that allow the symbiosis occurs involving taxonomically distinct groups of plants and AMF infective (Panstruga, 2003).

The objective of this study was to relate the morphology and root system development of plants of the rootstock cultivars of the peach trees Aldrighi and Okinawa with root colonization by AMF species and the influence of this relationship on nitrogen, phosphorus and potassium absorption and the vegetative development of the plants.

2. Material and methods

2.1. Execution area

The study was carried out under shading (Okinawa cultivar) and a greenhouse (Aldrighi cultivar) at the UFRGS Agronomic Experimental Station, county of Eldorado do Sul, RS, located at latitude 30° 05' South and longitude 51° 39' West from 2004 to 2005.

2.2. Plant and fungal material

Seeds from the two rootstock cultivars were stratified in sterilized sand and placed in a refrigerator at 4°C for 45 days to break the seed dormancy.

Afterwards the seeds were sown on a bed of sterilized sand in a greenhouse. When they were about 5 cm long, the seedlings were replicated to 5 liters black plastic bags containing substrate consisting of clay soil, sand with medium particle size and decomposed black acacia bark residue   (1:1:1, V:V.V). The substrate was previously disinfected with formaldehyde solution at 10%.

The AMF species tested were *Acaulospora sp.* (Trappe), *Glomus clarum* (Nicol. and Schenck) and *Glomus etunicatum* (Becker and Gerd) for the Okinawa cultivar and for the Aldrighi cultivar the same treatments were tested along with *Scutellospora heterogama* (Nicol. and Gerd.). The AMF species were inoculated by adding to each plastic bag 30 g of roots and rhyzospheric soil of oregano (*Origanum vulgare* Link) containing AMF structures. The inoculum was placed in a layer situated in the mid-part of the recipient.

A randomized block design was used, with 20 plants per plot and four replications, in a total of 320 plants for the Okinawa cultivar and 400 plants for the Aldrighi cultivar.

2.3. Determination of roots colonization and plant responses

When the plants had diameter for grafting (360 days for the Okinawa cultivar and 180 days for the Aldrighi cultivar) the height was assessed of the 20 plants in each plot, from the root-stem junction to the tip of the main stem, using a measuring tape, and the main stem diameter, at the root-stem junction and plant height using a pachymeter.

In addition, 5 plants were used from each replication of the treatments, for determination of leaf area, through the use of leaf area meter mark Li-Cor (LI - 3000). After, the shoot was dried and ground and where the fractions were removed for evaluation of plant tissue nitrogen, phosphorus and potassium content by digestion, distillation and spectrophotometry flames, following the methodology by Tedesco et al. (1995).

Five second order roots with similar length and diameter were collected from the root system to assess the root colonization rate (by the ratio number of infected segments/total analyzed). To determine the colonization rate the radicels were stained following methodology reported by Phillips and Hayman (1970).

2.4. Determination of reserve substances

Samples of the aerial part (leaves, stems and stem) and dried roots were ground in the mill, coupled with a sieve of 20 meshes per inch. Each sample was collected approximately one gram for determination of reserve substances.

A similar procedure was carried out with samples of roots. After each sample individually packaged in bags made of special screen for the filtration of food products and brought back to 65°C oven to constant weight, recording the weight of each bag, after, were digested in order to extract all components of plant tissue (carbohydrates, fats, fatty acids, etc.) that were not fibers (cellulose, hemicellulose and lignin), as conventionally known as reserve substances the method described by Priestley (1965).

The samples were placed in one liter Erlenmeyer flask containing an aqueous solution with 5% trichloroacetic acid (99%) and 35% methanol (99.8%) remained on heating gas burner, under a hood with hood, by eight hours. From the third hour to eight hours, distilled water was added to the solution, as it would evaporate in order to always maintain the same volume of liquid sufficient to maintain the samples immersed in the solution.

After the samples were rinsed with distilled water again and put in stove to dry at 65°C until constant weight. The difference in mass of the samples before and after digestion constisted substance content of the buffer that contained samples.

2.5. Histological studies

Secondary roots with similar diameter were used for the morphological studies, as shown in Tables 1 and 2. The histological studies followed the methods described by Johansen (1940), where 1 cm long samples were dehydrated and blocked in paraffin, and 10-15μm thick slices were made using a manual microtome.

The slices were placed on slides, removed the paraffin with xylol, rehydrated for later staining with aqueous Safranin (1%) and Toluidine Blue O (0.05%), and than dehydrated again and the preparations mounted in Canada balsam with a coverslip.

These sections were observed under a Leica DM microscope with 400X magnification. The images were captured with a Nikon CoolPix 990 digital camera (Photos of José Luis da Silva Nunes) and analyzed using the "WCIF Image J" software.

The morphometric parameters measured in the roots were area, diameter, number and perimeter of the tracheal element cells, regardless of the stage of ontogenetic development (primary or secondary) and, from the primary xylem, only the metaxylem was measured, because the protoxylem collapsed at the end of its differentiation (Figures 1 and 2).

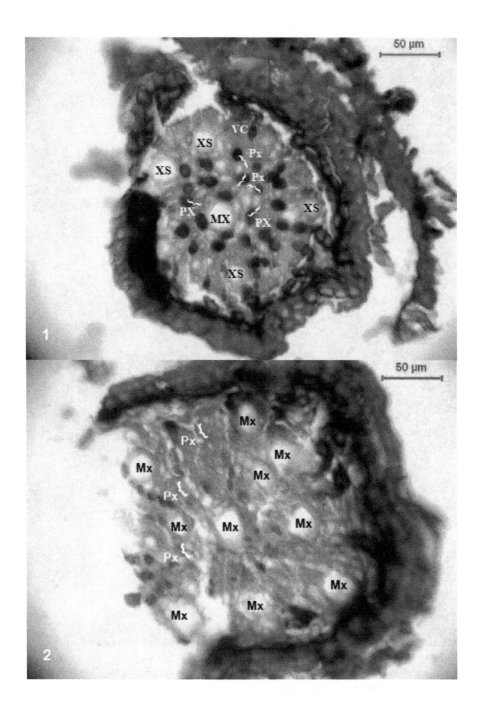

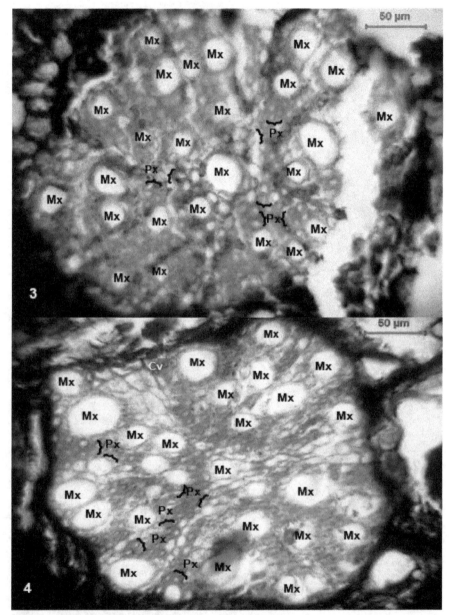

Figure 1. Cross sections of secondary roots of peach tree rootstock Okinawa without inoculation (1) and inoculated with AMF (2 – *G. clarum*, 3 – *G. etunicatum*, 4 – *Acaulospora sp.*). MX – metaxylem; PX – protoxylem; XS – secondary xylem. Scale 50 μm.

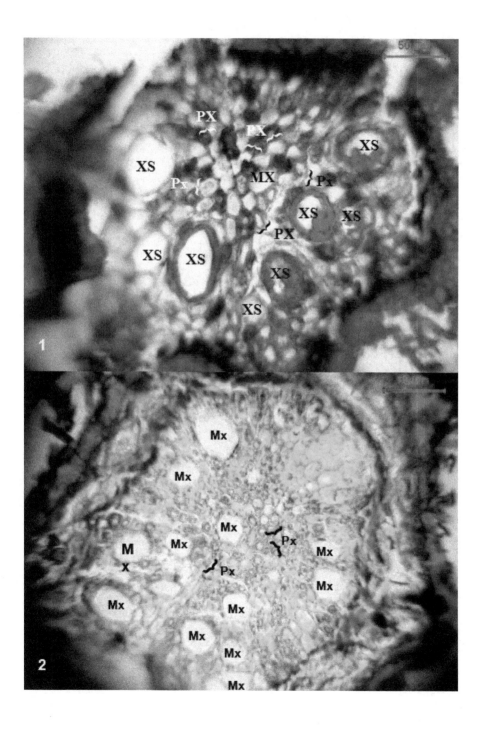

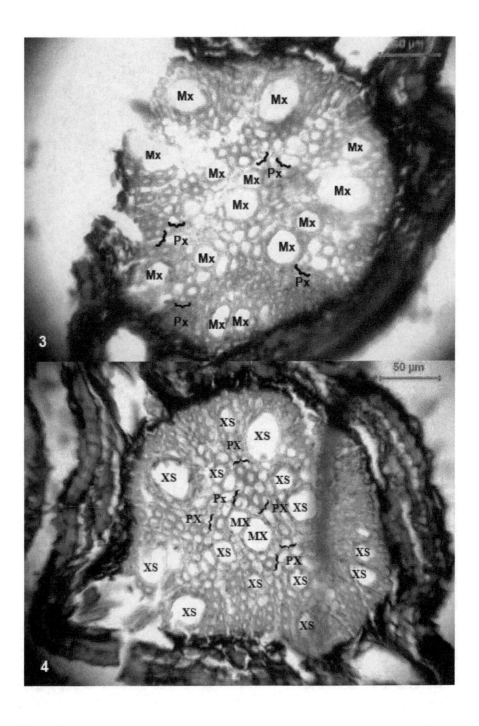

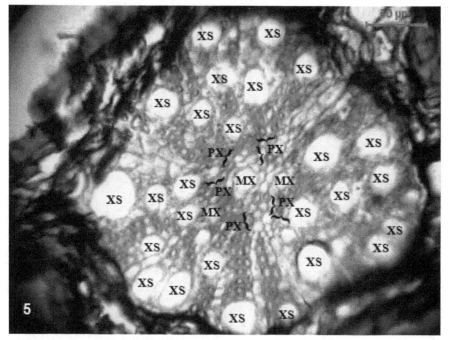

Figure 2. Cross sections of secondary roots of peach tree rootstock Aldrighi without inoculation (1) and inoculated with AMF (2 – *G. clarum*, 3 – *G. etunicatum*, 4 – *Acaulospora sp.*, 5 - *Scutellospora heterogama*). MX – metaxylem; PX – protoxylem; XS – secondary xylem. Scale 50 μm.

2.6. Statistics

The data were submitted to an analysis of variance by the SAS program and the measurements were compared by the Duncan test (Duncan, 1955) at the level of 5% significance.

3. Results

The results regarding the effect of the AMF on the conductor tissue of the roots of the Okinawa cultivar showed that the treatments with the *Acaulospora sp.* and *G. etunicatum* species performed similarly for the number and diameter of the metaxylem and secondary xylem cells, superior to the other treatments, while for cell area and perimeter, *Acaulospora sp.* was superior to the other treatments. The treatment with *G. etunicatum* presented cell area, perimeter and diameter similar to *G. clarum*, superior to the controls. The treatment with *G. clarum* presented the number of cells of these xylem classes similar to that of the controls. There was an inverse performance for cortical thickness, where the control plants presented the highest results, followed by the plants inoculated with *G. clarum* while those inoculated with *Acaulospora sp.* and *G. etunicatum* presented similar but lower results, compared to the control plants (Table 1).

Treatment	Diameter of root (μm)	Cortex thickness (μm)	Metaxylem e secondary xylem			
			Number of Cell	Cell diameter (μm)	Cell perimeter (μm)	Cell area (μm²)
Acaulospora sp.	957,14[ns]	118,12c	38,53a	5,50a	16,86a	58,55a
G. clarum	958,32[ns]	129,85b	26,00b	5,11b	15,61b	48,46b
G. etunicatum	961,35[ns]	119,31c	34,60a	5,26ab	15,70b	50,45b
Testemunha	952,61[ns]	139,39a	22,00b	4,04c	11,77c	41,14c
V. C. (%)	5,17	7,41	5,01	10,09	6,02	13,22

Table 1. Root morphology of secondary roots of the Okinawa cultivar rootstock inoculated with three AMF species (*Acaulospora sp.*, *G. clarum* and *G. etunicatum*), collected 360 days after sowing. Eldorado do Sul, RS, 2005. Means followed by the same letter in the column do not differ by the Duncan test at 5% significance. [ns]Non-significative.

For the Aldrighi cultivar, the treatment with the *S. heterogama* species presented better results than the other treatments for the parameters number of cells, cell area and perimeter of the metaxylem and the secondary xylem, but performed similarly to *G. etunicatum* for cell diameter, and was superior to the other treatments for this parameter (Table 2).

Treatment	Diameter of root (μm)	Cortex thickness (μm)	Metaxylem e secondary xylem			
			Number of Cell	Cell diameter (μm)	Cell perimeter (μm)	Cell area (μm²)
Acaulospora sp.	945,01[ns]	118,08c	33,40b	4,87b	12,95c	52,95c
G. clarum	939,11[ns]	128,85b	24,93c	4,93b	13,10c	53,10c
G. etunicatum	955,23[ns]	107,31d	30,20b	6,17a	16,86b	56,86b
S. heterogama	960,12[ns]	105,22d	37,00a	7,06a	22,70a	62,70a
Testemunha	940,04[ns]	138,39a	16,33d	3,87c	11,57d	41,57d
V. C. (%)	4,76	6,11	5,54	7,84	6,84	11,32

Table 2. Root morphology of secondary roots of the Aldrighi cultivar rootstock inoculated with four AMF species (*Acaulospora sp.*, *G. clarum*, *G. etunicatum* and *S. heterogama*) collected 180 days after sowing. Eldorado do Sul, RS, 2005. Means followed by the same letter in the column do not differ by the Duncan test at 5% significance. [ns]Non-significative.

G. etunicatum performed similarly to *Acaulospora sp.* regarding the number of cells that was better than *G. clarum* and the control. Furthermore, *G. etunicatum* performed better than *Acaulospora sp.*, *G. clarum* and the controls for the other parameters. The plants inoculated with *Acaulospora sp.* performed better than *G. clarum* for the number of cells and similarly to the other parameters. The control plants presented lower results for all the parameters in the assessment of the xylem classes. However, regarding cortical thickness, the control plants performed better than the other treatments, followed by *G. clarum* that was better than *Acaulospora sp.* that was better than the treatments with *S. heterogama* and *G. etunicatum* that presented similar results.

The inoculation with AMF species accelerated the growth of the plants of the Okinawa cultivar rootstock, inducing greater height, diameter, leaf area and greater nitrogen, phosphorus and potassium content, compared with the control. All presented root colonization rates were over 90%. *Acaulospora sp.* was the species that was most efficient among the AMF tested giving greatest height, diameter and nutritional state compared to the other species. The plants inoculated with *G. clarum* and *G. etunicatum* presented to mediate growth and nutritional state, and were similar (Table 3).

Treatment	H (cm)	D (mm)	L.A. (cm²/plant)	Nutrients (%)			Colonization (%)
				N	P	K	
Acaulospora sp.	136,46a	8,42a	197,01a	2,35a	0,16a	2,07a	97,00a
G. clarum	126,65b	7,79b	163,00b	2,23b	0,15b	1,74b	91,76b
G. etunicatum	129,04b	7,87b	167,00b	2,22b	0,15b	1,82b	92,62b
Control	119,23c	7,24c	142,03c	2,05c	0,14c	1,60c	00,00c
V. C. (%)	3,88	2,17	2,54	2,56	2,61	4,75	2,42

Table 3. Height (H), root-stem junction diameter (D), leaf area (L.A.), percentage of nitrogen (N), phosphorus (P) and potassium (K) in the plant tissue and root colonization (%) of the Okinawa cultivar rootstock inoculated with three AMF species (*Acaulospora sp.*, *G. clarum* and *G. etunicatum*), collected 360 days after sowing. Eldorado do Sul, RS, 2005. Means followed by the same letter in the column do not differ by the Duncan test at 5% significance.

For the plants of the Aldrighi cultivar, only the *S. heterogama* and *G. etunicatum* species were shown to be efficient for the height parameter (Table 4).

Treatment	H (cm)	D (mm)	L.A. (cm²/plant)	Nutrients (%)			Colonization (%)
				N	P	K	
Acaulospora sp.	130,94c	6,28c	125,00c	2,99c	0,16b	2,40b	30,00c
G. clarum	129,80c	6,24c	119,01c	2,96c	0,17b	2,44b	28,50c
G. etunicatum	138,36b	6,82b	137,00b	3,33b	0,20a	2,72a	91,50b
S. heterogama	143,97a	7,29a	173,10a	3,74a	0,22a	2,76a	97,75a
Control	129,70c	5,88d	105,00d	2,65d	0,16b	2,29c	00,00d
V. C. (%)	1,55	2,77	2,44	6,64	10,60	3,97	2,91

Table 4. Height (H), root-stem junction diameter (D), leaf area (L.A.), percentage of nitrogen (N), phosphorus (P) and potassium (K) in the plant tissue and root colonization (%) of the Aldrighi cultivar rootstock inoculated with four AMF species (*Acaulospora sp.*, *G. clarum*, *G. etunicatum* and *S. heterogama*), collected 180 days after sowing. Eldorado do Sul, RS, 2005. Means followed by the same letter in the column do not differ by the Duncan test at 5% significance.

These species were the only ones to present root colonization rates of over 90%. All of the AMF species were efficacious for the root-stem junction diameter and leaf area parameters and only varied in the response intensity. In all the assessments of plant growth and nutritional states, invariably *S. heterogama* induced greater growth compared to the other AMF species, while *G. etunicatum* induced intermediate performance and *Acaulospora sp.* and *G. clarum* presented similar performance.

Inoculation with AMF increased content of reserve substances to plants of cv. Okinawa, especially when inoculated with *Acaulospora sp.* In the shoots of rootstock plants inoculated with *G. etunicatum* and *G. clarum* showed intermediate levels. In roots, the plants inoculated with *G. etunicatum* also showed intermediate values, while those inoculated with *G. clarum* not differ from the controls (Table 5).

Treatment	Reserve substances (% in the plant)	
	Shoot	Roots
Acaulospora sp.	39,81a	28,38a
G. clarum	35,05b	21,02c
G. etunicatum	35,53b	24,28b
Control	27,29c	19,41c
V. C. (%)	5,24	2,58

Table 5. Reserve substances of shoots (leaves and stems) and roots of plants of the Okinawa cultivar rootstock inoculated with three AMF species (*Acaulospora sp.*, *G. clarum* and *G. etunicatum*), collected 360 days after sowing. Eldorado do Sul, RS, 2005. Means followed by the same letter in the column do not differ by the Duncan test at 5% significance.

In reviewing the data on the percentage of reserve substances from plants of cv. Aldrighi, present in the tissue of the shoot, it appears that the plants were inoculated with the AMF species had percentages higher than uninoculated plants (Table 6).

Treatment	Reserve substances (% in the plant)	
	Shoot	Roots
Acaulospora sp.	34,69b	22,03c
G. clarum	35,50a	22,99bc
G. etunicatum	36,61a	24,92b
S. heterogama	38,25a	28,27a
Control	28,57c	19,80c
V. C. (%)	2,76	3,26

Table 6. Reserve substances of shoots (leaves and stems) and roots of plants of the Okinawa cultivar rootstock inoculated with four AMF species (*Acaulospora sp.*, *G. clarum*, *G. etunicatum* and *S. heterogama*), collected 180 days after sowing. Eldorado do Sul, RS, 2005. Means followed by the same letter in the column do not differ by the Duncan test at 5% significance.

For the shoot, plants inoculated with G. *clarum*, G. *etunicatum* and S. *heterogama* showed percentages of reserve substances statistically similar but superior to *Acaulospora sp.*, which in turn was higher than the control. In roots, the plants were inoculated with S. *heterogama* presented the greatest results, while those inoculated with G. *etunicatum* showed intermediate levels, statistically similar to G. *clarum* that, in turn, was similar to those inoculated with *Acaulospora sp.* and the control.

4. Discussion

4.1. Anatomy and morphology changes in roots

It was observed that inoculation with AMF reduced the cortex thickness of inoculated plants in both cultivars, associated to increase in most of the morphological parameters of the root xylem assessed for the Okinawa cultivar and for all of those of the Aldrighi cultivar (Tables 1 and 2).

The main effect of the AMF occurred on the metaxylem, that is, one of the categories of the primary xylem, whose conductor cells differentiate later and are larger in diameter (Costa et al., 2003) and also on the secondary xylem cells. On the other hand, the AMF not did not seem to exercise effect on the protoxylem, that are conductor cells of the primary xylem that differentiate first, that is, they acquire secondary lignin walls early (Apezzato-da-Glória & Hayashi, 2003) that reduce the possibility of the AMF acting on the growth of this category of cells of the primary xylem.

The decrease in the cortex area seems to be directly linked to the increase in the number of cells in the metaxylem and the secondary xylem of the plants inoculated with AMF. The control plants presented a smaller number of metaxylem and secondary xylem cells that were smaller in diameter compared to the cells of the inoculated plants, especially in the case of the treatments with the species *Acaulospora sp.* (cv. Okinawa) and S. *heterogama* (cv. Aldrighi).

The mycelia of endomycorrhizal fungi were extracted from roots of *Ophrys lutea* (Orchidaceae) and placed in culture medium kept and in the dark for three weeks (Barroso et al., 1986). Then the derivatives released in the culture medium were extracted, whose greatest concentrations consisted of indol-3-acetic acid and indol-3-ethanol acid, showing the ability of these fungi to synthesize hormones. The authors concluded that the identification of these compounds in the mycelia extract suggested the transference of these compounds from the fungi to the host plants in the phase when symbiosis was established.

Roots colonized by AMF presented an increase in auxin and cytokinin production that are involved in the increase or continuity of the growth of the conductor tissue cells, especially in the size and number of the cells of the metaxylem and the secondary xylem (Hirsch et al., 1997). According to the same authors, the establishment of symbiosis would lead to the production of biochemical signals that would activate genes involved in the production of these plant hormones, and thus the same signals would be responsible for the formation of the root nodes on legumes colonized by *Rhizobium sp*. These authors showed that in roots of the MN 1008 alfalfa mutant cultivar, that carries out the transcription of these signals in the absence of the symbiants, the responses of the root tissue were identical to those of roots of plants colonized both by AMF and *Rhizobium meliloti*.

Thus it can be inferred that the presence of AMF would favor the constant differentiation of the xylem tracheal elements, that coincides with the results obtained in this study for both the root stock cultivars.

There appear to be possible variable effects on root morphology, according to the AMF species and the plant species involved in the symbiosis that also influences the size and growth of the xylem cells, that was also observed in this study, because some species presented variable performance in increasing the size and number of cells, in function of the cultivar used, and in function of the AMF species used for the same cultivar (Atkinson et al., 1994). The species *G. fasciculatum* and *G. etunicatum* induced modification in the roots of the Elsanta and Cambridge Favorite strawberry cultivars, but did not cause any alteration in the morphology of the roots of the Rhapsody cultivar, shedding the variable effect of these AMF species on different cultivars of the same plant species (Norman et al., 1996).

Moreover, roots of plants were colonized by AMF may or may not show increases in longevity, depending on plant species and the fungi involved in symbiosis (Atkinson et al, 2003; Eissenstat et al., 2000; Hodge et al., 2000). However, the morphological attributes of the roots that may be affected by the AMF, such as roots and branches of the diameter of the conducting tissue, has a direct influence on increasing the longevity of roots (Wells et al., 2002). In addition to increasing longevity, root colonization by AMF provides a quick renewal of the root system, increasing the rate of substitution of roots that have collapsed.

4.2. Acquisition of nutrients and benefits

The increase in the absorption and transport volume of nutrients such as nitrogen, that is a constituent of proteins (Tedesco et al., 1995), phosphorus that is essential for a cell division and photosynthesis metabolism and potassium that acts on the electric equilibrium of the cells and on the stomata opening and closing (Tedesco et al., 1995), is vital for plant growth. This contributed to greater responses of the inoculated plants in terms of plant development that was observed in this study for both the cultivars, especially in the plants where there were the highest percentages of root colonization (Tables 3 and 4).

The AMF obtain carbohydrates from their host plants and provide nutrients, especially phosphate. In the case of phosphate, depending on the combination plant-fungus, the acquisition can be performed wholly or partly by the fungus (Smith et al., 2003). The metabolic pathway of nutrient acquisition starts with the uptake by hyphae-soil interface (Benedetto et al., 2005). In hyphae, the nutrient is transported to structures of the fungus in the roots (Ohtomo et al., 2005), where it is transferred to the plant via arbúsculo (Nagy et al.,

2005). The route of transfer of carbohydrates from the plant to the AMF follows the opposite direction (Nagy et al., 2005; Ohtomo et al., 2005).

The benefits given by the AMF on xylem development is associated to many action mechanisms of these fungi, that act directly or indirectly on the plants (Souza et al., 2000). One of the positive effects of the AMF is in function of the presence of the external mycelia, which play an important role in slow diffusion nutrient absorption, such as phosphorus and potassium (Minhoni & Auler, 2003; Souza et al., 2000; Tobar et al., 1994), increasing the nutritional content of the plants (An et al., 1993; Barea, 1991). Associated to this, the modifications caused by the AMF in the xylem structure, such as increase in the number and diameter of the metaxylem and secondary xylem cells, permitted a greater flow of nutrient absorption, such as nitrogen, phosphorus and potassium, translocated to the upper part of the plant, culminating in accelerated growth (Souza, 2000; Souza et al., 2000).

The fact that the AMF species induce major development parameters such as height, diameter and leaf area per plant provides greater photosynthesis and, consequently, a higher level of production of assimilates (Nunes et al., 2006). This report confirms the data obtained in this study with respect to the reserve substances of shoots of both cultivars, for all species used provided an increase in leaf area compared to control (Tables 3 and 4). There is also agreement with other authors, who found higher levels of reserve substances in the tissues of plants inoculated with AMF (Theodore et al., 2003; Sena et al., 2004, Souza et al., 2005).

Another fact to be noted is that only the AMF species that provided the greatest results for height, diameter and leaf area for both cultivars (*Acaulospora sp.* Okinawa cultivar, *S. heterogama* for Aldrighi cultivar and *G. etunicatum* for both cultivars) yielded significant differences in plant reserves, both to the tissues of shoots and roots to the tissues, compared to other treatments. Plants with greater height and leaf area, are more light-gathering capacity and production of assimilates, which allows a higher flow of carbohydrates into the root system where one part would be used by the AMF in its nutrition and accumulation in structures buffer (vesicles, where *Acaulospora sp.* and *G. etunicatum*), and the rest would be accumulated in the storage tissue of the plant in the form of reserve substances (Souza et al., 1999; Scatena & Scremin-Dias, 2003). Moreover, the larger diameter provided by the AMF, would increase the upward flow of water and nutrients, and sap formulated in the downward (Mazzoni-Viveiros & Trufem, 2004).

5. Conclusion

Plants inoculated with AMF have changes in the morphological structure of the roots, such as reduction of the cortex and increased the number and size of cells of the metaxylem, which provides greater volume of water and nutrients translocated to the top of the plant. This benefits plants, accelerating its vegetative growth, improving the content of macronutrients and allowing the production and accumulation of assimilates.

Author details

José Luis da Silva Nunes
BADESUL Desenvolvimento, Brasil

Paulo Vitor Dutra de Souza, Gilmar Arduino Bettio Marodin
and Jorge Ernesto de Araújo Mariath
Universidade Federal do Rio Grande do Sul, Brasil

José Carlos Fachinello
Universidade Federal de Pelotas, Brasil

Acknowledgement

To Ministério da Agricultura, Pecuária e Abastecimento (MAPA) and Conselho Nacional de Desenvolvimento Científico e Tecnológico (CNPq) for research support and grants of the authors.

6. References

An Z.Q., Shein T., Wang H.G. (1993). Mycorrhizal fungi in relation to growth and mineral nutrition of apple seedlings. *Scientia Horticulturae*, Vol.54, Nº 4 (July 1993), pp. 275 – 285, ISSN 0304-4238

Appezzato-da-Glória B., Hayashi A.H. (2003). Raiz. In: *Anatomia Vegetal*, Appezzato-da-Glória B., Carmello-Guerreiro S.M. (eds), pp. 267 – 287, UFV, ISBN 85-7269-240-1, Viçosa, Brasil

Atkinson D., Berta G., Hooker J.E. (1994). Impact of mycorrhizal colonization on root archicture, roots longevity and the formation of growth regulators. In: *Impact of Arbuscular Mycorrhizas on Sustainable Agriculture and Natural Ecosystems*, Gianinazzi S., Schüepp H. (eds), pp. 89 – 99, Birkhäuser Verlag, ISBN 3-7643-5000-8, Basel, Switzerland

Atkinson, D., Blanck, K.E., Forbes, P.J., Hooker, J.E., Baddeley, J.A., Watson, C.A. (2003). The influence of arbuscular mycorrhizal colonization and environment on root development in soil., Vol.54, Nº 4 (December 2003), pp. 751 – 757, ISSN 1365-2389

Barea J.M. (1991). Vesicular-arbuscular mycorrhizae as modifiers of soil fertility. In: *Advances in Soil Science*, STEWART, B.S. (ed), pp. 01 – 40, Springer-Verlag, ISBN 0-3879-7354-0, New York, USA

Barroso J., Neves H.C., Pais M.S. (1986). Production of indole-3-ethanol and indole-3-acetic acid by the mycorrhizal fungus of *Ophrys lutea* (Orchidaceae). *New Phytologist*, Vol.103, Nº 4 (December 1986), pp. 745 – 749, ISSN 1469-8137

Benedetto, A., Magurno, F., Bonfante, P., Lanfranco, L. (2005). Expression profiles of a phosphate transporter gene (GmosPT) from the endomycorrhizal fungus *Glomus mosseae*. *Mycorrhiza*, Vol.15, Nº 8 (December 2005), pp. 620 – 627, ISSN: 0940-6360

Berta G., Trotta A., Fusconi A., Hooker J.E., Munro M., Atkinson P., Giovannetti M., Morini S., Fortuna P., Tisseranti B., Gianinazzi-Pearson V., Gianinazzi S. (1995). Arbuscular mycorrhizal induced changes to plant growth and root system morphology in *Prunus cerasifera*. *Tree Physiology*, Vol.15, N⁰ 5 (May 1995), pp. 281 – 293, ISSN 0829-318X

Bressan W., Vasconcellos C.A . (2002). Alterações morfológicas no sistema radicular do milho induzidas por fungos micorrízicos e fósforo. *Pesquisa Agropecuária Brasileira*, Vol.37, N⁰ 4 (Abril 2002), pp. 509 – 517, ISSN 0100-204X

Cooper K.M. (1984). Physiology of VA Mycorrhizae associations. In: *VA Mycorrhiza*. Powel CL, Bagyaraj J (eds), pp. 155 – 186, CRC, ISBN 08-493-569-46, Boca Raton, USA

Costa C.G., Callado C.H., Coradin V.T.R., Carmello-Guerreiro S.M. (2003). Xilema. In: *Anatomia Vegetal*, Appezzato-da-Glória B., Carmello-Guerreiro S.M. (eds), pp. 129 – 154, UFV, ISBN 85-7269-240-1, Viçosa, Brasil

Dodd, J. C. (2000) The role of arbuscular mycorrizal fungi in agro – and natural ecosystems. *Outlook on Agriculture*, Vol.29, N⁰ 1 (March 2000), p. 55 – 62, ISSN 0030-7270

Duncan, D.B. (1955). Multiple range and multiple F tests. *Biometrics*, Vol.11, N⁰ 1 (March 1955), pp. 1- 42, ISSN 1947-2006

Eissenstat, D.M., Wells, C.E., Yanai, R.D., Whitbeck, V.L. (2000) Building roots in a changing environment: implications for root longevity. *New Phytologist*, Cambridge, Vol.147, N⁰ 1 (July 2000), pp. 33 – 42, ISSN 1469-8137

Hirsch A.M., Fang Y., Asad S., Kapulnik Y. (1997). The role of phytohormones in plant-microbe symbioses. *Plant and soil*, Vol.194, N⁰ 2 (January 1997), pp. 171 – 184, ISSN 0032-079X

Hodge, A., Robinson, D., Fitter, A.H. (2000). An arbuscular mycorrhizal inoculum enhances root proliferation in, but not nitrogen capture from, nutrient-rich patches in soil. *New Phytologist*, Cambridge, Vol.147, N⁰ 3 (September 2000), pp. 575 - 584, ISSN 1469-8137

Johansen D.A. (1940). *Plant microtechnique*. McGraw-Hill, ISBN 007592, New York, USA

Kothari B.K., Maschner, H., George, E. (1990) Effect of VA mycorrhizal fungi and rhizosphere microorganisms on root and shoot morphology, growth and water-relations in maize. *New Phytologist*, Vol.116, N⁰ 2 (October 1990), p. 303 – 311, ISSN 1469-8137

Marschner, H. (1995) *Mineral nutrition of higher plants*. Academic Press, ISBN 978-0124735415, San Diego, USA

Mazzoni-Viveiros, S.C., Trufem, S.F.B. (2004) Efeitos da poluição aérea e edáfica no sistema radicular de *Tibouchina pulchra* Cogn. (Melastomataceae) em área de mata Atlântica: associações micorrízicas e morfologia. *Revista Brasileira de Botânica*, Vol.27, N⁰ 2 (Abril/ Junho 2004), pp. 337 – 348, ISSN 0100-8404

Minhoni M.T.A., Auler P.A.M. (2003) Efeito do fósforo, fumigação do substrato e fungo micorrízico arbuscular sobre o crescimento de plantas de mamoeiro. *Revista Brasileira de Ciência do Solo*, Vol.27, N⁰ 5 (Outubro 2003), pp. 841 – 847, ISSN 0100-0683

Moreira F.M.S., Siqueira J.O. (2002) *Microbiologia e bioquímica do solo*. Editora UFLA, ISBN 85-8769-233-X, Lavras, Brasil

Nagy, R., Karandashov, V., Chague, V., Kalinkevich, K.; Tamasloukht, M., Xu, G., Jakobsen, I., Levy, A.A., Amrhein, N., Bucher, M. (2005). The characterization of novel mycorrhiza-specific phosphate transporters from *Lycopersicon esculentum* and *Solanum tuberosum* uncovers functional redundancy in symbiotic phosphate transport in solanaceous species. *Plant Journal*, Vol.42, Nº 2 (April 2005), pp. 236 – 250, ISSN 1365-313X

Norman J.R., Atkinson D., Hooker J.E. (1996). Arbuscular mycorrhizal-fungal-induced alteration to root architecture in strawberry and induced resistance to the pathogen *Phytophthora fragariae*. *Plant and Soil*, Vol.185, Nº 2 (September 1996), pp. 191 – 198, ISSN 0032-079X

Nunes M.S., Soares, A.C.F., Soares Filho, W.S., Lêdo, C.A.S. (2006). Colonização micorrízica natural de porta-enxertos de citros em campo. *Pesquisa Agropecuária Brasileira*, Vol.41, Nº 3 (Março 2006), pp. 525 – 528, ISSN 0100-204X

Ohtomo, R., Saito, M. (2005). Polyphosphate dynamics in mycorrhizal roots during colonization of an arbuscular mycorrhizal fungus. *New Phytologist*, Vol. 167, Nº 2 (August 2005), pp. 571 – 578, ISSN 1469-8137

Panstruga, R. (2003). Establishing compatibility between plants and obligate biotrophic pathogens. *Current Opinions in Plant Biology*, Vol.6, Nº 4 (August 2003), p. 320-326, ISSN: 1369-5266

Paszkowski, U. (2006). Mutualism and parasitism: the yin and yang of plant symbioses. *Current Opinions in Plant Biology*, Vol.9, Nº 4 (August 2006), p.364 – 370, ISSN: 1369-5266

Phillips J.M., Hayman D.S. (1970). Improved procedures for clearing roots and staining parasitic and vesicular-arbuscular mycorrhizal fungi for rapid assessment of infection. *Transactions of the British Mycological Society*, Vol.55, Nº 1 (January 1970), pp. 157-160, ISSN 0007- 1536

Priestley, G.A. (1965) New method for the estimation of the resources of apple tress. *Journal of the Science of Food and Agriculture*, Vol.16, Nº 12 (December 1965), pp. 717 – 721, ISSN 1097-0010

Scatena, V.L., Scremin-Dias, E. (2003). Parênquima, Colênquima e Esclerênquima. In: *Anatomia Vegetal*, Appezzato-da-Glória B., Carmello-Guerreiro S.M. (eds), pp. 109 – 127, UFV, ISBN 85-7269-240-1, Viçosa, Brasil

Sena, J.O.A., Labate, C.A., Cardoso, E.J.B.N. (2004) Caracterização fisiológica da redução de crescimento de mudas de citros micorrizadas em altas doses de fósforo. *Revista Brasileira da Ciência do Solo*, Vol.28, Nº 5 (Setembro/ Outubro 2004), pp. 827 – 832, ISSN 0100-0683

Siqueira J.O., Lambais M.R., Stürmer S.L. (2002) Fungos micorrízicos arbusculares. *Biotecnologia, Ciência & Desenvolvimento*, Vol.25 (Março/Abril 2002), pp. 12 – 21, ISSN 1414-4522

Silva, L.M.S.; Alquini, Y.; Cavallet, V.J. (2005) Inter-relações entre a anatomia vegetal e a produção vegetal. *Acta Botanica Brasilica*, Vol.19, Nº 1 (Janeiro/ Março 2005), p. 183 – 194, ISSN 0102-3306

Souza, F.A., Trufem, S.F.B., Almeida, D.L., Silva, E.M.R., Guerra, J.G.M. (1999) Efeito de pré-cultivos sobre o potencial de inoculo de Fungos Micorrízicos Arbusculares e produção de mandioca. *Pesquisa Agropecuária Brasileira*, Vol.34, Nº 10 (Outubro 1999), pp. 1913 – 1923, ISSN 0100-204X

Souza P.V.D., Agustí M., Abad M., Almela V. (2000). Desenvolvimento vegetativo e morfologia radicular de Citrange Carrizo afetado por ácido indolbutírico e micorrizas arbusculares. *Ciência Rural*, Vol.30, Nº 2 (Março/Abril 2000), pp. 249 – 255, ISSN 0103-8478

Souza P.V.D. (2000). Interação entre micorrizas arbusculares e ácido giberélico no desenvolvimento vegetativo de plantas de Citrange Carrizo. *Ciência Rural*, Vol.30, Nº 5 (Setembro /Outubro 2000), pp. 783 – 787, ISSN 0103-8478

Souza, P.V.D., Carniel, E., Schimitz, J.A.K., Silveira, S.V. (2005) Influência de substratos e fungos micorrízicos arbusculares no desenvolvimento do porta-enxerto Flying Dragon (*Poncirus trifoliata*, var. monstruosa Swing.). *Revista Brasileira de Fruticultura*, Vol.27, Nº 2 (Agosto 2005), pp. 285 – 287, ISSN 0100-2945

Smith, S.E., smith, F.A., Jakobsen, I. (2003) Mycorrhizal fungi can dominate phosphate supply to plants irrespective of growth responses. *Plant Physiology*, Vol.133, Nº 2 (October 2003), pp. 16–20, ISSN 0032-0889

Sylvia, D. M., Chellemi, D. O. (2001) Interations among root-inhabiting fungi and their implications for biological control of root pathogens. *Advances in Agronomy*, Vol.73, Nº 1 (April 2001), pp. 1 – 33, ISBN 978-0-12-000773-8

Tedesco M.J., Gianello C., Bissani C.A., Bohnen H., Volkweiss S.J. (1995). *Análises de solo, plantas e outros materiais (Boletim Técnico, 5)*, UFRGS/Departamento de solos, ISBN 000-148837, Porto Alegre, Brasil

Theodoro, V.C.A., Alvarenga, M.I.N., Guimarães, J., Mourão Junior, M. (2003) Carbono da biomassa microbiana e micorriza em solo sob mata nativa e agroecossistemas cafeeiros. *Acta Scientiarum: Agronomy*, Vol.25, Nº 1 (Maio 2003), pp. 147 – 153, ISSN 1679-9275

Tobar R., Azcón R., Barea J.M. (1994) Inproved nitrogen uptake and transport from ¹⁵N-labelled nitrate by external hyphae of arbuscular mycorrhiza under water stressed condictions. *New Phytologist*, Vol.126, Nº 1 (January 1994), pp. 119 – 122, ISSN 1469-8137

Tokeshi, H. (2000) Doenças e pragas agrícolas geradas e multiplicadas pelos agrotóxicos. *Fitopatologia Brasileira*, Vol.25 (Janeiro 2000), pp. 264-270, ISSN 0100-4158

Wells, C.E., Glenn, D.M., Eissenstat, D.M. Changes in the risk of fine-root mortality with age: a case study in peach, *Prunus persica* (Rosaceae). *American Journal of Botany*, Vol.89, Nº 1 (January 2002), pp. 79 – 87, ISSN 0002-9122

Wilcox, H. E. (2002). Mycorrhizae. In: *Plants roots*, Waisel, Y., Eshel, A., Kafkafi, U. (eds), Marcel Dekker, ISBN 0-8247-0631-5, New York, USA

Environmental and Genetic Variation for Water Soluble Carbohydrate Content in Cool Season Forage Grasses

Ali Ashraf Jafari

Additional information is available at the end of the chapter

1. Introduction

Forage dry matter has been divided into two fractions on the basis of nutritional availability. One fraction (non-structural) corresponds to the cell content and is composed of WSC (water soluble carbohydrates) (fructans and starches), lipids, most of the protein, nucleic acid and minerals. Fructan, a polymer of fructose, is accumulated as an energy reserve mainly in the leaf sheaths of temperate grasses which are made in the cytoplasm during periods of photosynthesis (Pollock and Cairns, 1991). The other dry matter fraction corresponds to the plant cell wall and consists mostly of structural carbohydrates (cellulase, hemicellulase and lignin) (Gill et al, 1989). Chemical composition of grasses changes with advancing maturity. As grass matures the proportion of the cell wall increases and the cell content fraction decreases. Together, non-structural and structural carbohydrates, depending on maturity stage, make up approximately 50 to 80% of dry matter of forages (Gill et al, 1989).

The most important traits that affecting the feeding value of herbage are digestibility (DMD), the ratios of crude protein, water-soluble carbohydrate, fiber, and the concentration of alkaloid toxins (Smith, et al., 1997; Wilkins and Humphreys, 2003). High WSC content is an important breeding goal for milk production and liveweight gain. Over a range of species, Grimes et al (1967) found strong positive correlation between WSC content and intake and liveweight gain of lambs. In the last decade, few improved ryegrass varieties had been bred for increase WSC concentration, in UK (Wilkins and Lovatt 2007; Wilkins et al. 2010), and New Zealand (Rasmussen et al., 2009). Most published data suggest that animals consuming grasses with high WSC concentrations were able to utilize the protein in their diet more efficiently, resulting in increased liveweight gain, milk production and lower loss of N (Lee et al., 2001; Evans et al., 2011; Miller et al., 2001). Similarly, Mayland et al (2001) and Smit (2006) suggested that forage with higher WSC concentration is related to preference. In contrast, few trials that animal grazing of pure swards with high versus low concentrations of WSC, showed low (Cosgrove et al. 2010) or non significant animal performance advantages than control (Allsop et al., 2009; Parsons et al. 2010).

WSC are completely digestible and have an important role in animal nutrition, as they are a primary source of the readily available energy necessary for efficient microbial fermentation in the rumen. Fermentation of structural carbohydrates also provides energy directly to the microbial population and indirectly to the animal. Increased microbial metabolism is a prerequisite to improving forage intake and nutrient utilisation but this only occurs if the interrelationships of carbohydrate and protein are optimized (Carlier, 1994; Evans et al., 2011). One of the major problems in grasses and legumes consumption is reduced efficiency of protein utilisation and unbalanced supplies of carbohydrate and protein to support microbial metabolism (Wilkins and Humphreys, 2003). A large proportion of the protein may be lost during rumen fermentation, mainly caused by a deficiency in available carbohydrates. To increase the carbohydrate/protein ratio in the diet, either grass composition should be changed or grass should be supplemented with feeds of readily available carbohydrate. Beever and Reynolds (1994) suggested that breeding for yield and digestibility alone is no longer adequate, and that forage breeding programs should concentrate on improved nutrient value of forages, such as improved carbohydrate availability and more controlled protein degradation. Beerepoot and Agnew (1997) advised against simple selection for increased carbohydrate levels alone because of possible negative effects upon rumen pH and cell wall degradation and suggested that selection for highly digestible cell walls is preferable.

Success in conserving legumes and grasses as silage often depends on the amount of readily available fermentable carbohydrate present in herbage. If the concentration is sufficiently high, conditions are more favorable for establishment and growth of lactic acid bacteria which leads to better fermentation and preservation (Humphreys, 1994). A minimum of 3.7% WSC in fresh grass (i.e. about 15% WSC in dry weight) is considered necessary to produce good quality silage without additives (Haigh, 1990).

2. Environmental and genetic effects on WSC content

2.1. Environmental effects

WSC in forage grasses appears to be associated with good tiller survival and sward persistency (Thomas and Norris, 1981). Stored WSC provides the energy and structural carbohydrate required for growth when demand cannot be met by contemporary photosynthesis e.g. regrowth after defoliation, recovery from drought and persistency during winter (Humphreys, 1989d, 1994).

The WSC level in grasses depend on a wide range of factors, including the plant part, ploidy level, plant maturity, diurnal and seasonal effects, temperature, light intensity, growth rate, endophyte, water status, as well as the inherent differences between cultivars (Stewart and Hayes, 2011). Accumulation of WSC in tiller bases of cocksfoot is highly correlated with summer survival and rapid recovery after drought (Volaire and Gandoin, 1996). The amount of WSC maintained through the winter appears critical. Factors which reduce carbohydrate reserves such as autumn disease, autumn growth forced by high nitrogen application, late cutting, or uncompensated respiratory loss through growth at low light intensity, can all seriously reduce winter hardiness. If carbohydrate levels can be maintained during periods of winter growth a decline in hardiness may not occur (Humphreys, 1989d).

Radojevic et al (1994) found significant variation for WSC and DMD among four European and New Zealand varieties of perennial ryegrass under Australian conditions. A variation of 20% for WSC concentration among six diploid ryegrasses was found to be consistent over several years (Smit, 2006), the largest variation among cultivars was found during summer, in July and August. Tetraploid cultivars generally had a higher WSC concentration than diploid cultivars (Smith et al. 2001; Gilliland et al. 2002). DMD concentration declined during summer due to the influence of higher temperatures, however, high WSC lines exhibited fewer declines in DMD than the lines that did not accumulate as much WSC.

Intake depends, to some extent, on digestibility but other factors such as palatability (animal preference) are important. Jones and Roberts (1991), in a survey comparing four varieties of perennial ryegrass with the same digestibility but different WSC contents, found that varieties with high WSC content were more palatable than varieties with moderate content of WSC.

WSC content is sensitive to temperature and light intensity. Low levels of WSC are found in Western Europe country as Ireland and UK due to cloudy weather conditions (i.e. low light intensity) (Jung et al, 1996). WSC generally decline during autumn. Fulkerson and Donaghy (2001) suggested that the lower WSC concentration in autumn has several causes. The reduced solar radiation will decrease photosynthetic activity, which results in a reduced primary production of non-structural carbohydrates and WSC concentrations are closely related with hours of sunlight. The higher night temperatures in autumn will induce respiration during night time (McGrath 1988). The altering source–sink relations deplete WSC reserves, and hence lower the WSC concentration in the herbage.

There is seasonal variation in WSC content. Dent and Aldrich (1963) reported WSC content was high in spring and fell to the lowest point in summer, rising again in autumn, to about half of spring level. McGrath (1988) reported that average concentration of WSC in perennial ryegrass was 20% (mean of three cuts per year) with maximum level in late April each year. WSC content was at least 50% higher in stem than in leaf. He also reported that, on average, fructans accounted for 70% of soluble carbohydrate, the remainder being fructose, glucose and sucrose. For these reasons (environmental effects) WSC level in forage grasses vary considerably ranging from less than 5% (Waite and Boyd, 1953) to up to 50% (Bugge, 1978).

Moisture stress retards growth of grasses and is accompanied by an increase in WSC (Brown and Blaser, 1970). WSC contents can be influenced by management practices such as defoliation and nitrogen application level. Application of N fertilizer allows more rapid growth, and thus tends to result in lower levels of WSC accumulation. WSC is greatest when swards are allowed to accumulate higher herbage mass, such as silage crops, with levels lower under very frequent grazing (Rasmussen et al. 2008). Cutting of herbage invariably causes a decrease in the amount of WSC in the remaining plant parts and this decrease reaches a minimum about one week after defoliation and is then followed by an increase associated with increasing photosynthetic activity. If cutting is repeated before the original level is restored, the successive herbage yields will be smaller and eventually such a treatment will lead to plant death. Based on this, Alberda (1966) concluded that WSC are of greater importance for the regrowth and recovery of sward after cutting. McGrath (1992) reported that application of nitrogen fertilizer increased DM yield and invariably increased CP content and decreased WSC content. However, he found no effect of nitrogen on total WSC yield.

Sowing density has a major effect on the development of individual plants. Smouter et al (1995) in a study on the effect of three tiller densities (100, 500, and 5000 plant m^{-2}) on WSC concentration of *Lolium rigidum*, found that the concentration of WSC varied in response to treatment. Yield of WSC in the low-density sward was 30 to 50% higher than other swards at anthesis. In addition, the digestibility of low density was higher than of the high-density sward.

Maturity and herbage age generally have a greater influence on forage quality than environmental factors. As plants advance in maturity, cell wall concentration within stems and leaves, increases and the proportion of cell soluble content decreases. The rate of decline in digestibility of herbage is greatest during reproductive growth (Buxton et al. 1996). Breese and Thomas (1967) showed that small differences in maturity of cocksfoot could greatly affect digestibility. Rezaeifard, et al (2010) in cocksfoot grown in Iran, found that both WSC and acid detergent fiber (ADF) values increased by plant growth, however, WSC increases were slower than for ADF by advancing maturity from vegetative to milky stage. Contents of WSC were higher in stem than in leaf, so by advance of plant maturity WSC contents increase. These results are expected because with plant maturity and reducing the ratio of leaf to stem, the WSC content will increase in stem. In general, DM yield, WSC and ADF values dramatically increased by advancing of plant maturity, in contrast, DMD, CP and ash values decreased (Rezaeifard, et al., 2010). Similarly, Jafari et al (2010) in tall wheatgrass *Agropyron elongatum* and Jafari and Rezaeifard (2010) in tall fescue found the higher values of WSC-yield in pollination and milky stage under conservation management, respectively.

Humphreys (1989c) found considerable genetic variation in WSC content of ryegrass during both vegetative and reproductive growth phases. The range of variation for WSC (13.9 to 28.8%) was greater than that found for CP or DMD and WSC showed generally high heritability. Frandsen (1986) found significant variation for WSC among 36 HS families (22.9 to 44.9%) in perennial ryegrass. Wilkins and Davies (1994) also reported significant differences for WSC among four varieties of perennial ryegrass under conservation management. But, the range was relatively small (30.5 to 33.6%). Dent and Aldrich (1963), using perennial ryegrass varieties under sward condition, reported a range of 14.1 to 19.3 and 17.6 to 26.3 WSC% for frequent and conservation management, respectively.

More recently, significant variation for WSC in five phenological stages (vegetative, heading, pollination, milky and dough seed stage), were obtained in by Jafari et al (2010) in tall wheatgrass (11.2 to 14.4%), Jafari and Rezaeifard (2010) in tall fescue (8.38 to 12.69) and by Rezaeifard, et al., (2010) in cocksfoot (7.76 to 11.37), respectively.

2.2. Gene action and heritability

By comparison to DM yield and DMD there is little information on genetic control of WSC and the extent of genetic variation in forage grasses. Estimates of three heritability estimates broad sense heritability ($h^2{}_b$), narrow sense heritability ($h^2{}_n$), and heritability, offspring/parent regression ($h^2{}_{op}$) are summarized in Table 1. Cooper (1962) reported $h^2{}_{op}$ of 0.11 to 0.62 in perennial ryegrass (Table 1). Based on a diallel analysis, he estimated high $h^2{}_n$=0.84 and concluded that gene effects were additive (Cooper, 1973).

Contrasting results were reported by Humphreys (1989a, b). From an analysis of crosses between the early heading ryegrass, Aurora, and five late heading varieties, he found different degrees of directional dominance and over-dominance (heterosis) and concluded that WSC behaved as a complex polygenic trait, which was controlled by mainly non-additive gene effects. Genetic variation within crosses was relatively small and estimated F2 family heritability was 0.08 to 0.14 and 0.20 to 0.38 under frequent cutting and conservation management, respectively (Table 1). In other experiments Humphreys (1995) found much higher estimates of $h^2{}_b$ while the values for conservation management was again higher than that for frequent cutting (Table 1). While assessing winter hardiness of 86 accessions of perennial ryegrass under spaced plants condition, Humphreys (1989d) found significant genetic variation for WSC and estimated $h^2{}_b$ from 0.38 to 0.60 for March and October harvests, respectively. In another survey he estimated $h^2{}_b$ in the range of 0.27 to 0.70 for different cuts over two years in 81 accessions of perennial ryegrass (Table 1) (Humphreys, 1991).

Jafari (1998) using a polycross progeny test in ryegrass under spaced plants and swards, estimated relatively high, ($h^2{}_b$) and ($h^2{}_n$) for WSC ($h^2{}_b$=0.60-0.80), ($h^2{}_n$=0.50-0.54), respectively. The genetic analysis indicated additive genetic variances were more important for WSC and other quality traits. In another survey using full sib families of ryegrass he estimated relatively high $h^2{}_b$=0.60 and 0.80 for conservation and frequent cutting management, respectively (Table 1) (Jafari et al., 2003).

In other grass species some estimates of genetic control of WSC have been published (Table 1). Cooper (1962), in cocksfoot, estimated relatively high $h^2{}_{op}$ which ranged from 0.78 to 0.56 for July and August cuts, respectively. Buckner et al (1981), using a polycross progeny test in *Lolium-festuca* hybrid derivatives, estimated $h^2{}_b$ of 0.55 and $h^2{}_{op}$ of 0.01 to 0.47 for WSC. Burner et al (1983) analyzed variance of tall fescue and obtained $h^2{}_b$ =0.55. Grusea and Oprea (1994) found $h^2{}_b$ estimates for WSC of 0.58 in cocksfoot. Jafari and Javarsineh (2005) in tall fescue and Jafari and Naseri (2007) in cocksfoot estimated relatively moderate to high values of $h^2{}_b$, while $h^2{}_n$ estimates were low suggests that genetic variation in this trait is controlled by both additive and non additive gene action. Sanada, et al (2007) in cocksfoot found significant genetic variation and moderate $h^2{}_n$ estimates, and suggested that forage quality of cocksfoot was influenced by an additive gene effect and could be improved genetically by recurrent and phenotypic-genotypic selection (Table 1).

Author (s)	Management/ Environment	Grass species	Basis of estimation	Heritability		
				h^2_b	h^2_n	h^2_{op}
Cooper (1962)	C/pots/ 2 cuts	*L.perenne*	FS-families			0.11 to 0.62
Cooper (1973)	C/SP/1st cut	*L.perenne*	Diallel		0.84	
Humphreys (1989d)	C/SP/2 cuts	*L.perenne*	Populations	0.38 to 0.60		
Humphreys (1991)	C/SP/2 years	*L.perenne*	Populations	0.27 to 0.70		
Humphreys (1989b)	C/Sward/2 yr.	*L.perenne*	FS-families	0.08 to 0.38		
Humphreys (1995)	C/Sward/2nd yr.	*L.perenne*	HS-families	0.34 to 0.71		
Jafari (1998)	C/SP/2 years	*L.perenne*	HS-families	0.75	0.50	0.26
Jafari (1998)	C/Sward/ 2 yr.	*L.perenne*	HS-families		0.54	0.17
Jafari (1998)	C/ Pots/3 yr.	*L.perenne*	HS-families	0.67	0.32	0.35
Jafari et al (2003)	C/Sward/ 2 yr.	*L.perenne*	FS-families	0.60		
Jafari et al (2003)	F/Sward/ 2 yr.	*L.perenne*	FS-families	0.80		
Jafari & Javarsineh (2005)	C/SP/ 2 yr.	*F.arundinacea*	HS-families	0.50	0.11	0.00
Burner et al (1983)	C/SP/ 2 cuts	*F.arundinacea*	Clones	0.55		
Jafari and Naseri (2007)	C/SP/2 yr.	*D.glomerata*	HS-families	0.50	0.23	0.62
Cooper (1962)	C/pots/ 2 cuts	*D.glomerata*	FS-families			0.56 to 0.78
Sanada et al (2007)	C/SP/	*D.glomerata*	HS-families	0.10	0.53	0.78
Sanada et al (2007)	C/Sward/	*D.glomerata*	Clones	0.50	0.59	
Grusea & Oprea (1994)	C/Sward/ 2 yr.	*D.glomerata*	FS-families	0.58		
Buckner et al (1981)	C/SP/ 2nd cut	*Festolulium*	Clones	0.55		
Buckner et al (1981)	C/Sward/2ndcut	*Festolulium*	HS-families		0.39 to 0.49	0.01 to 0.47

C, F, SP, HS, FS, Conservation management, frequent cutting management, spaced plants, Half-sib, Full-sib, respectively

Table 1. Heritability estimates for water soluble carbohydrates (WSC) in some cool-season grass species

2.3. Genotype × Environment interactions

There is little evidence for GE interaction of WSC. Burner et al (1983) reported that interaction between tall fescue genotypes and environments for WSC concentration were minor. Humphreys (1989b) reported that although WSC is a trait which is subject to large environmental fluctuations, genetic differences could remain fairly stable without large GE interaction. He also found relative WSC contents in parents and hybrids were consistent over generations under both spaced plants and sward conditions. In contrast, Buckner et al (1981), using a polycross progeny test in *Lolium-festuca* hybrid derivatives, reported when parent and HS families were evaluated in the same environment h^2_{op} were much higher than when they were grown in different environments. They suggested that a minimum of two locations with widely different environments was necessary to develop varieties for forage quality. Conaghan et al (2008) for DM yield came to same similar conclusions. They suggested at least three locations and 2 sowing years is necessary for evaluation of perennial ryegrass trials.

Sanada, et al (2007) in cocksfoot reported a significant genotype × year interaction for WSC and suggests that the evaluation of forage quality traits should be carried out by divergent selection under multiple environments, especially for the evaluation of parental clones. Jafari and Rezaeifard (2010) in tall fescue found non significant genotype × phenological stage interaction effect for WSC. Similarly, Rezaeifard et al (2010) in cocksfoot reported that genotypes by environments interaction for WSC concentration were minor. This finding were in agreement of Buxton and Casler (1993), that in a review, concluded that most environment stresses have a greater effect on DM yield than on quality traits and G x E interactions should be smaller for forage quality than for yield.

In conclusion, despite great fluctuations in WSC due to environmental and plant developmental factors, there is genetic variation within and between forage grasses populations for this character which could be exploited by selection.

3. Correlation between WSC and yield/quality traits

3.1. Correlation between WSC and Digestibility (DMD)

An understanding of the interrelationships between quality traits is important in the development of a selection program. Table 2 summarizes some of the results published for the relationships between DMD and WSC.

A broad pattern of positive phenotypic correlation between digestibility and WSC has been found for forage grasses under both conservation and frequent cutting management (Table 2). Since WSC is completely digestible a positive correlation between these two parameters is expected. The summary data for confirms that these characters show a moderate/strong positive correlation. Humphreys (1989c) found large positive correlation between DMD and WSC for spaced plants. However, under sward conditions with frequent cutting management only three of eight cuts showed significant correlation and over all cuts it was non significant (Table 2). Significant positive correlations between digestibility and WSC concentration in tall fescue forage have been reported (Burns and Smith, 1980). Bugge (1978), in Italian ryegrass, and Clements (1969), using canarygrass, found no significant correlation between DMD and WSC. Strong positive phenotypic and genotypic correlation between WSC and DMD were obtained by Jafari (1998), Jafari et al (2003) in ryegrass, Jafari and Naseri (2007) in cocksfoot, Jafari and Javarsineh (2005) in tall fescue (Table 2).

Source	Management / Nursery/ harvests	Grass species	WSC/ DMD	
			Phenotypic	Genotypic
Frandsen (1986)	C/SP/1st Cut	*Lolium perenne*	0.76**	
Humphreys (1989c)	C/SP/annual	*Lolium perenne*	0.70**	
Dent & Aldrich (1963)	C/Sward /2 Cuts	*Lolium perenne*	0.29 to 0.51**	
Jafari (1998)	C/SP/ 2 years	*Lolium perenne*	0.89**	0.94
Jafari (1998)	C/Sward 2years	*Lolium perenne*	0.69**	0.82
Jafari (1998)	C/ Pots/3 generation	*Lolium perenne*	0.55	0.67
Jafari et al (2003)	C/Sward/ 2 years	*Lolium perenne*	0.61**	0.58
Dent & Aldrich (1963)	F/Sward	*Lolium perenne*	0.02 to 0.56**	
Grimes et al (1967)	F/Sward/annual	*Lolium perenne*	0.87**	
Humphreys (1989c)	F/Sward/ 8Cuts	*Lolium perenne*	-0.20 to 0.90**	
Jafari et al (2003)	F/Sward/ 2 years	*Lolium perenne*	0.62**	0.67
Clements (1969)	C/ Pots/3 generation	*Phalaris tuberosa*	0.16	
Buckner et al (1981)	C/SP/ 2nd Cut	*Festololium*	0.17 to 0.57**	
Thomson & Rogers (1971)	C/Sward/2 years	*Phleum pratense*	0.12 to 0.49**	
Jafari and Naseri (2007)	C/SP/ 2 years	*Dactylis glomerata*	0.79**	0.98
Jafari and Javarsineh (2005)	C/SP/ 2 years	*Festuca arundinacea*	0.71**	0.77

*, **, Significant at 5%, 1%, respectively, respectively
SP, C, F, Spaced plant, Conservation and Frequent cutting management, respectively.

Table 2. Correlation coefficients (r) between dry matter digestibility (DMD) and water soluble carbohydrates (WSC) in some cool-season grass species.

3.2. Correlation between WSC and Crude protein (CP)

Phenotypic correlations between WSC and CP are consistent and negative across all species (Table 3). Cooper (1962) reported negative genetic correlation between WSC and CP and concluded that the limits are sufficiently wide to allow selection of a large range of protein/carbohydrate ratios. This negative relationship is present over a wide range of environment for all species. Apart from data in Table 3, there are other results which confirm this conclusion. Vose and Breese (1964), in a glasshouse study, reported high negative phenotypic relationship between WSC and CP in ryegrass. However, using covariance analysis, they suggested there is sufficient degree of genetic independence to allow simultaneous selection for both characters. Radojevic et al (1994) investigated the relationship between WSC and CP using a multiple regression analysis based on three variables (harvest date, genotype and WSC concentration). The analysis showed that the largest influences on CP concentration were time of year (72%) and genotype (5%). WSC did not explain a significant proportion of the variance of CP concentration and they concluded that, although WSC and CP were negatively correlated, this was due mainly to divergent seasonal variation in these components of herbage. Humphreys (1989c) suggested that as growth increases with rapid uptake of nitrogen fertilizer, increase of CP and decrease of WSC content are environmentally induced effects.

Strong negative phenotypic and genotypic correlation between WSC and CP were obtained by Jafari (1998), Jafari et al (2003) in ryegrass, Jafari and Naseri (2007) in cocksfoot, Jafari and Javarsineh (2005) in tall fescue. However, Sanada et al (2007) obtained weak negative phenotypic and genotypic correlation between two traits in cocksfoot (Table 3).

Source	Management / Nursery/ harvests	Species	WSC/CP Phenotypic	WSC/CP Genotypic
Cooper (1962)	C/Pots/2 Cuts	*Lolium perenne*		-0.20 to -0.70**
Humphreys (1989c)	C/SP/annual	*Lolium perenne*	-0.20*	
Humphreys (1989d)	C/SP/Oct. Cut	*Lolium perenne*		-0.70**
Valentine & Charles (1979)	C/Sward/3rd Cut	*Lolium perenne*	-0.61**	
Bugge (1978)	C/SP/1st Cut	*Lolium perenne*	-0.50**	
Jafari (1998)	C/SP 2 years	*Lolium perenne*	-0.75**	*-0.84*
Jafari (1998)	C/Sward 2years	*Lolium perenne*	-0.69**	*-0.80*
Jafari et al (2003)	C/Sward/ 2 years	*Lolium perenne*	-0.65**	-0.60**
Dent & Aldrich (1963)	F/Sward	*Lolium perenne*	-0.44** to -0.69**	
Grimes et al (1967)	F/Sward/annual	*Lolium perenne*	-0.65**	
Humphreys (1989c)	F/Sward/ 8Cuts	*Lolium perenne*	-0.30 to -0.90**	
Jafari et al (2003)	F/Sward/ 2 years	*Lolium perenne*	-0.51**	-0.66**
Marais et al (1993)	F/SP/annual	*Lolium multiflorum*	-0.17 to -0.26*	-0.56**
Cooper (1962)	C/Pots/2 Cuts	*Dactylis glomerata*		-0.48* to -0.57**
Sanada et al (2007)	C/Sward/	*Dactylis glomerata*	-0.34	-0.33
Sanada et al (2007)	C/SP/	*Dactylis glomerata*	-0.14	-0.16
Jafari and Naseri (2007)	C/SP/ 2 years	*Dactylis glomerata*	–0.53*	–0.32
Clements (1969)	C/Pots/3generation	*Phalaris tuberosa*	-0.13	
Thomson & Rogers (1971)	C/Sward/2 years	*Phleum pratense*	-0.56** to -0.68**	
Jafari &Javarsineh (2005)	C/SP/ 2 years	*Festuca arundinacea*	-0.77**	–0.83

*, **, Significant at 5%, 1%, respectively, respectively
SP, C, F, Spaced plant, Conservation and Frequent cutting management, respectively.

Table 3. Correlation coefficients (r) between water soluble carbohydrates (WSC) and crude protein (CP) in some cool-season grass species.

In summary, data in Table 3 show a consistent negative correlation between WSC and CP at both phenotypic and genotypic levels and over various herbage species. It could be concluded that, due to genetic correlation, selection for one component alone is likely to have negative effect on the other component.

3.3. Correlation between WSC and DM yield

Genotypic and phenotypic correlations between DM yield and WSC in perennial ryegrass and in other forage species are inconsistent although significant correlations are generally positive (Humphreys, 1989c). In a glasshouse study, Vose and Breese (1964) found a low and non-significant correlation between DM yield and WSC. However, Valentine and Charles (1979) obtained a highly positive correlation between DM yield and WSC particularly at low nitrogen level (Table 4).

Jafari (1998) in ryegrass found, positive and significant correlation between WSC and DM yield under spaced plant. For sward experiments their relationships were positively non significant (Jafari, 1998; Jafari et al., 2003). In contrast, Jafari and Naseri (2007) in cocksfoot, found negative and significant relationships between two traits. Jafari and Javarsineh (2005) in tall fescue and Sanada et al (2007) in cocksfoot reported low and inconsistent values between traits. The overall pattern of results suggests that DM yield and WSC are independent or show a weak positive correlation (Table4).

Source	Management /	Grass Species	WSC vs. DM yield	
	Nursery/harvests		Phenotypic	Genotypic
Valentine & Charles (1979)	C/Sward/3rd Cut	*Lolium perenne*	0.24 to 0.78**	
Humphreys (1989d)	C/SP/Oct. Cut	*Lolium perenne*		0.60**
Vose & Breese (1964)	C/Pots/4 Cuts	*Lolium perenne*		0.16
Jafari (1998)	C/SP 1 years	*Lolium perenne*	0.36*	0.61**
Jafari (1998)	C/Sward 2years	*Lolium perenne*	0.25	0.17
Jafari et al (2003)	C/Sward/ 2 years	*Lolium perenne*	0.39	0.04
Humphreys (1989c)	F/Sward/8 Cuts	*Lolium perenne*	-0.40 to 0.60**	
Jafari et al (2003)	F/Sward/ 2 years	*Lolium perenne*	0.36	0.08
Clements (1969)	C/Sward /3 years	*Phalaris Tuberosa*	0.20	
Brown & Blaser (1970)	C/Sward / 1st Cut	*Dactylis glomerata*	0.38 to 0.82**	
Sanada et al (2007)	C/Sward/	*Dactylis glomerata*	-0.11	-0.10
Jafari & Naseri (2007)	C/SP/ 2years	*Dactylis glomerata*	-0.45**	-0.51**
Jafari & Javarsineh (2005)	C/SP/ 2years	*Festuca arundinacea*	0.04	0.06

*, **, Significant at 5%, 1%, respectively, respectively.
SP, C, F, Spaced plant, Conservation and Frequent cutting management, respectively.

Table 4. Correlation coefficients (r) between DM yield and water soluble carbohydrates (WSC) in some cool-season grass species.

4. Conclusion and grass breeding strategy

Under rotational grazing and cutting there is frequently an excess of CP in most of grasses in terms of animal requirement, particularly in autumn cuts. This excess of CP leads to high levels of rumen degradable protein (RDP). In the absence of a readily available supply of fermentable carbohydrate this can result in elevated ruminal ammonia concentrations (Beever and Siddons, 1986; Van Vuuren et al, 1986).   There is also evidence that this imbalance between CP and carbohydrate has a negative influence on reproductive behavior in grazing animals (Visek, 1984; Canfield et al, 1990). Given the relationship between WSC, CP, DM yield and DMD, it is tempting to suggest that selection for high WSC is a means to improve quality in general. Beerepoot and Agnew (1997) have argued that this simple approach may not result in improved herbage quality because of possible negative effects on rumen pH. There is, however, indirect evidence that higher WSC in ryegrass may result in improved animal performance (Lee et al., 2001; Evans et al., 2011; Miller et al., 2001). Tetraploid ryegrass varieties have higher levels of WSC compared to diploids (Jung et al, 1996; Smith et al. 2001; Gilliland et al. 2002). There are two large scale animal production trials in which tetraploid and diploid varieties were compared (Castle and Watson, 1971; Connolly et al, 1977). The results from both experiments indicated that tetraploids were superior to diploids of comparable DM yield. It is possible, that this superiority was due to the slightly higher WSC content of tetraploid varieties (Smith et al. 2001; Gilliland et al. 2002). Grimes et al (1967) also found strong positive correlation between WSC content and intake and liveweight gain of lambs. On the basis of published data, it is suggested that increased WSC, particularly in the vegetative leaf and in the summer/autumn period when CP is frequently in excess, would improve herbage quality. The data also indicate that response to combined selection for both WSC and DM yield should be possible.

Abbreviations

WSC	Water soluble carbohydrates	h^2_b	Broad sense heritability
DMD	Dry matter digestibility	h^2_n	Narrow sense heritability
DM yield	Dry matter yield	h^2_{op}	Heritability, offspring/parent regression
CP	Crude protein	r_p	Phenotypic correlation
HS families	Half -sib families	r_g	Genotypic correlation

Author details

Ali Ashraf Jafari
Plant Breeding, Gene Bank Division, Research Institute of Forests and Rangelands, Tehran, Iran

5. References

Alberda, T. 1966. The influence of reserve substances on dry matter production after defoliation. *Proceeding X International Grassland Congress* pages 140-147.

Allsop, A.R.C., Nicol, A.M. and Edwards, G.R. 2009. Effect of water soluble carbohydrate concentration of ryegrass on the partial preference of sheep for clover. Proceedings of the New Zealand Animal Production Society 69: 20–23.

Beerepoot, L. J. and Agnew, R. E. 1997. Breeding for improved herbage quality in perennial ryegrass. In. *"Seeds of Progress"* (ed. Weddell, J. R.) *Occasional Symposium of the British Grassland Society*. 32: 135-145.

Beever, D. E. and Reynolds, C. K. 1994. Forage quality, feeding value and animal performance. In: *"Grassland and society"* (eds. tMannetje and Frame), *Proceeding of the 15th EUCARPIA general Meeting of the European Grassland Federation,* Wageningen, Netherlands, pages 48-60.

Beever, D. E. and Siddons, R. C. 1986. Digestion and metabolism in the grazing ruminant. In control of digestion and metabolism in ruminants (eds. Milling, et al). Englewood Cliffs, NJ: Prentice-Hall, pages 479-497.

Breese, E. L. and Thomas, A. C. 1967. *In vitro* digestibility in cocksfoot. Report of the Welsh Plant Breeding Station for 1966, UK, pp. 35-41.

Brown, R. H. and Blaser, R. E. 1970. Soil moisture and temperature effects on growth and soluble carbohydrates of orchardgrass (*Dactylis glomerata*). *Crop Science* 10: 213-216.

Buckner, R. C., Burrus, P. B., Cornelius, L. P., Bush, L. P. and Leggett, J. E. 1981. Genetic variability and heritability of certain forage quality and mineral constituents in *Lolium-Festuca* hybrid derivatives. *Crop Science* 21: 419-423.

Bugge, G. 1978. Genetic variability in chemical composition of Italian ryegrass ecotypes. *Z. Pflanzenzuchtg* 81: 235-240.

Burner, D. M., Balasko, J. A. and Thayne, W. V. 1983. Genetic and environmental variance of water soluble carbohydrate concentration, yield and disease in tall fescue. *Crop Science* 23: 760-763.

Burns, J. C. and Smith, D. (1980). Non-structural carbohydrates residue, neutral detergent fiber, and *in vitro* dry matter disappearance of forages. *Agronomy Journal* 72: 276-281.

Buxton, D. R. and Casler, M. D. 1993. Environmental and genetic effects on cell wall composition and digestibility. In: *"Forage cell wall structure and digestibility"* (eds. Jung, et al), ASA, CSSA, and SSSA, Madison, USA, pages 685-714.

Buxton, D. R., Mertens, D. R. and Fisher, D. S. 1996. Forage quality and ruminant utilisation. In: *"Cool-season forage grasses"* (eds. Moser et al) ASA, CSSA, and SSSA, Madison, USA pages 229-266.

Canfield, R. W., Sniffen, C. J. and Butler, R. W. 1990. Effects of excess degradable protein on postpartum reproduction and energy blance in dairy cattle. *Journal of Dairy Science* 73: 2342-2349.

Carlier, l. 1994. Breeding, forage quality, feeding value and animal performance. *Proceeding of the 19th EUCARPIA Fodder Crops Section Meeting,* Brugge, Belgium, pages 25-27.

Castle, M. E. and Watson, J. N. 1971. A comparison between a diploid and a tetraploid perennial ryegrass for milk production. *Journal of agricultural Science Cambridge* 77: 69-76.

Clements, R. J. 1969. Selection for crude protein content in *Phalaris tuberosa* L. I. Response to selection and preliminary studies on correlated response. *Australian Journal of Agricultural Research* 20: 643-652.

Conaghan, P., M. D. Casler, D. A. McGilloway, P. O'Kiely and L. J. Dowley. 2008. Genotype x environment interactions for herbage yield of perennial ryegrass sward plots in Ireland *Grass and Forage Science* 63: 107–120.

Connolly, V., Ribeiro, M. do Valle and Crowley, J.G. 1977. Potential of grass and legume cultivars under Irish conditions. *Proceedings International Meeting on animal production from Temperate Grassland*, Dublin, pages 23-28.

Cooper, J. P. 1962. Selection for nutritive value. *Report of the Welsh Plant Breeding Station for 1961*, UK, pages 145-156.

Cooper, J. P. 1973. Genetic variation in herbage constituents. In: *"Chemistry and biochemistry of herbage"* (eds. Butler and Bailey), Vol. II. Academic press, London pages 379-417.

Cosgrove, G.P., Koolaard, J., Potter, F., Luo, D., Burke, J.L. and Pacheco, D. 2010. Milk solids yield from high water soluble carbohydrate ryegrasses: a combined analysis of multiyear data. In: "Proceedings of the 4th Australian Dairy Science Symposium", 31st Aug.2nd Sept. Lincoln University, New Zealand, pages 334–338.

Dent, J. W. and Aldrich, D. T. A. 1963. The inter relationship between heading date, yield chemical composition and digestibility in varieties of perennial ryegrass, timothy, cocksfoot and meadow fescue. *Journal of the National Institute of Agriculture Botany* 9: 261-281.

Evans, J. G., M. D. Fraser, I. Owen and D. A. Davies. 2011. An evaluation of two perennial ryegrass cultivars (AberDart and Fennema) for sheep production in the uplands. *Journal of Agricultural Science* 149: 235-248.

Frandsen, K. J. 1986. Variability and inheritance of digestibility in perennial ryegrass (*Lolium perenne*), meadow fescue (*Festuca pratensis*), and cocksfoot (*Dactylis glomerata*). II. F1 and F2 progeny. *Acta Agricuturae Scandinavica* 36: 241-263.

Fulkerson, W.J. and Donaghy, D.J., 2001. Plant-soluble carbohydrate reserves and senescence: key criteria for developing an effective grazing management system for ryegrass-based pastures: a review. *Australian Journal of Experimental Agriculture* 41 (2): 261-275.

Gill, M., Beever, D. E. and Osbourn, D. F. 1989. The feeding values of grass and grass products. In: *"Grass, its production and utilisation"* (ed. Holmes, W.), Blackwell Scientific Publications, London, pages 89- 129.

Gilliland, T.J., Barrett, P.D., Mann, R.L., et al., 2002. Canopy morphology and nutritional quality traits as potential grazing value indicators for *Lolium perenne* varieties. *Journal of Agricultural Science*, 139(3): 257-273.

Grimes, R. C., Watkin, B. R. and Gallagher, J. R. 1967. The growth of lambs on perennial ryegrass, tall fescue and cocksfoot, with and without white clover, as related to the botanical and chemical composition of the pasture and pattern of fermentation in rumen. *Journal of Agricultural Science Cambridge* 68: 11-21.

Grusea, A. and Oprea, G. 1994. Variation and inheritance of quality of *Dactylis glomerata* varieties which were obtained by different breeding methods. *Proceeding of the 19th EUCARPIA Fodder Crops Section Meeting* Brugge, Belgium, pages 145-149.

Haigh, P. M. 1990. Effect of herbage water-soluble carbohydrate content and weather conditions at ensilage on the fermentation of grass silage made on commercial farms. *Grass and Forage Science* 45: 263-271.

Humphreys, M. O. 1989a. Water-soluble carbohydrates in perennial ryegrass breeding. I. Genetic differences among cultivars and hybrid progeny grown as spaced plants. *Grass and Forage Science* 44: 231-236.

Humphreys, M. O. 1989b. Water-soluble carbohydrates in perennial ryegrass breeding. II. Cultivar and hybrid progeny performance in cut plot. *Grass and Forage Science* 44: 237-244.

Humphreys, M. O. 1989c. Water-soluble carbohydrates in perennial ryegrass breeding. III. Relationships with herbage production, digestibility and crude protein content. *Grass and Forage Science* 44: 423-430.

Humphreys, M. O. 1989d. Assessment of perennial ryegrass (*Lolium perenne* L.) for breeding. II. Components of winter hardiness. *Heredity* 41: 99-106.

Humphreys, M. O. 1991. A genetic approach to the multivariate differentiation of perennial ryegrass (*Lolium perenne* L.) populations. *Heredity* 66: 437-443.

Humphreys, M. O. 1994. Variation in the carbohydrate and protein content of ryegrass: Potential for genetic manipulation. *Proceeding of the 19ᵗʰ EUCARPIA Fodder Crops Section Meeting* Brugge, Belgium, pages 165-171.

Humphreys, M. O. 1995. Multitrait response to selection in *Lolium perenne* L. (perennial ryegrass) populations. *Heredity* 74: 510-517.

Jafari A. A. and Javarsineh S. 2005. Estimates of heritability and predicted of genetic gain for yield and quality traits in parents and half sib families of tall fescue (*festuca arundinacea* schreb). *Iranian Journal of Rangelands Forests Plant Breeding And Genetic Research* 13(1): 99-124. ISSN 1735-0891 (Abstract in English).

Jafari A. A. and Rezaeifard, M. 2010. Effects of maturity on yield and quality traits in tall fescue (*Festuca arundinace* Schreb) American-Eurasian J. Agric. & Environ. Sci. 9 (1): 98-104

Jafari, A. A. 1998. Genetic analysis of yield and quality in perennial ryegrass (*Lolium perenne* L.). Ph.D. Thesis, Department of Crop Science, Horticulture and Forestry. University College Dublin, Ireland.

Jafari, A. A. and Naseri, H. 2007. Genetic variation and correlation among yield and quality traits in cocksfoot (*Dactylis glomerata* L). *Journal of Agricultural Science, Cambridge*. 145: 599-610.

Jafari, A. A., V. Connolly and E.K. Walsh. 2003. Genetic analysis of yield and quality in full-sib families of perennial ryegrass (*Lolium perenne*. L) Under two cutting managements. *Irish Journal of Agricultural and Food Research* 42: 275-292.

Jafari, A.A., Anvari, H., Nakhjavan, S. and Rahmani, E. 2010. Effects of phenological stages on yield and quality traits in 22 populations of tall wheatgrass grown in Lorestan, Iran. *Journal of Rangeland Science* 1(2): 13-20

Jones, E. L. and Roberts, J. E. 1991. A note on the relationship between palatability and water-soluble carbohydrates content in perennial ryegrass. *Irish Journal of Agricultural and food Research* 30: 163-167.

Jung, G. A., Van Wijk, J. A. P., Hunt. W. F. and Watson, C. E. 1996. Ryegrass. In: "*Cool-season forage grasses*" (eds. Moser et al.), ASA, CSSA, and SSSA, Madison, USA, pages 605- 641.

Lee, M. R. F. , Jones, E. L. ,Moorby, J. M., Humphreys, M. O., Theodorou, M. K., MacRae, J. C., Scollan, N. D. 2001. Production responses from lambs grazed on *Lolium perenne* selected for an elevated water-soluble carbohydrate concentration Animal Research 50(6): 441-449.

Marais, J. P., de Figueiredo, M. and Goodenough, D. C. W. 1993. Dry mater and non-structural carbohydrate content as quality parameters in *Lolium multiflorum* breeding program. *African Journal Forage Science* 10: 118-123.

Mayland H. F., Shewmaker G.F., Harrison P. A. and Chatterton N. J. 2001. Nonstructural carbohydrate in tall fescue cultivars; relationship to animal preference. *Agronomy Journal*, 92: 1203-1206.

McGrath, D. 1988. Seasonal variation in the water-soluble carbohydrates of perennial and Italian ryegrass under cutting conditions. *Irish Journal of Agricultural and food Research* 27: 131-139.

McGrath, D. 1992. A note on the influence of nitrogen application and time of cutting on water soluble carbohydrate production by Italian ryegrass. *Irish Journal of Agricultural and food Research* 31: 189-192.

Miller, L. A., Moorby, J. M., Davies, D. R., Humphreys, M. O., Scollan, N. D., Macrae, J. C., Theodorou, M. K. 2001. Increased concentration of water-soluble carbohydrate in perennial ryegrass (*Lolium perenne* L.). Milk production from late-lactation dairy cows. *Grass and Forage Science* 56 (4): 383-394.

Parsons, A. J., Edwards, G.R., Newton, P. C. D., Chapman, D. F., Caradus, J. R., Rasmussen, S. and Rowarth, J.S. 2010. Past lessons and future prospects: plant breeding for cool temperate pastures. In: "Proceedings of the 4th Australasian Dairy Science Symposium", Lincoln University, New Zealand, pages 272–291.

Pollock C.J. and Cairns A.J. 1991. Fructan metabolism in grasses and cereals. Annual Review of Plant Physiology and Plant Molecular Biology 42:77-101.

Radojevic, I., Simpson, R. J., John, J. A. St. and Humphreys, M. O. 1994. Chemical composition and *in vitro* digestibility of lines of *Lolium perenne* selected for high concentrations of water-soluble carbohydrate. *Australian Journal of Agriculture Research* 45: 901-912.

Rasmussen, S., Parsons, A.J., Fraser, K., Xue, H. and Newman, J. 2008. Metabolic profiles of *Lolium perenne* are differentially affected by nitrogen supply, carbohydrate content, and fungal endophyte infection. Plant Physiology 146: 1440–1453.

Rasmussen, S., Parsons, A.J., Xue, H. and Newman, J.A. 2009. High sugar grasses harnessing the benefits of new cultivars through growth management. Proceedings of the New Zealand Grassland Association 71: 167–175.

Rezaeifard M., Jafari, A. A. and Assareh, M. H. 2010. Effects of phenological stages on forage yield quality traits in cocksfoot (*Dactylis glomerata*). *Journal of Food, Agriculture & Environment* 8(2): 365-369.

Sanada, Y., Takai, T. and Yamada, T. 2007. Inheritance of the concentration of water-soluble carbohydrates and its relationship with the concentrations of fiber and crude protein in herbage of cocksfoot (*Dactylis glomerata* L.). *Grass and Forage Science* 62 (3): 322–331.

Smit, H. J. 2006. Cultivar effects of perennial ryegrass on herbage intake by grazing dairy cows. In: A. Elgersma, J. Dijkstra and S. Tamminga (eds.), Fresh Herbage for Dairy Cattle, 45-62.

Smith K. F., Reed K.F.M. and Foot J.Z. 1997. An assessment of the relative importance of specific traits for the genetic improvement of nutritive value in dairy pasture. Grass and Forage Science, 52: 167-175.

Smith, K.F., Simpson, R.J., Culvenor, R.A., et al., 2001. The effects of ploidy and a phenotype conferring a high water-soluble carbohydrate concentration on carbohydrate accumulation, nutritive value and morphology of perennial ryegrass (*Lolium perenne* L.). *Journal of Agricultural Science*, 136 (1): 65-74.

Smouter, H. Simpson, R. J. and Pearce, G. R. 1995. Water-soluble carbohydrates and *in vitro* digestibility of annual ryegrass (*Lolium rigidum* Gaudin) sown at varying densities. *Australian Journal of Agricultural Research* 46: 611-625.

Stewart A. and R. Hayes. 2011. Ryegrass breeding-balancing trait priorities. *Irish Journal of Agricultural and Food Research* 50: 31–46.

Thomas, H. and Norris I. B. 1981. The effect of light and temperature during winter on growth and death in simulated swards of *Lolium perenne*. *Grass and Forage Science* 36: 107-116.

Thomson, A. J. and Rogers, H. H. 1971. The interrelationship of some components of forage quality. *Journal of Agricultural Science Cambridge* 76: 283-293.

Valentine, J. and Charles, A. H. 1979. The associations of dry matter yield with nitrogen and soluble carbohydrate concentration in perennial ryegrass (*Lolium perenne* L.). *Journal of Agricultural Science, Cambridge* 93: 657-667.

Van Vuuren, A. M., Van Der Koelen, C. J. and Vroons-De-Bruin, J. 1986. Influence of level and composition of concentrate supplements on rumen fermentation patterns of grazing dairy cows. *Netherlands Journal of Agricultural Science* 34: 457-467.

Visek, W. J. 1984. Ammonia: its effects on biological system, metabolic hormones, and reproduction. *Journal of Dairy Science* 67: 481-489.

Volaire, F. and Gandoin, J. M. 1996. The effect of age of the sward on the relationship between water-soluble carbohydrate accumulation and drought survival in two contrasted populations of cocksfoot (*Dactylis glomerata* L.). *Grass and Forage Science* 51: 190-198.

Vose, P. B. and Breese, E. L. 1964. Genetic variation in utilization of nitrogen by ryegrass species *Lolium perenne* and *Lolium multiflorum*. *Annals of botany* 110: 251-270.

Waite, R. and Boyd, J. 1953. The water-soluble carbohydrates of grasses. II. Grasses cut at grazing height several times during growing season. *Journal Science Food Agriculture* 4: 257-261.

Wilkins P. W. and Humphreys M. O. 2003. Progress in breeding perennial forage grasses for temperate agriculture. Journal of Agricultural Science, Cambridge, 140: 129-150.

Wilkins, P. W. and Davies, R. W. 1994. Progress in combining high dry matter yield with reduced flowering intensity and improved digestibility in perennial ryegrass. *Proceeding of the 19th EUCARPIA Fodder Crops Section Meeting* Brugge, Belgium, pages 247-250.

Wilkins, P.W. and Lovatt, J.A. 2007. "AberMagic (Ba13582)", IBERS Internal Reference 1340 4213. Institute of Biological, Environmental and Rural Sciences, University of Aberystwyth, UK

Wilkins, P.W., Lovatt, J. A., Hayes, R.C. and Thomas, G.L. 2010. "AberGreen (Ba13926)", IBERS Internal Reference. Institute of Biological, Environmental and Rural Sciences, University of Aberystwyth, UK.

Biotechnology

Chapter 9

Carbohydrates from Biomass: Sources and Transformation by Microbial Enzymes

Luis Henrique Souza Guimarães

Additional information is available at the end of the chapter

1. Introduction

In the last decades the destination of biomass, especially agro-industrial residues, is an important world problem that has been target of many researches. Accumulation of agro-industrial residues in the environment can cause serious ecological problems. On the other hand, these kinds of rich carbohydrate materials can aggregate economical value to different biotechnological process as for example in the microbial fermentative processes. According to this, the proposal of this chapter is to describe and discuss the utilization of biomass from agro-industrial residues and products and its transformation by microbial enzymes to obtain products (saccharides) with industrial interest. This is a subject that has attracted the attention of many researches and industrial sectors. To organize the information concerning this subject in a chapter is very interesting to qualify the state of the art on the utilization and importance of carbohydrates from the agro-industrial residues and products. The importance of microorganism for the transformation of biomass is another important aspect that will be highlighted.

Microorganisms, as bacteria and fungi, are able to use a great variety of inorganic and organic compounds as nutrients, reflecting an interesting metabolic diversity. Among these nutrients, nitrogen and carbon sources are indispensible for a primary metabolism. Others nutrients are required at low concentration, as vitamins. According to the growing, microorganisms are able to produce many enzymes that can show interesting biochemical properties for biotechnological application (Guimarães et al., 2006). Among these enzymes, some are constitutive while others are inducible. The induction of enzyme production by microorganisms can be obtained by use of properly biomass as carbon sources. Microorganisms are able to produce a diversity of enzymes as, for instance, the carbohydrate-active enzymes (figure 1). The glycoside hydrolases are enzymes able to acts on disaccharides, oligosaccharides and polysaccharides where can be found important enzymes as cellulases, amylases, inulinases and invertases (Table 1). Carbohydrate esterase is involved in the removal of O-(ester) and N-Acetyl moieties from carbohydrates. The polysaccharide lyase catalyzes the β-elimination reaction on uronic acid glucosides while the glycosyltransferase acts forming glycosidic bonds using activated sugar donors.

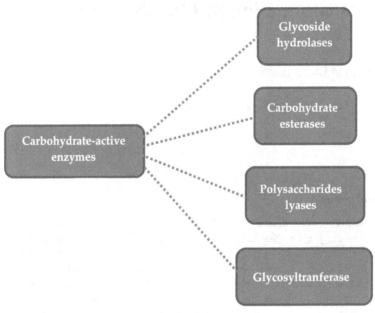

Figure 1. Classes of carbohydrate-active enzymes.

Microorganisms	Enzymes				
	Cellulase*	Xylanase	Invertase	Inulinase	Amylase
Bacteria					
Acremonium cellulolyticus	◆	◆			
Arthrobacter sp.				◆	
Bacillus amyloliquefaciens	◆				◆
Bacillus cellulyticus	◆				
Bacillus circulans		◆			◆
Bacillus licheniformis		◆			◆
Bacillus subtilis	◆	◆			◆
Bifidobacterium sp.			◆		
Cellulomonas sp.	◆	◆			
Clostridium cellulolyticum	◆				
Clostridium thermocellum	◆	◆			
Lactobacillus sp.			◆		◆
Pseudoalteromonas sp.	◆				
Streptomyces sp.		◆			

Microorganisms	Enzymes				
	Cellulase*	Xylanase	Invertase	Inulinase	Amylase
Fungi					
Aspergillus aculeatus	◆	◆			
Aspergillus caespitosus		◆	◆		
Aspergillus japonicus		◆	◆		
Aspergillus niger	◆	◆	◆	◆	◆
Aspergillus ochraceus		◆	◆		
Aspergillus oryzae	◆	◆	◆		◆
Aspergillus phoenicis		◆	◆		
Aspergillus terreus	◆	◆			
Chaetomium thermophilum		◆			◆
Emericella nidulans	◆	◆	◆		
Fusarium oxyporum		◆			
Humicola grisea	◆				◆
Humicola insolens	◆				◆
Neurospora crassa	◆	◆	◆		
Penicillium sp.	◆	◆		◆	◆
Trichoderma viride	◆	◆			

*Including endo-1,4- β-glucanases, exo-1,4- β-glucanases or cellobiohydrolases and 1,4- β-glucosidases. Data obtained from BRENDA (The Comprehensive Enzyme Information System).

Table 1. Some microbial (bacteria and fungi) sources of the enzymes involved in the utilization of carbohydrates found in the plant biomass.

2. Microbial cultivation using biomass

Different kinds of biomass have been used as carbon sources in the microbial cultivations under submerged and solid-state fermentations. Agro-industrial residues and products as, for example, rice straw, fruit peels, sugar cane bagasse and oat meal are important alternatives of carbon sources for both kinds of fermentation. The solid-sate fermentation is characterized as a system constituted by solid material in absence of free water where microorganisms are able to grown. This condition is more similar than that found by microorganisms in the environment if compared to the submerged condition. In addition, some other advantages for use of solid-state fermentation has been mentioned as: i) higher yields of products; ii) similar or higher yield if compared to submerged fermentation; iii) uniform dispersion of spore suspension; iv) higher levels of aeration and v) reduction of problems with contamination by bacteria and yeast. It is also important to consider that the medium for solid-state fermentation is simple and low cost substrates as agro-industrial residues can be used.

According to the substrate nature, two processes can be used for solid-state fermentation. The solid substrate, in the first case, is used as both support and nutrient source. These substrates are obtained from agriculture activity or from by-products from food industry.

Generally, they are heterogeneous and water insoluble. When substrates with amylaceous or lignocellulosic nature will be used, a pre-treatment is required to convert raw substrate into a suitable substrate. After, the liquid medium containing nutrients necessaries to the microbial growth can be used to moisten the inert support. The microbial growth and the product synthesis in solid state fermentation is influenced by environmental factors, such as water activity, moisture content of the substrate, mass transfer processes, temperature and pH. The control of these factors is not easy, configuring a negative aspect from solid-state fermentation. However, under economic view, solid-sate fermentation can be applied in different sectors for biotransformation of crop residues, food additives, biofuels, bioactive products, production of organic acids, detoxification of agro-industrial wastes, bioremediation, biodegradation and enzyme production (Pérez-Guerra et al., 2003).

The use of agro-industrial residues and/or products as substrates/ carbon source for SSF should be considered under some aspects. According to the biomass characteristics a pre-treatment step is necessary as cited above. To transform the raw material to the available form for microbial utilization it is necessary, many times, to reduce the size of the material using for example grinding among others. Other possibility is to promote damages on the superficial substrate layers using cracking, grinding or pearling. The utilization of chemical or enzymatic pre-treatments, cooking or vapor treatment and elimination of contaminants can be also utilized. According to the nutritional exigency of the microorganism, supplementation with phosphorus and nitrogen sources and salts can improve the microbial growth and the product yield. On the other hand, the influence of environmental factors on the SSF system also deserves consideration. The microbial growth as well as the obtainment of the products in SSF is directly affected by the moisture content. Excessive or reduced moisture content is prejudicial to the microorganism and, consequently, to the product recovery. According to this, the moisture content should be adjusted for each microorganism used in process considering the nature of the matrix used as substrate, and it has been used water content of substrate from 30% to 70% (Pérez-Guerra et al., 2003).

In SSF the gases and nutrients diffusion are severally affected by the matrix structure and also by the liquid phase in the system. The aeration permits an effective supplement of oxygen that can be used for aerobic metabolism and, at the same time, it promotes the removing of CO_2 and water vapour as well as the heat and volatiles compounds produced by the microbial metabolism. The temperature in the SSF system is a consequence of the microbial metabolism if the heat is not removed. The acquirement of nutrients depends on both hydrolysis of the polymeric structure to obtain monomers and after diffusion through the cell membrane from outside to inside the cell. The pH is another factor that affects the SSF system but its control is difficult (for review, see Pérez-Guerra et al., 2007).

Advantages for enzyme and secondary metabolites productions have been reported for both fermentations. In addition, agro-industrial residues and products are excellent alternatives as substrates for solid-state fermentation. These substrates, and consequently their carbohydrate content, can be transformed by action of a set of enzymes instead the chemical conversion. Enzymatic technology is a clear and secure process minimizing the environment problems while chemical process can generate pollulents.

Sugar cane bagasse is one of the most important agro-industrial residues accumulating biomass in the environment (figure 2) that can be used for microbial transformation. In Brazil, around 80 million of ton of sugar cane bagasse is produced per year. This raw material is constituted by 26-47% cellulose, 19-33% hemicelluloses, 14-23% lignin and 1-5% ashes. Part of this biomass as used for electric energy generation, but the most part is accumulated without

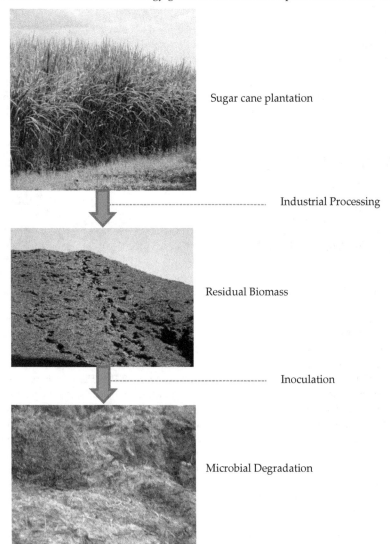

Sugar cane plantation

Industrial Processing

Residual Biomass

Inoculation

Microbial Degradation

Figure 2. Utilization of sugar-cane bagasse as biomass source for microbial activity (photos by Guimarães L.H.S).

destination. So, sugar cane bagasse can be used for many other purposes as paper production, fertilizer, feeding for ruminant animals and ethanol production, among others. It is an interesting carbon source/substrate for microbial cultivation and enzyme production as verified for many filamentous fungi. High levels of β-fructofuranosidase were obtained using sugar cane bagasse as carbon source in submerged cultivation from *Aspergillus niveus* (Guimarães et al., 2009) and *Aspergillus ochraceus* (Guimarães et al., 2007).

Many others agro-industrial residues can be used for microbial cultivation such as wheat bran, which was used for invertase production by *Aspergillus caespitosus* under submerged and solid-state fermentation (Alegre et al., 2009).

Although the submerged fermentation and solid state fermentation as good option for microbial cultivation, a new kind of fermentation has been proposed. The biofilm fermentation (BF) is characterized by the fungal growing on the inert support as can be observed in the figure 3.

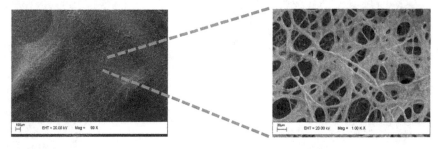

Figure 3. The *Aspergillus phoenicis* biofilm on the polyethylene as inert support.

The adhesion on surface is a natural process observed for the filamentous fungi in the environment. This complex process involves the production of adhesive compounds that fix the spore to the substrate, germ tube formation, hyphae elongation and, finally, surface colonization. These events can be observed in the *Aspergillus niger* biofilm formation on polyester cloth as reported by Villena and Gutiérrez-Correa (2007). These authors observed that the morphological pattern for *A. niger* growth attached to the surfaces is similar to that found in microbial biofilms with micro colonies development, extracellular matrix production and formation of pores and channels. The fungal morphology is an important factor to the enzyme production and, compared to the submerged fermentation, biofilms are more productive and more efficient if considered the metabolism associated with the specific enzymes which act on biomass, as for instance lignocellulosic enzymes. Recently, the use of *Aspergillus phoenicis* biofilms was reported for fructooligosaccharides production by one simple step (Aziani et al., 2012).

3. Biomass conversion by microbial enzymes

Residues and products of plant origin are recognized by their carbohydrate composition. According to this composition, the action of different enzymes on polysaccharides permits the obtainment of a variety of mono- and oligosaccharides which can be used by different sectors including biofuel, food, beverage and pharmaceutical among others.

Nowadays, the future of our energy sources and consequently the life in the planet is target of discussion around the world with participation of different sectors of the society as researches, politicians, undertakers and third sector. It has been noted that there is an increasing interest on biomass utilization as renewable energy source since that there is a conscience that the fossil fuels are restricted. In addition, this kind of fuel is a determinant factor of pollution to the atmosphere, where the CO_2 concentration has been increased. Hence, the utilization of biomass from plant residues for biofuel production is pointed-out as an important alternative for reduction of the energetic and environment problems. Brazil and USA are the main producer countries of ethanol to be used as fuel, the former using sugar cane and the later using the corn. For example, ton of sugar cane bagasse is generated as residue from the ethanol production in Brazil, which could be used for obtainment of fermentable sugars by enzymatic hydrolysis. In the next step, these sugars can be used for fermentation process to obtain ethanol.

3.1. Cellulases and lignocelullosic biomass

Considering the plant biomass, the main component from plant cell wall is the cellulose, the more abundant carbohydrate found in the planet. Structurally, this saccharide is constituted by glucopyranose monomers linked by β-1,4 glycosidic bonds with two distinct regions, the crystalline and amorphous regions. For the complete cellulose hydrolysis, an enzymatic complex (known as cellulases) constituted by endo-1,4- β-glucanases (EC 3.2.1.4), exo-1,4- β-glucanases or cellobiohydrolases (EC 3.2.1.91), and 1,4- β-glucosidases (EC 3.2.1.21), is necessary. Cellulases are modular enzymes included in the GH family (glycoside hydrolases). These enzymes have a complex structure with different modules as one or more catalytic domain and/or CBD module in the same protein. The CBD module is able to modify the catalytic domain and, consequently the cellulase properties, facilitating the interaction catalytic domain/crystalline cellulose. Cellulases can act using two main catalytic mechanisms, inversion or retention of the anomeric carbon. Two catalytic carboxylate residues are involved in both mechanisms and they are responsible for the acid-base catalysis in the reaction. Endoglucanases (EG; carboxymethiylcellulases, CMCase) catalyze random cleavage of cellulose internal bonds at amorphous region. Exoglucanases, also known as cellobiohydrolases (CBH) act at the chains ends (CBHI at the reduncing end and CBHII at non-reducing end), releasing cellobiose that can act as competitive inhibitor, while β-glucosidases (BGL) convert short cellooligosaccharides and cellobiose to glucose monomers. It is important to detach that the BGL activity is competitively inhibited by glucose. The GH1 family includes BGL obtained from bacteria, plant and mammalian and the GH3 family includes BGL from bacteria, fungi and plants. However, the full hydrolysis of cellulose depends on the previous hydrolysis of the other cell wall compounds, *i.e.* hemicelluloses and lignin (Dashtban et al., 2009; Bayer et al., 1998).

Hemicelluloses are polymeric molecules constituted by pentoses (as xylose and arabinose), hexoses (as mannose, glucose, galactose) and sugar acids. Because their heterogeneity, the hemicelluloses hydrolysis is only obtained by the action of different enzymes called hemicellulases. The most important hemicellulase is the enzyme that catalyzes the breakdown of β-1,4 linkages in the xylan, a polymer constituted by monomers of xylose. The oligomers obtained from this reaction are now substrates to the reaction catalyzed by β-xylosidase to obtain xylose (Dashtban et al., 2009).

Lignin is a heterogeneous aromatic polymer constituted by non-phenolic and phenolic structures that is able to link both cellulose and hemicelluloses making difficult the access of the enzymatic preparations to the cellulose and hemicelluloses. The enzyme able to catalyze the lignin hydrolysis are generically named as ligninases, which can be divided in two main families, phenol oxidase (laccase) and peroxidases that includes manganese peroxidase (MnP) and lignin peroxidase (LiP) (Dashtban et al., 2009).

The conversion of the lignocelluloses biomass from different sources to ethanol as can be observed in the figure 4 should take in account different steps as pre-treatment of the material, hydrolysis of the cellulose and hemicelluloses, fermentation and, finally distillation and evaporation. The pre-treatment will facilitate the asses of the hydrolases to the polyssacharides (they will be cleaved by cellulases and hemicellulases) through the lignin breakdown using physical-chemical or enzymatic process. The separated lignin can be used as matrix for energy production, as electricity. The hydrolysis and fermentation steps can be conducted separately or through of simultaneous saccharification and fermentation as shown in SSF square. Some considerations that will be done in next lines on the microorganism selection can be used for the others enzymes discussed in the next pages.

All enzymes from the cellulolitic complex, hemicellulases and lignin hydrolase can be obtained from microorganism as filamentous fungi. In the nature, filamentous fungi are able to produce and secrete different enzymes to the extracellular medium to hydrolyze polymeric compounds to obtain monomers that can be used as nutritional source.

3.2. Amylases, starch sources and structure

Another interesting carbohydrate found in the plant biomass is the starch, the main form carbohydrate reserve in these organisms. The starch is the result of interaction of two structure, one linear structure formed by glucose monomers linked by α-1,4 glycosidic bonds (amylose) with molar mass of 10^1 e 10^2 Kg/mol and another branched structure with α-1,6 bonds (amylopectin) with molar mass of 10^4 and 10^6 Kg/mol. The units of amylopectin can be classified in tree groups, A, B and C. The type A is simple and characterized by the α-1,6 linkage to the amylopectin structure. The type B is subdivided in B_1, B_2, B_3 e B_4 according to the size and the group formation. The type C is a mixture of A and B types. The starch granule is formed by 25% and 75% of amylase and amylopectin, respectively. The last one is responsible for the granule crystalinity. According to the crystallographic structure the native starch can be classified as cereal starch (type A), tuber starch (Type B) and leguminosea starch (type C), that corresponds to the amylopectin groups. The more stable structure for the α-1,4 chains, as observed for starch, is the helix with high degree of spiralization with intra chain hydrogen bridge. The helicoidal structure has six residues per loop where each glucose residue forming an angle of 60° with the next residue (Yoshimoto et al., 2000; Ritte et al., 2006).

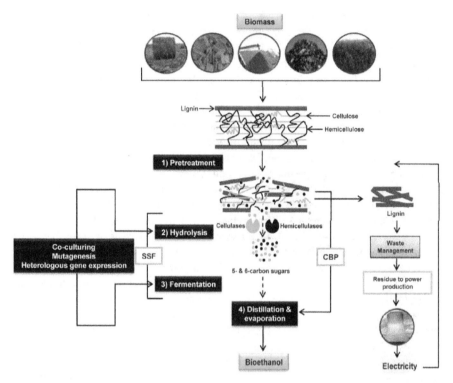

Figure 4. Schematic picture for the conversion of lignocellulosic biomass to ethanol, including the major steps (Original figure from Dashtban M., Schraft H., Qin W. Fungal Bioconversion of Lignocellulosic Residues; Opportunities & Perspectives. Int. J. Biol. Sci. 2009; 5(6):578-595. Available from http://www.biolsci.org/v05p0578.htm)

The starch synthesis is realized in the plastids and it is characterized, at the first step, by the conversion of glucose-1-phosphate and ATP to ADP-glucose and Pi by the ADP-glucose phosphorylase. After, the ADP-glucose can be used as a donor of glucose to different starch synthases that are able to elongate the glucan chain in the α-1,4 positions for both amylase and amylopectin. It is important to highlight the participation of the branching enzymes that are responsible to add α-1,6 linkages by re-organization of linear pre-existent chains. Some introduced branching can be removed by other enzymes to finalize the starch structure. The starch is highly hydrated because there are many hydroxyl groups permitting the interaction with the water. The starch is accumulated in the leaves during the day and it is used at night to maintain the respiration, the sucrose export and the grown. This transitory starch can be use through two ways, i) hydrolytic way to obtain maltose and ii) phosphorolytic way to provide carbon to the reactions in the chloroplast during the light phase.

Similar to the cellulose, the starch cleavage occurs under the action of an enzymatic complex (figure 5). The enzymes found in this complex are organized in four main groups: debranching enzymes; endoamylases; exoamylases; and transferases. The debraching enzymes are able to act exclusively on the α-1,6 glycosidic bonds and they are separated in isoamylase (EC 3.2.1.68) and pullulanase (EC 3.2.1.41). The former hydrolyze this kind of bond exclusively in amylopectin while the later hydrolyzes the α-1,6 bond from amylopectin and pullulan. Endoamylases acts on α-1,4 glycosidic bonds inside the amylose and amylopectin structure. In this group are located the α-amylases that release oligosaccharides with different length from their substrates. They can be found in many microorganisms as bacteria and fungi. Exoamylases are enzymes that hydrolyze the external bonds of amylase and amylopectin to release only glucose or maltose and β-limiting dextrin. Three main hydrolytic characteristics can be recognized for the exomylases: the specific breakdown of α-1,4 glycosidic bonds catalyzed by β-amylases (EC 3.2.1.2) and the breakdown of both α-1,4 and α-1,6 bonds catalyzed by amyloglucosidase (glucoamylase; EC 3.2.1.3) and α-glucosidase (EC 3.2.1.20). β-amylases as well as glucoamylase are able also to convert the anomeric configuration from α to β in the released maltose. In addition, glucoamylases act better on long-chain polysaccharide while α-glucosidases have preference on maltooligosaccharides. The last group of starch-converting enzymes, i.e. transferases, acts on α-1,4 glycosidic bonds of a donor molecule transferring part of this molecule to a glycosidic acceptor producing a new glycosidic bond. In this group are found the enzyme known as amylomaltase (EC 2.4.1.25), cyclodextrin glycosyltranferase (EC 2.4.1.19) and branching enzymes (EC 2.4.1.18). Amylomaltase and cyclodextrin glycosyltranferase have similar mechanism of reaction. Although the reduced hydrolytic activity from these enzymes, they are able to catalyze the transglycosylation reaction to obtain cyclodextrins by breakdown of α-1,4 glycosidic bonds and linkage of the reducing to the non-reducing end. However the product obtained by amylomaltase activity is linear while the cyclic product is obtained by cyclodextrin glycosyltranferase action (van der Maarel et al., 2002).

Many of these enzymes are involved in the complete hydrolysis of the starch. First, disbranching enzymes should act on the α-1,6 bonds to expose the linear structure that can be hydrolyzed by α-amylase and β-amylase (figure 5). Many authors have demonstrated the importance of obtainment of maltooligosaccharides from complex carbohydrate using amylolytic enzymes. Maltooligosaccharides can be used in different industrial sectors as food. Fungi are able to produce amylases. In addition, other important application of the starch is related with its use as source of renewable energy in the production of bioethanol as an alternative to the fossil fuels.

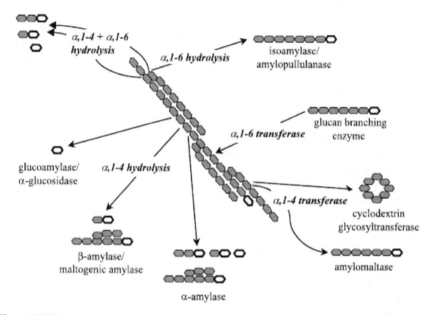

Figure 5. Different enzymes involve in the degradation of the starch. The open ring structure represents the reducing ends of a polyglucose molecule (Original figure from van der Maarel MJEC, van der Veen B, Uitdehaag JCM, Leemhuis H, Dijkhuisen L. Properties and applications of starch-converting enzymes of the α-amylase family. J. Biotechnol. 2002; 94: 137-155).

3.3. Fructosidases and other carbohydrates from biomass

Other saccharides can be also obtained from plant sources as inulin and sucrose, which are substrates for fructofuranosidase action, as β-fructofuranosidases and inulinases, which can be produced by microorganism as for example yeast and filamentous fungi. These enzymes have an important role in the microbial nutrition since monomers can be obtained and used in the metabolism. On the other hand, the enzymes involved in this process can be used with biotechnological goal.

3.3.1. The inulin and its utilization

Inulin, a polymer constituted by linear chain of β-2,1 fructufuranose residues terminated by a glucose residue, can be obtained from plant sources, especially from tubers and roots, as for example chicory, dahlia, yacon and Jerusalen artichoke. The plant sources of inulin have been considered as renewable raw material for many applications such as ethanol, obtainment of fructose syrup and fructooligosaccharides (FOS) production. This carbohydrate can be hydrolyzed by action of inulinases (2,1 β-D-fructan fructanohydrolase; EC 3.2.1.7), which can be classified into endoinulinase that hydrolyze the internal linkages from inulin to obtain inulotriose, inulotetraose and inulopentaose as end products, and exoinulinase that acts removing the terminal fructose from the non-reducing end from the inulin until the last linkage to release glucose (Ricca et al., 2007). It is important to observe that the type of enzymatic action depends on the microbial source. The most of fungal inulinases acts using the exo-mechanism. However, it was demonstrated that *Aspergillus ficuum* was able to produce endo- and exo-inulinases with different properties. The mixture of both enzymes can be considered as a good strategy to increase the conversion of the inulin to fructose (for review, see Ricca et al., 2007). Despite the action and affinity similarities for sucrose as substrate, with fructosidases as invertase, inulinases has been separated since invertase has reduced activity on high molecular mass substrates as inulin. The relation S/I has been used to separate inulinase from invertase. The S/I values depends on the inulin sources and also on the methodology used to determine the enzyme activity. However, kinetic studies are good methodologies that can be help the differentiation of these enzymes as for example considering the substrate affinity and catalytic efficiency. On the other hand, the enzymes recognized as true invertases have no activity on inulin.

The inulinases obtained from yeast are enzymes that can be linked to the cell membrane and partially secrete to the extracellular environment. In addition the synthesis of these enzymes is subject to the catabolic repression. In addition, inulinases are recognized as inducible enzymes and they are encoded by *INU* genes. The enzymes obtained from filamentous fungi has demonstrated optimum of pH activity from 4.5 to 6.0 differing than that observed for some bacterial inulinases with higher pH of activity. The optimum of temperature for activity for the most inulinases is from 30°C to 55°C but higher temperatures can be also found.

3.3.2. The sucrose and its utilization

Sucrose, a disaccharide constituted by D-glucose and D-fructose linked by α-1,2 glycosidic bond, is the main carbohydrate produced by plants using photosynthesis to generate ATP and NADPH, which will be used in the Calvin Cycle to fix CO_2 in the dark step. Two main enzymes are involved in the sucrose synthesis, the sucrose-phosphate syntase (EC 2.4.1.14) and the sucrose-phosphate phosphatase (EC 3.1.3.24) (Winter and Huber, 2000). After the synthesis, the sucrose produced in the photosynthetic leaves is distributed to the other plant organs and tissues. The ethanol production in Brazil is performed using rich-sucrose sugar cane juice. However, after the sucrose extraction, the residual sugar cane bagasse also has residual sucrose that can be used to obtain monosacharides by microbial action. These monomers can be used together with other monosaccharides obtained from the hydrolysis of polysaccharides present in the sugar cane bagasse, as cellulose, to fermentation process. The sucrose hydrolysis (figure 6) is catalyzed by the β-fructofuranosidases (invertases; EC 3.2.1.26), which are found in many microorganisms. The product obtained is an equimolar (1:1) mixture of the monosaccharides and residual sucrose known as invert sugar with wide application in the food and beverage industries. The fructose is much more attractive for application since it can has liquid and non-crystalizable constitution. The β-fructofuranosidases are located in the GH32 family of the glycosil hydrolases and grouped in different isoforms according to their pH of actuation as acid, alkaline and neutral enzymes (Vargas et al., 2003).

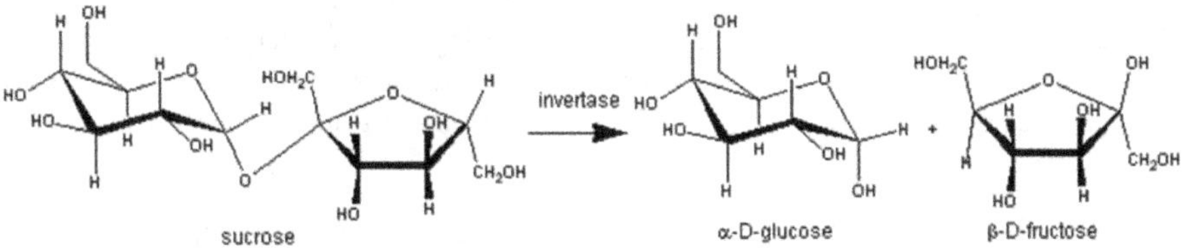

Figure 6. Hydrolysis of sucrose by invertase.

The production of β-fructofuranosidases (FFases) by microorganisms has been characterized, especially for the yeast *Saccharomyces cerevisiae*. In this microorganism it was observed the synthesis of two isoforms of FFases where one is gylcosylated and another non-glycosylated. Both enzymes are result from the two mRNA (1.8 and 1.9 kb) encoded by the same gene *SUC2*. The glycosylated enzyme is found in the periplasmic space while the non-glycosylated is found in the citosol (Belcarz *et al.*, 2002). In *S. cerevisiae* the sucrose metabolism occurs throughout two main ways: i) the sucrose is hydrolyzed in the extracellular environment by extracellular invertases to liberate glucose and fructose, which can be transported to inside of the cell by hexose transporters and ii) the sucrose is actively transported to inside of the cell by proton-symport mechanism and after hydrolyzed by intracellular invertase. The expression of the *SUC2* gene that encodes both enzymes is severally regulated by glucose (Basso et al., 2011).

The production of FFases by other microorganism, especially filamentous fungi as *Aspergillus* genera, among others, has been studied. In this situation, different fermentation processes are used, which many times the residual biomass are used as carbon sources (submerged fermentation) and/or substrates (solid-sate fermentation). Some microorganisms are able to produce multiple β-fructofuranosidases as observed for *Aureobasidium pullulans* (Yoshikawa et al., 2006). In this situation the authors observed the presence of five FFases (I, II, III, IV and V) with high FFase I activity at the initial times of culture and reduced FFase II-V activities. After initial times the FFase II-V activities are increased. In addition, the multiple FFases produced by *A. pullulans* have distinct properties as suggested by authors. FFase IV has high hydrolytic activity acting as FOS-degrading enzyme at the FOS-degrading period while the participation of FFases II, III and V is uncertain, since they have significant transfructosylating activity and they are present in the FOS-degrading periods (Yoshikawa et al., 2006).

The most of β-fructofuranosidases are dimmers but monomers also can be found. The optimum of temperature and pH of reaction considering all microbial sources are variable. This carbohydrate has a negative influence on the invertase synthesis by *A. niger* as well as fructose. Only β-fructofuranoside saccharides were able to induce the invertase synthesis

(Rubio & Navarro, 2006). It was observed that the *A. niger* is able to produce two β-fructosidases known as SUC1 and SUC2. Both enzymes catalyzed the sucrose hydrolysis but only SUC 2 was able to act on inulin. Other fungal strains have been used for invertase production as *Aspergillus ochraceus* (Guimarães et al., 2007), *Aspergillus niveus* (Guimarães et al., 2009), *Aspergillus caespitosus* (Alegre et al., 2009), *Aspergillus phoenicis* (Rustiguel et al., 2011) and *Paecylomyces variotii* (Giraldo et al., 2012) using both submerged and solid-state fermentation with agro-industrial residues as carbon source/substrate. Thermostable FFases has been obtained by cultivation of *A. ochraceus* and *A. niveus* using sugar cane bagasse as carbon source.

At high sucrose concentration, some β-fructofuranosidases are able to catalyze transfructosylation reaction to obtain fructooligosaccharides (FOS) as 1-kestose (GF2), 1-nystose (GF3) and fructofuranosyl nystose (GF4). The molecular structure of these FOS can be observed in the figure 7. The GF2 is constituted by two molecules of fructose binding to the D-glycosyl unit at the non reducing end while the GF3 and GF4 by three and four fructose residues, respectively. These oligosaccharides have functional properties that have attracted the attention of different sectors. FOS are no caloric sugars that can be used by diabetic peoples with security since they are not metabolized by the organism. In addition, FOS can also stimulate the bifidobacteria development in the intestine and minimize the colon tumor. It has been demonstrated that some components of plant sources (biomass) used in pet foods exhibit FOS concentration of GF2, GF3 and GF4, as for example wheat bran, peanut hulls and barley, among others. The hydrolysis of FOS by microbial sources as bacteria using enzymes that act on these saccharides was demonstrated in some reports. Hence, enzymes that are able to act on FOS can be used to obtain saccharides as glucose from the biomass containing GF2, GF3 and/or GF4. On the other hand, different approaches have been used to obtain FOS as the utilization of immobilized enzymes on lignocellulosic materials and by substrate and enzyme engineering. Recently, the one-step FOS production was obtained using *Aspergillus phoenicis* biofilms in rich-sucrose medium as demonstrated in our laboratory (Aziani et al., 2012).

In the same way, the fuctosyltransferases as levansucrases (sucrose:2,6-β-fructans: 6-β-D-fructosyltransferase; EC 2.4.1.10), inulosucrases (sucrose:2,1-β-D-fructan: 1-β-D-fructosyltransferase; EC 2.4.1.9) and fructosyltransferase (sucrose:2,6-β-fructan:6-β-D-fructosyltranferase; EC 2.4.1.10) should be considered. The former is responsible by the synthesis of microbial levans using glucose or levan as acceptor to the β-D-fructosyl residues while the inulosucrases are able to catalyze the transference of the β-D-fructosyl residues to the sucrose or inulin as acceptors. The later are involved in the levan synthesis but it is not able to catalyze hydrolysis or exchange reactions as observed for the levansucrase (Velázques-Hernandez et al., 2009). These enzymes that catalyze the fructans synthesis are inserted in the GH68 family of glycoside hydrolases. In general, five main domains are recognized in the fructosyltransferases of microorganisms: a signal peptide; an N-terminal domain with variable length; a catalytic domain with around of 500 amino acids; a cell wall binding domain; and a C-terminal domain with variable length (Van Hijum et al., 2006).

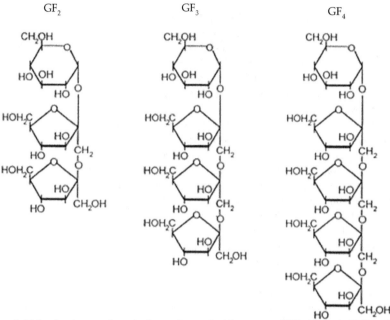

Figure 7. Molecular structure from the fructooligosaccharides nystose (GF2), 1-kestose (GF3) and fructosyl nystose (GF4).

4. Conclusion

In conclusion, the biomass that has been accumulated around the world as residue can be widely used for different applications considering its carbohydrate composition which can be accessed by microbial activity according to the enzymatic potential of each one. Microorganisms show metabolic versatility permitting the carbohydrate utilization and transformation from biomass since they are important sources of enzymes with biotechnological potential. According to this, different products can be obtained from biomass and applied in different industrial sectors. The view of the biomass as an important renewable energy source is very important to the future of the life in our planet, especially if considered the agro-industrial residues. In addition, the environment problems of bioaccumulation of residues can be reduced. Future studies to improve the biomass utilization are important as well as on the carbohydrate-active enzymes produced by microorganisms to optimize this process.

Author details

Luis Henrique Souza Guimarães

Faculdade de Filosofia, Ciências e Letras de Ribeirão Preto, University of São Paulo, Brasil

5. References

Alegre, A.C.P., Polizeli, M.L.T.M., Terenzi, H.F., Jorge, J.A. & Guimarães, L.H.S. (2009) Production of thermostable invertases by *Aspergillus caespitosus* under submerged or solid state fermentation using agro-industrial residues as carbon source. *Brazilian Journal of Microbiology*, Vol. 40, pp. 612-622.

Aranda, C., Robledo, A., Loera, O., Contreras-Esquiavel, J., Rodríguez, R. & Aguilar, C.N. (2006). Fungal invertase expression in solid-state fermentation. *Food Technology and Biotechnology*, Vol. 44, pp. 229-233.

Aziani, G., Terenzi, H.F., Jorge, J.A., Guimarães, L.H.S. (2012) Production of fructooligosaccharides by *Aspergillus phoenicis* biofilm on polyethylene as inert support. *Food Technology and Biotechnology*, Vol. 50, pp. 40-45.

Basso, T.O., de Kok, S., Dario, M., Espirito-Santo, J.C., Müller, G., Schlölg, P.S., Silva, C.P., Tonso, A., Daran, J.M., Gombert, A.K., van Maris, A.J., Pronk, J.T. & Stambuk, B.U. (2011). Engineering topology and kinetics of sucrose metabolism in *Saccharomyces cerevisiae* for improved ethanol yield. *Metabolism Engineering*, Vol. 13, pp. 694-703.

Bayer, E.A., Chanzy, H., Lamed, R. & Shoham, Y. (1998). Cellulose, cellulases and cellulosomes. *Current Opinion in Structural Biology* Vol. 8, pp. 548-557.

Belcarz, A., Ginalska, G., Lobarzewski, J. & Penel, C. (2002). The novel non-glycosilated invertase from *Candida utilis* (the properties and the conditions of production and putification). *Biochemistry and Biophysics Acta*, Vol. 1594, pp. 40-53.

Chi, Z., Chi, Z., Zhang, T., Liu, G. & Yue, L. (2009). Inulinase-expressing microorganisms and applications of inulinases. *Applied Microbiology and Biotechnology*, Vol. 82, pp. 211-220.

Dashtban, M., Schraft, H. & Qin, W. (2009). Fungal bioconversion of lignocellulosic residues: opportunities & perspectives. *International Journal of Biological Sciences*, Vol. 5, pp. 578-595.

Davies, G.J., Gloster, T.M. & Henrissat, B. (2005). Recent structural insights into the expanding world of carbohydrate-active enzymes. *Current Opinion in Structural Biology*, Vol. 15, pp. 637-645.

Fettke, J., Hejazi, M., Smirnova, J., Höchel, E., Stage, M. & Steup, M. (2009). Eukaryotic starch degradation: integration of plastidial and cytosolic pathways. *Journal of Experimental Botany*, Vol. 60, pp. 2907-2922.

Geigenberger, P. (2011). Regulation of starch biosynthesis in response to a fluctuating environment. *Plant Physiology*, Vol.155, pp. 1566–1577.

Gilbert, R.G. (2011). Size-separation characterization of starch and glycogen for biosynthesis–structure–property relationships. *Analytical and Bioanalytical Chemistry*, Vol.399, pp.1425–1438.

Giraldo, M.A., Silva, T.M., Salvato, F., Terenzi, H.F., Jorge, J.A. & Guimarães, L.H.S. (2012). Thermostable invertases from *Paecylomyces variotii* produced under submerged and solid-state fermentation using agroindustrail residues. *World Journal of Microbiology and Biotechnology*, Vol 28, pp. 463-472.

Guimarães, L.H.S., Somera, A.F., Terenzi, H.F., Polizeli, M.L.T.M. & Jorge, J.A. (2009). Production of β-fructofuranosidases by *Aspergillus niveus* using agroindustrail residues as carbon sources: characterization of an intracellular enzyme accumulated in the presence of glucose. *Process Biochemistry*, Vol. 44, pp. 237-241.

Guimarães, L.H.S., Terenzi, H.F., Polizeli, M.L.T.M. & Jorge, J.A. (2007). Production and characterization of a thermostable β-D-fructofuranosidase produced by *Aspergillus ochraceus* with agro-industrial residues as carbon source. *Enzyme and Microbial Technology*, Vol. 42, pp. 52-57.

Guzmán-Maldonado, H. & Paredes-López, O. (1995). Amylolytic enzymes and products derived from starch: A review. *Critical Review in Food Science and Nutrition*, Vol. 35, pp. 373-403.

Hizukuri, S. (1996). Starch analytical aspects. In: *Carbohydrates in Foods*. pp. 347–429. Eliasson, A.C., Ed., Marcel Dekker Inc., New York.

Hoover, R. (2010). The impact of heat-moisture treatment on molecular structures and properties of starches isolated from different botanical sources. *Critical Reviews in Food Science and Nutrition*, Vol. 50, pp. 835–847.

Jobling, S. (2004). Improving starch for food and industrial applications. *Current Opinion in Plant Biology*, Vol. 7, pp. 210-218.

Mischnick, P. & Momcilovic, D. (2005). Chemical structure analysis of starch and cellulose derivatives. *Advances in Carbohydrate Chemistry and Biochemistry*, Vol. 64, pp. 118-210.

Morrison,W.R. & Karkalas, J. (1990). Starch. In: Methods in Plant Biochemistry, Vol. 2. pp. 323–352. Academic Press, Inc., New York.

Parada, J. & Aguilera, J.M. (2011). Review: starch matrices and the glycemic response. *Food Science Technology International*, Vol. 17, pp. 187-204.

Pérez-Guerra, N., Torrado-Agrasar, A., López-Macias, C. & Pastrana, L. (2003). Main characteristics and applications of solid substrate fermentation. *Electronic Journal of Environmental, Agricultural and Food Chemistry*, Vol. 2, pp. 343-350.

Ricca, E., Calabrò, V., Curcio, S. & Iorio, G. (2007). The state of the art in the production of fructose from inulin enzymatic hydrolysis. Critical Reviews in Biotechnology, Vol. 27, pp. 129-145.

Ritte, G., Heydenreich, M., Mahlow, S., Haebel, S., Kötting, O. & Steup, M. (2006). Phosphorylation of C6- and C3-positions of glucosyl residues in starch is catalysed by distinct dikinases. *FEBS Letters*, Vol. 580, N° 20, pp. 4872-4876.

Rubio, M.C. & Navarro, A.R. (2006). Regulation of invertase synthesis in *Aspergillus niger*. *Enzyme and Microbial Technology*, Vol. 39, pp. 601-606.

Rustiguel, C.B., Oliveira, A.H.C., Terenzi, H.F., Jorge, J.A. & Guimarães, L.H.S. (2011). Biochemical properties of an extracellular β-D-fructofuranosidase II produced by *Aspergillus phoenicis* under solid-state fermentation using soy bran as substrate. *Electronic Journal of Biotechnology*, Vol. 14, N° 2 (doi: 10.2225/vol14-issue2-fulltext-1).

Szydlowski, N., Ragel, P., Hennen-Bierwagen, T.A., Planchot, V., Myers, A.M., Mérida, A., d'Hulst, C. & Wattebled, F. (2011). Integrated functions among multiple starch synthases determine both amylopectin chain and branch linkage location in *Arabidopsis* leaf starch. *Journal of Experimental Botany*, Vol. 62, N° 13, pp. 4547-4559.

Szydlowski, N., Ragel, P., Raynaud, S., Lucas, M., Roldan, I., Montero, M., Muñoz, F.J., Ovecka, M., Bahaji, A., Planchot, V., et al. (2009). Starch granule initiation in *Arabidopsis* requires the presence of either class IV or class III starch synthases. *Plant Cell*, Vol. 21, pp. 2443-2457.

Teixeira, F.A., Pires, A.V. & Nascimento, P.V.N. (2007). Sugarcane pulp in the feeding of bovine. *Revista Eletrônica de Veterinária*, Vol. 8, pp. 1-9.

van der Maarel, M.J.E.C., van der Veen, B., Uitdehaag, J.C.M., Leemhuis, H. & Dijkhuisen, L. (2002). Properties and applications of starch-converting enzymes of the α-amylase family. *Journal of Biotechnology*, Vol. 94, pp. 137-155.

Van Hijum, S.A.F.T., Kralj, S., Ozimek, L.K., Dijkhuizen, L. & van Geel-Schutten, I.G.H. (2006). Structure-function relationships og glucasucrase and fructansucrase enzymes from lactic acid bacteria. *Microbiology and Molecular Biology Review*, Vol. 70, pp. 157-176.

Vargas, W., Cumino, A. & Salerno, G.L. (2003). Cyanobacterial alkaline/neutral invertases. Origin of sucrose hydrolysis in the plant cytosol? *Planta*, Vol. 216, 951-960.

Velázquez-Hernández, M.L.; Baizabal-Aguirre, V.M., Bravo-Patiño, A., Cajero-Juárez, M., Chávez-Moctezuma, M.P. & Valdez-Alarcón, J.J. (2009). Microbial fructosyltransferases and the role of fructans. *Journal of Applied Microbiology*, Vol. 106, N°6, pp. 1763-1768.

Vijayaraghavan, K., Yamini, D., Ambika, V. & Sowdamini, N.S. (2009). Trends in inulinase production – a review. *Critical Reviews in Biotechnology*, Vol. 29, pp. 67–77.

Villena, G.K. & Gutiérrez-Correa, M. (2007). Morphological patterns of *Aspergillus niger* biofilms and pellets related to lignocellulosic enzyme productivities. *Letters in Applied Microbiology*, Vol. 45, pp. 231-237.

Weise, S.E., van Wijk, K.J. & Sharkey, T.D. (2011). The role of transitory starch in C(3), CAM, and C(4) metabolism and opportunities for engineering leaf starch accumulation. *Journal of Experimental Botany*, Vol. 62, pp. 3109-3118.

Winter, H. & Huber, S.C. (2000). Regulation of sucrose metabolism in higher plants: localization and regulation of activity of key enzymes. *Critical Review in Plant Science*, Vol. 19, pp. 31-67.

Yoshimoto, Y., Tashiron, J., Takenouchi, T. & Takeda, Y. (2000). Molecular structure and some physicochemical properties of high amylose barley starch. *Cereal Chemistry*, Vol. 77, pp. 279–285.

Chapter 10

Starch and Microbial α-Amylases:
From Concepts to Biotechnological Applications

Amira El-Fallal, Mohammed Abou Dobara, Ahmed El-Sayed and Noha Omar

Additional information is available at the end of the chapter

1. Introduction

Starch is a polymer of glucose linked to one another through the C1 oxygen, known as the glycosidic bond. Amylases are capable of digesting these glycosidic linkages found in starch. Amylases have been isolated from diversified sources including plants, animals, and microbes, where they play a dominant role in carbohydrate metabolism. In spite of the wide distribution of α-amylase, microbial sources are used for the industrial production. This is due to their advantages such as cost effectiveness, consistency, less time and space required for production as well as ease of process modification and optimization.

In the present day scenario, α-amylases have applications in all the industrial processes such as in food, detergents, textiles and paper industry, for the hydrolysis of starch. They can also be of potential use in the pharmaceutical and fine chemical industries. In this light, microbial α-amylases have completely replaced chemical hydrolysis in the starch processing industry. Despite this, interest in new and improved α-amylase is growing and consequently, the research is intensified as well to meet requirements set by specific applications.

2. Starch

Starch and starch-containing substrates are wide spread in nature and also in industrial praxis. They can predominantly find their application in many industrial processes.

2.1. Sources and utilization

Starch occurs mainly in the seeds, roots and tubers of higher plants. Some algae produce a similar reserve polysaccharide called phytoglycogen. Plants synthesize starch as a result of photosynthesis. It is synthesized in plastids as a storage compound for respiration during dark periods. It is also synthesized in amyloplasts found in tubers, seeds, and roots as a long-term storage compound. In these latter organelles, large amounts of starch accumulate as water-insoluble granules. The shape and diameter of these granules depend on the botanical origin. Regarding to commercial starch sources, the granule sizes range from 2–30 μm (maize starch) to 5–100 μm (potato starch) (Robyt, 1998). A variety of different enzymes are involved in the synthesis of starch. Sucrose is the starting point of starch synthesis. It is converted into the nucleotide sugar ADP-glucose that forms the actual starter molecule for starch formation. Subsequently, enzymes such as soluble starch synthase and branching enzyme synthesize the amylopectin and amylose molecules (Smith, 1999).

Starch-containing crops form an important constituent of the human diet. Besides the direct use of starch-containing plant parts as a food source, starch is harvested and chemically or enzymatically processed into a variety of different products such as starch hydrolysates, glucose syrups, fructose, starch or maltodextrin derivatives, or cyclodextrins. In spite of the large number of plants able to produce starch, only a few plants are important for industrial starch processing. The major industrial sources are maize, tapioca, potato, and wheat.

2.2. Structure and properties

Starch is a polymer of glucose linked to one another through the C1 oxygen by a glycosidic bond. This glycosidic bond is stable at high pH but hydrolyzes at low pH. At the end of the polymeric chain, a latent aldehyde group is present. This group is known as the reducing end. Two types of glucose polymers are present in starch: (i) amylose and (ii) amylopectin. While amylopectin is soluble in water, amylose and the starch granule itself are insoluble in cold water.

Amylose is a linear polymer consisting of up to 6000 glucose units with α, 1-4 glycosidic bonds (Fig. 1a.). The number of glucose residues, also indicated with the term DP (degree of polymerization), varies with the origin. The relative content of amylose and amylopectin varies with the source of starch. The average amylose content in most common starches, e.g. in barley, corn and potato, is 20-30% (Marc et al., 2002).

Amylopectin consists of short α, 1-4 linked linear chains of 10–60 glucose units and α,1-6 linked side chains with 15–45 glucose units (Fig. 1b.). The average number of branching points in amylopectin is 5% (Thompson, 2000), but varies with the botanical origin. The complete amylopectin molecule contains about 2 000,000 glucose units, thereby being one of the largest molecules in nature (Marc et al., 2002). The most commonly accepted model of the structure of amylopectin is the cluster model, in which the side chains are ordered in clusters on the longer backbone chains (Bertoft, 2007; Thompson, 2000). In general, Zhu et al. (2011) suggested that the internal part of amylopectin is critical to the physical behavior of granular starch.

The diameter of starch granules ranges from 2 to 100μm (Whistler & Daniel, 1985) depending on its source. The orientation of the starch chains is thought to be perpendicular to the granule surface (French, 1984). Native starch is partly crystalline. The crystallinity of native starch varies between 15 and 45% depending on the origin and pretreatment (French, 1984). According to the currently accepted concept, amylopectin forms the crystalline component whereas amylose exists mainly in the amorphous form (Hanashiro et al., 1996; Marc et al., 2002; Zobel, 1992). Structural studies have shown that native starch has crystalline polymorphism. In x-ray diffraction, cereal starch typically gives A-type patterns of monoclinic symmetry, and tuber starch gives B-type patterns of hexagonal symmetry (Gerard et al., 2000; Imberty et al., 1991). The crystal lattice of B-type starch contains more water molecules than the A- structures, which is proposed to be the reason for higher stability of the A- structure. Both structures' molecular conformations are practically identical. They have left-handed double helices with parallel strands. Double helices contain six glucose units per turn in each chain and the glucose units are in a chair conformation. With in the double helix, there are inter-chain but no intra-chain hydrogen bonds. In additional, parallelly packed double helices are connected through a hydrogen bonding network.

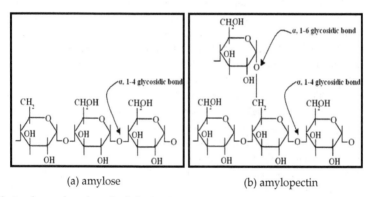

(a) amylose   (b) amylopectin

Figure 1. Amylose and amylopectin chain structure

3. Amylases

Amylases are a class of enzymes that are capable of digesting these glycosidic linkages found in starches. Amylases can be derived from a variety of sources. They are present in all living organisms, but the enzymes vary in activity, specificity and requirements from species to species and even from tissue to tissue in the same organism. Raw-starch digesting amylases are produced by a variety of living organisms, ranging from microorganisms including fungi, yeast, and bacteria to plants and humans.

3.1. Microbial sources of amylases

Several amylase-producing bacteria, fungi and other microrganisms have been isolated and characterized over many decades. Bacteria and fungi secrete amylases outside their cells to carry out extra-cellular digestion.

Among mold species producing high levels of amylase, those of *Aspergillus niger, Aspergillus oryzae* (Aunstrup, 1979), *Thermomyces lanuginosus* (Arnesen et al., 1998) and *Penicillium expansum* (Doyle et al., 1989) in addition to many species of the genus *Mucor* (Domsch et al., 1995; Petruccioli & Federici, 1992; Zare-Maivan & Shearer, 1988). It was reported that four species of *Ganoderma* mushrooms could produce relatively weak amylase in sawdust medium (Y.W. Wang & Y. Wang, 1990). Amylolytic yeasts differ strongly with regard to amylase secretion and the extent of starch hydrolysis (De Mot et al., 1984a, 1984b). Strains of *Filobasidiuim capsuligenum* are capable of extensive starch hydrolysis (De Mot et al., 1984c; McCann& Barnett, 1984).

Regarding to bacteria, *Bacillus* spp and the related genera produce a large variety of extracellular enzymes, of which amylases are of particular significance to the industry e.g., *B. cereus* (Rhodes et al., 1987), *B. circulans* (Siggens, 1987), *B. subtilis* (El-Banna et al., 2007), *B. licheniformis* (El-Banna et al., 2008) and *Clostridium thermosulfurogenes* (Hyun & Zeikus, 1985a). Bacteria belonging mainly to the genus *Bacillus* have been widely used for the commercial production of thermostable α-amylase (Tonkova, 2006). However, most of the *Bacillus* liquefying amylases, such as the enzymes from *B. amyloliquefaciens* and *B. stearothermophilus* have pH optima of between 5 and 7.5 (Yamamoto, 1988). Many alkaline amylases have been found in cultures of *Bacillus* sp. (Hayashi et al., 1988; Kim et al., 1995). This alkaline amylases are all of the saccharifying type, except for the enzymes from *Bacillus* sp. strain 707 (Kimura et al., 1988) and *B. licheniformis* TCRDC-B13 (P. Bajpai and P.K. Bajpai, 1989). Thermostable β-amylases have been isolated from *Bacillus* species (Shinke et al., 1974; Takasaki, 1976). Also, *Lactobacillus plantarum* strain A6 was selected for its ability to synthesize large amounts of extracellular α-amylase (Giraud et al., 1991). Furthermore, a variety of ruminal bacteria exhibit the ability to utilize starch as a growth substrate and are present in the rumen in sufficient numbers to be of quantitative significance in the fermentation of this substrate. These species include *Bacteroides ruminicola, Ruminobacter amylophilus, Butyrivibrio fibrisolvens, Selenomonas ruminantium,* and *Streptococcus bovis* (Russell, 1984).

Genes encoding intracellular α-amylases have been reported for *Escherichia coli* and *Streptococcus bovis* (Satoh et al., 1997; Whitehead & Cotta, 1995). Although there has been some characterization of these activities, no clear physiological role for intracellular α-amylase has been established for either *E. coli* or *Streptococcus bovis*. However, it is postulated that it plays an important role in rapid cell growth in *Streptococcus bovis* (Brooker & McCarthy, 1997).

Many hyperthermophilic microorganisms possess starch-hydrolyzing enzymes in their genomes even though they live in environments where starch is rare (Sambrook et al., 1989). Among the polysaccharide-degrading enzymes of *Thermotoga maritime* described so far are two α-amylases, one is an extracellular putative lipoprotein (AmyA) (Liebl et al., 1997) and one is located in the cytoplasm (AmyB) (Lim et al., 2003). *Geobacillus thermoleovorans* has been found to produce hyperthermostable, high maltose-forming and Ca^{2+} independent α-amylase (Malhotra et al. 2000; Narang & Satyanarayana 2001). Numerous hyperthermophilic Archaea, especially deep-sea *Thermococcale* and *Sulfolobus* species have been reported to produce α-amylases (Leuschner and Antranikian, 1995; Sunna et al., 1997).

The industrial potential of high-maltose forming α-amylases from *Thermomonospora curvata* (Collins et al., 1993) is limited by their moderate thermostability and Ca^{2+} requirement.

α-Amylases are secreted by several species of *Streptomyces*, for example *S. albus* (Andrews & Ward, 1987), *S. griseus* IMRU3570 (Vigal et al., 1991), *S. thermocyaneoviolaceus* (Hang et al., 1996). Gene encoding extracellular α-amylase has been cloned from many *Streptomyces* species (Bahri & Ward, 1990; Virolle et al., 1988). In addition, α-amylase activity of *Thermoactinomyces* species was first reported by Kuo & Hartman (1966). After that, several α-amylases with different characters were found in other studies (Obi & Odibo, 1984; Omar et al., 2011; Shimizu et al., 1978; Uguru et al., 1997). Within actinomycetes, available reports on β-amylase production are scanty and refer mainly to nonthermostable enzyme (Shinke et al., 1974).

3.2. Amylases types

Enzymes belonging to amylases, endoamylases and exoamylases, are able to hydrolyse starch. These enzymes are classified according to the manner in which the glycosidic bond is attacked. The starch degrading enzymes are found in the numerous glycoside hydrolase (GH) families (13, 14 and 15), mainly in GH family 13 (Coutinho & Henrissat, 1999; Henrissat, 1991).

Endoamylases are able to cleave α,1-4 glycosidic bonds present in the inner part (endo-) of the amylose or amylopectin chain. α-amylase (EC 3.2.1.1) is a well-known endoamylase. It is found in a wide variety of microorganisms, belonging to the Archaea as well as the Bacteria (Pandey et al., 2000). The end products of α-amylase action are oligosaccharides with varying length with α-configuration and α-limit dextrins, which constitute branched oligosaccharides. α-amylases are often divided into two categories according to the degree of hydrolysis of the substrate (Fukumoto & Okada, 1963). Saccharifying α-amylases hydrolyze 50 to 60% and liquefying α-amylases cleave about 30 to 40% of the glycosidic linkages of starch.

Enzymes belonging to the second group, the exoamylases, either exclusively cleave α,1-4 glycosidic bonds such as β-amylase (EC 3.2.1.2) or cleave both α,1-4 and α,1-6 glycosidic bonds like amyloglucosidase or glucoamylase (EC 3.2.1.3) and α-glucosidase (EC 3.2.1.20). Exoamylases act on the external glucose residues of amylose or amylopectin and thus produce only glucose (glucoamylase and α-glucosidase), or maltose and β-limit dextrin. β-amylase and glucoamylase also convert the anomeric configuration of the liberated maltose from α to β. Glucoamylase and α-glucosidase differ in their substrate preference: α-glucosidase acts best on short maltooligosaccharides and liberates glucose with α-configuration while glucoamylase hydrolyzes long-chain polysaccharides best. β-amylases and glucoamylases have also been found in a large variety of microorganisms (Pandey et al., 2000).

3.3. α-amylases actions and structure

3.3.1. Mode of action

In general, it is believed that α-amylases are endo-acting amylases which hydrolyze α-(1-4) glycosidic bonds of the starch polymers internally. Several models for amylase action pattern have been proposed, such as the random action and the multiple attack action. Random action has also been referred to as a single attack or multi-chain attack action (Azhari & Lotan, 1991). In the former, the polymer molecule is successively hydrolysed completely before dissociation of the enzyme-substrate complex. While, in the latter, only one bond is hydrolysed per effective encounter. The multiple attack action is an intermediate between the single-chain and the multi-chain action (Bijttebier et al., 2008) where the enzyme cleaves several glycosidic bonds successively after the first (random) hydrolytic attack before dissociating from the substrate.

In short, it can clearly be seen that the multiple attack action is generally an accepted concept to explain the differences in action pattern of amylases (Kramhøft et al. 2005; Svensson et al. 2002). However, most of the endoamylases have a low to very low level of multiple attack action (Bijttebier et al., 2008). Although only few reports deal with the influence of pH and temperature on the action pattern of amylases, this influence was confirmed. Bijttebier et al. (2007) showed that the level of multiple attack of several endoamylases increased with temperature to a degree depending on the amylase itself.

3.3.2. Molecular weight

Despite wide difference of microbial α-amylases characters, their molecular weights are usually in the same range 40-70 kDa (Gupta et al., 2003). Ratanakhanokchai et al. (1992) reported the highest molecular weight of α-amylases, 210 kDa, for *Chloroflexus aurantiacus*. Whereas, 10 kDa of *Bacillus caldolyticus* α-amylase was reported to be the lowest value (Gupta et al., 2003).

This molecular weight may be raised due to glycosylation as in the case of *T. vulgaris* α-amylase that reach 140 kDa (Omar et al., 2011). In contrast, proteolysis may lead to decrease in the molecular weight. For example, α-amylase of *T. vulganis* 94-2A (AmyTV1) is a protein of 53 kDa and smaller peptides of 33 and 18 kDa that have been shown to be products of limited AmyTV1 proteolysis (Hofemeister et al., 1994).

3.3.3. Modular structure

α-amylases from different organisms share about 30% amino acid sequence identity and all belong to the same glycosyl hydrolase family 13 (Henrissat & Bairoch, 1993). The three dimensional (3D) structures of α-amylases have revealed monomeric, calcium-containing enzymes, with a single polypeptide chain folded into three domains (A-C).

The most conserved domain in α-amylase family enzymes, the A-domain, consists of a highly symmetrical fold of eight parallel β-strands arranged in a barrel encircled by eight α-helices.

The highly conserved amino acid residues of the α-amylase family involved in catalysis and substrate binding are located in loops at the C-termini of β-strands in this domain. This is typical to all enzymes belonging to the α/β –barrel protein family (Farber & Petsko, 1990).

α-amylases have a B-domain that protrudes between β-sheet no 3 and α-helix no. 3. It ranges from 44 to 133 amino acid residues and plays a role in substrate or Ca^{2+} binding (Marc et al., 2002). The sequence of this domain varies most; in *Bacillus* α-amylases it is relatively long and folds into a more complex structure of β-strands (Machius et al., 1995), whereas in barley α-amylase there is an irregularly structured domain of 64 residues (Kadziola et al., 1994).

All known α-amylases, with a few exceptions, contain a conserved Ca^{2+} binding site which is located at the interface between domains A and B (Linden et al., 2003; Prakash & Jaiswal, 2010). In addition, α-amylase produced by *Bacillus thermooleovorans* was found to contain a chloride ion binding site in their active site (Malhotra et al., 2000), which has been shown to enhance the catalytic efficiency of the enzyme, presumably by elevating the pKa of the hydrogen-donating residue in the active site (Prakash & Jaiswal, 2010).

α-amylases have a domain C which is relatively conserved and folds into an antiparallel β-barrel. The orientation of domain C relative to domain A varies depending on the type and source of amylase (Bayer et al., 1995). The function of this domain is unknown.

Structural studies have confirmed that the active sites of glycosyl hydrolases are composed of multiple binding sites, or subsites, for the sugar units of polymeric substrates. The open active site cleft is formed between domains A and B, so that residues from domain B participate in substrate binding. The substrate binding sites are commonly lined with aromatic residues which make hydrophobic stacking interactions with the sugar rings. In addition, the active sites contain many residues which form hydrogen bonds to the substrate either directly or via water molecules (Aleshin et al., 1994; Svensson & Sogaard, 1993).

In Taka-amylase A, the first examined protein α-amylase by X-ray crystallography, three acidic residues, i.e., one glutamic and two aspartic acids were found at the centre of the active site (Matsuura et al., 1984), and subsequent mutational studies have shown that these residues are essential for catalysis (Janecek, 1997; Svensson, 1994). The glutamic acid residue is now believed to be the proton donor, while the first of the two conserved aspartic acids appearing in the amino acid sequence of an α-amylase family member is thought to act as the nucleophile. The role of the second aspartic acid is less certain, but it has been suggested to be involved in stabilising the oxocarbenium ion-like transition state and also in maintaining the glutamic acid in the correct state of protonation for activity (Uitdehaag et al., 1999). These residues occur near the ends of strands 3, 4, 5 and 7 of the α/β -barrel and are found in four short sequences, long-recognised as being conserved in α-amylase family enzymes.

3.3.4. Glycosylation

Glycosylation is one of the major post-translation modifications that affect a variety of enzyme functions including secretion, stability, and folding (Barros et al., 2009; Shental-

Bechor & Levy, 2009). Oligosaccharides are usually linked to asparagine side chains (N-linked glycosylation) or to serine and threonine hydroxyl side chains (O-linked glycosylation) (Shental-Bechor & Levy, 2009).

Glycoproteins have been detected in α-amylases of *A. oryzae* (Eriksen et al., 1998), *B. stearothermophilus* (Srivastava, 1984) and *B. subtilis* strains (Matsuzaki et al., 1974; Yamane et al., 1973). Generally, this is about 10 % for most α-amylases (Vihinen & Mantsala, 1989). These carbohydrate moieties are thought to be responsible for high molecular weight of some α-amylases. A carbohydrate content as high as 56 % has been reported in *S. castelii* (Sills et al., 1984). Also, the high molecular weight α-amylase of 140 kDa produced by *T. vulgaris* (Abou Dobara et al., 2011) is a good example of highly glycosylated α-amylase (Omar et al., 2011). Using SDS-PAGE, glycoproteins can be detected by initial oxidation of carbohydrates by periodic acid and subsequent staining with cationic dyes such as alcian blue (Wardi & Michos, 1972).

3.4. Production of microbial α-amylases

The major advantage of using microorganisms for the production of amylases is the economical bulk production capacity and easy manipulation of microbes to obtain enzymes of desired characteristics (Lonsane & Ramesh, 1990). Screening for the α-amylase producers is a key step for production. Starch hydrolysis is usually detected directly on plates as clear zones surrounding the colonies. The diameter of the area of hydrolysis, with in limits, was always related to the potency of the amylase (Dhawale et al., 1982).

3.4.1. Factors affecting production

The production and stability of α-amylase in the medium is affected by a variety of physicochemical factors. In spite of expression's possibility under a wide range of culturing conditions, α-amylase could be denatured under some conditions. Many proteins easily aggregate into so-called inclusion bodies during expression in bacterial systems (Espargaro et al., 2008). Inhibition of protein aggregation during fermentation/expression can be achieved by adjusting the production conditions (Bahrami et al., 2009; Hao et al., 2007).

Regarding to the incubation period, many investigators have found that extracellular α-amylase production is growth associated (Abou Dobara et al., 2011; Asoodeh et al., 2010; Murthy et al., 2009). The changes in productivity of extracellular enzymes can be attributed to the differences in the timing of induction of separate components of the enzyme system, the inhibition by products of substrate hydrolysis and differential inactivation by proteases and/or variation in the pH during cultivation conditions (Tuohy & Coughlan, 1992; J.P. Wang et al., 1993). The accumulation of sugars over a critical concentration in the medium is well documented to inhibit the enzyme production (Dona et al., 2010; J.P. Wang et al., 2006).

Among the physical parameters, the temperature and pH of the medium play an important role in α-amylase production and stability. Generally, the influence of temperature on amylase production is related to the growth of the organism. Hence, the optimum temperature depends on whether the culture is mesophilic, thermophilic or psycrophilic. Among the fungi and actinomycetes, most amylase production studies achieved the optimum yields within the range 25°C- 40°C (Gupta et al., 2003). However, thermophilic fungi ,such as *Thermomyces lanuginosus* (Mishra & Maheshwari, 1996), and actinomycetes, namely; *Thermomonospora fusca* (Busch & Stutzenberger, 1997) and *Thermoactinomyces vulgaris* (Abou Dobara et al., 2011) have been reported to produce α-amylase optimally at 50 °C, 55 °C and 55 °C, respectively. On the other hand, it has been produced at a wider range of optimal temperature by bacteria reaching to 90 °C in *Thermococcale* and *Sulfolobus* species (Leuschner & Antranikian, 1995; Sunna et al., 1997). Also, the pH values were reported to serve as an indicator of the initiation and end of enzyme synthesis (Friedrich et al., 1989) because the change in pH affects α-amylase stability in the medium (Calamai et al., 2005). It is worth noting that the α-amylase active site consists of a large number of charged groups (Lawson et al., 1994; Strokopytov et al., 1996; Uitdehaag et al., 1999) which explain the fact that most α-amylases had optimum pH in the acidic to neutral range (Bozic et al., 2011; Pandey et al., 2000; Sun et al., 2010).

In general, amylase activity is connected with the substrate utilization. The inducibility nature of α-amylase has been assured in different microorganisms (Abou Dobara et al., 2011; Aiyer, 2004; Asoodeh et al., 2010; Ryan et al., 2006). α-amylase production is also appeared to be subjected to catabolite repression by maltose and glucose, like most other inducible enzymes that are affected by substrate hydrolytic products (Bhella & Altosaar, 1988; Morkeberg et al., 1995). However, α-amylase synthesis by *Bacillus* strains was reported to not subject to catabolite repression by monosaccharides (Kalishwaralal et al., 2010). Gupta et al. (2003) have classified xylose and fructose as strongly repressive to α-amylase synthesis. Addition of starch to the medium has normally been employed for the production of α-amylase from various microorganisms as reported in the literature.

Nitrogen source as a basal component of the medium is a major factor affecting α-amylase production. Its effect was not only as a nitrogen source but also as a metal ion source and a pH controller as well. Many investigators had recorded that organic nitrogen sources supported maximum α-amylase production by various bacteria (Abou Dobara et al., 2011; Aqeel & Umar, 2010; Mrudula & Kokila, 2010; Saxena et al., 2007). The increased α-amylase production by organic nitrogen sources could be attributed to the high nutritional amino acids and vitamins content. However, various inorganic salts have been reported to support better production in fungi (Gupta et al., 2003). As a metal ion source, ammonium chloride was found to enhance the production of the α-amylase by *T. vulgaris*, where chloride is a stabilizer, over that of other ammonium salts (Abou Dobara et al., 2011). In addition, the same authors also reported different productivity of α-amylase by using sodium nitrate from potassium nitrate.

3.4.2. Activity measurement of enzyme

The diversity and heterogeneity of natural substrates coupled with the mixed specificities of individual enzymes presents a problem in the characterization of amylases. Furthermore, the enzymatic degradation of native insoluble substrates involves steps and mechanisms which are not yet understood at the molecular level. Therefore biochemical studies always use starch in some modified form to simplify analyses. There are basically four different types of substrates used for activity measurements: purified insoluble substrates approximated to a native substrate, modified insoluble substrates, soluble modified polysaccharides and soluble oligosaccharides. Catalytic activity is usually measured by quantifying formed soluble saccharides or chromophoric aglycon. The action of enzyme on insoluble substrates can also be assayed by other means. For example, a viscosimetric method has been used to measure α-amylase activity on starch pastes (Marciniak & Kula, 1982).

The measurement of soluble products from insoluble or soluble polymeric substrates often means assaying the formed reducing sugars. One of the simplest and most widely used is the 3, 5-dinitrosalisylic acid (DNS) method (Miller, 1959). However, the colour development in the reaction is not strictly proportional to the number of reducing sugars present, but also to the length of the oligosaccharides, leading to higher apparent reducing values with longer sugars (Robyt & Whelan, 1972). DNS itself also breaks down the substrate. Several other reducing sugar determination methods have also been developed. In some cases dye groups have been attached to the polymeric substrate, e.g. dyed amylose-and amylopectin (Klein et al., 1970) and dyed and cross-linked starch (Cesk et al., 1969). The enzymatic assay is based on colour released from the substrate.

Starch forms a deep blue complex with iodine and with progressive hydrolysis of the starch, it changes to red brown. Several procedures have been described for the quantitative determination of amylase based on the reduction in blue colour intensity resulting from enzyme hydrolysis of starch (Swain et al., 2006). This method determines the dextrinising activity of α-amylase in terms of decrease in the iodine colour reaction. Also, the coupled assay methods have been used for amylases, in which the concentration of released glucose is determined either by glucose oxidase/peroxidase (Kunst et al., 1984) or by hexokinase/glucose-6-phosphate dehydrogenase method (Rauscher, 1984).

Generally, various available methods for the determination of α-amylase activity are based on decrease in starch–iodine colour intensity, increase in reducing sugars, degradation of colour-complexed substrate and decrease in viscosity of the starch suspension.

3.4.3. Purification of enzyme

Purification is a key step in the enzymes production where residual cell proteins and other contaminants are removed. Different techniques have been developed for purification of enzymes based on their properties, prior to their characterization or use in biotechnological and industrial processes. The commercial use of α-amylase generally does not require purification of the enzyme but enzyme applications in food industries, pharmaceutical and clinical sectors require high purity amylases. The enzyme in purified form is also a prerequisite in studies of structure-function relationships and biochemical properties.

The used methods to purify amylases can vary considerably, but most purification protocols involve a series of steps (Sun et al., 2010). The choice of purification protocol naturally depends on the intended use, the highest purity usually being required for basic purposes in which even separation of isozymes may be important. The purity and the yield attained depend on the number of steps and separation techniques employed.

The purification of α-amylases from microbial sources in most cases has involved classical purification methods. These methods involve separation of the culture from the fermentation broth, selective concentration by precipitation using ammonium sulphate or organic solvents. The crude enzyme is then subjected to chromatography. The most commonly used techniques are usually affinity chromatography, ion exchange, and/or gel filtration. Cross-linked starch or starch derivatives are useful affinity adsorbents for the isolation of bacterial α-amylases (Somers et al., 1995). Primarini & Ohta (2000) isolated and separated two pure α-amylases from Streptomyces sp. using starch adsorption, α-CD Sepharose 6B and DEAE-Toyopearl 650M. Adsorption of α-amylase, from Streptomyces sp. E-2248; on starch followed by separation on DEAE-Toyopearl and Toyopearl-HW55S gave the highest purification (2130-fold) (Kaneko et al., 2005). Also, α-amylase from Bacillus licheniformis has been purified 6-fold with a yield of 38% using by two gel filtration chromatography steps on Sephadex G-100 and Superose 12 column (Bozic et al., 2011).

In addition to the classical chromatographic techniques, immunoaffinity chromatography has been applied for the preparation of highly purified amylases (Jang et al., 1994). Recent advances in the understanding of the physical and functional properties of amylases, and of the selectivity and capacity of the adsorbents, have led to greater rationality in the design of separation methods. However, the potential of the methods for the separation of amylases has not been fully exploited.

3.4.4. Industrial desirable aspects

The stability of biocatalysts is often a limiting factor in the selection of enzymes for industrial applications due to the elevated temperature or extreme pH of many biotechnological processes. Therefore, there is a continuing demand to improve the stability of the enzymes and thus meet the requirements set by specific applications.

As an example, the problem with traditional detergent enzymes is that they have to function in a washing machine under conditions that are very unfavorable for the stability of the enzyme. The pH is highly alkaline in washing conditions. The high temperature (55–60°C) in a dishwasher requires thermostable enzymes. In addition, it is preferred to be resistant to various detergent ingredients, such as surfactants, chelating and oxidative agents (bleach).

In general, temperature has a complex effect on protein either directly or indirectly for both physical and chemical induced aggregation processes (Y.W. Wang et al., 2010). Therefore, it is the most critical environmental factor for consideration when proteins are handled during the entire development and commercialization processes. The advantages for using thermostable α-amylases in industrial processes include the decreased risk of contamination, the increased diffusion rate and the decreased cost of external cooling. In short, almost all industries need thermostable enzymes. Besides thermostability and other factors such as activity with high concentrations of starch, i.e. more than 30% dry solids, or the protein yields of the industrial fermentation are important criteria for commercialization (Schäfer et al., 2000). Also, α-amylases with wide pH range is desired to satisfy all applications either acidic as glucose syrup production or alkali as detergents industries.

However, there is a recent trend to use intermediate temperature stable (ITS) α-amylases (Ahuja et al., 1998, as cited in Gupta et al., 2003). Olesen (1991) found that this feature render the enzyme to be useful for baking industry through avoiding stickiness in bread. Also, a modern trend among consumers is to use colder temperatures for doing the laundry or dishwashing. At these lower temperatures, detergents with α-amylases optimally working at moderate temperatures and alkaline pH would be favourable (Marc et al., 2002). Although a wide variety of microbial α-amylases is known, α-amylase with 'ITS' property has been reported from only a few microorganisms (Gigras et al., 2002).

Another important desirable feature is calcium independency. Most known α-amylases, with a few exceptions, contain a conserved Ca^{2+} binding site (Linden et al., 2003; Prakash & Jaiswal, 2010) which make calcium be important to the enzyme activity. In manufacture of fructose syrup, the Ca^{2+} ions inhibit the glucose isomerase enzyme used in the final step of the process (Tonkova, 2006) and may lead to the formation of inorganic precipitates which have deleterious effects on fermentation and downstream processing (Kelly et al., 2009). Because the removal of these metal ions is both cost and time consuming to the overall industrial process (Kelly et al., 2009), the use of stable and functional α-amylases in the absence of Ca^{2+} ions at high temperatures would be highly favored.

4. Biodegradation of starch

The degradation of starch occurs mainly through the action of microorganisms in plant litter and soil. Since the native substrate is water-insoluble and cannot penetrate into cells, the biodegradation of starch occurs extracellularly. Amylases are mainly secreted into the medium or are found membrane-bound. Some microbial strains are known to produce intracellular amylases; the reason for this is unknown (Vihinen & Mantsala, 1989).

4.1. Enzymatic degradation of starch

The effective hydrolysis of starch demands the action of many enzymes due to its complexity, although a prolonged incubation with one particular enzyme can lead to (almost) complete hydrolysis. Few microorganisms produce a complete set of enzymes capable of degrading starch efficiently. There are basically four groups of starch-converting enzymes: (i) endoamylases; (ii) exoamylases; (iii) debranching enzymes; and (iv) transferases.

Endoamylases are able to cleave α,1-4 glycosidic bonds present in the inner part (endo-) of the amylose or amylopectin chain. Exoamylases act on the external glucose residues of amylose or amylopectin and thus produce only glucose (glucoamylase and α-glucosidase), or maltose and β-limit dextrin (β-amylase).

The third group of starch-converting enzymes is the debranching enzymes that exclusively hydrolyze α,1-6 glycosidic bonds: isoamylase (EC 3.2.1.68) and pullulanase type I (EC 3.2.1.41). These enzymes exclusively degrade amylopectin, thus leaving long linear polysaccharides. There are also a number of pullulanase type enzymes that hydrolyze both α, 1-4 and α,1-6 glycosidic bonds. These belong to the group II pullulanase and are referred to as α-amylase–pullulanase or amylopullulanase. The main degradation products are maltose and maltotriose.

The fourth group of starch-converting enzymes are transferases that cleave an α,1-4 glycosidic bond of the donor molecule and transfer part of the donor to a glycosidic acceptor with the formation of a new glycosidic bond. Enzymes such as amylomaltase (EC 2.4.1.25) and cyclodextrin glycosyltransferase (EC 2.4.1.19) form a new α, 1-4 glycosidic bond while branching enzyme (EC 2.4.1.18) forms a new α,1-6 glycosidic bond. Cyclodextrin glycosyltransferases have a very low hydrolytic activity and make cyclic oligosaccharides with 6, 7, or 8 glucose residues and highly branched high molecular weight dextrins, the cyclodextrin glycosyltransferase limit dextrins. Amylomaltases are very similar to cyclodextrin glycosyltransferases with respect to the type of enzymatic reaction. The major difference is that amylomaltase performs a transglycosylation reaction resulting in a linear product while cyclodextrin glycosyltransferase gives a cyclic product.

Depending on the relative location of the bond under attack as counted from the end of the chain, the products of this digestive process are dextrin, maltotriose, maltose, and glucose, etc. Dextrins are shorter, broken starch segments that form as the result of the random hydrolysis of internal glucosidic bonds. A molecule of maltotriose is formed if the third bond from the end of a starch molecule is cleaved; a molecule of maltose is formed if the point of attack is the second bond; a molecule of glucose results if the bond being cleaved is the terminal one; and so on.

Most of the enzymes that convert starch belong to one family based on the amino acid sequence homology: the α-amylase family or family 13 glycosyl hydrolases according to the classification of Henrissat (1991). Other little enzymes that convert starch don't belong to family 13 glycosyl hydrolases like β-amylases that belong to family 14 glycosyl hydrolases (Henrissat & Bairoch, 1993); and glucoamylases which belong to family 15 glycosyl hydrolases (Aleshin et al., 1992).

4.2. Catalytic mechanism and substrate binding

The α-glycosidic bond is very stable having a spontaneous rate of hydrolysis of approximately 2×10^{-15} s^{-1} at room temperature (Wolfenden et al., 1998). Members of the α-amylase family enhance this rate so enormously that they can be considered to belong to the most efficient enzymes known.

The α-amylase family enzymes always carry strictly conserved three essential catalytic residues (Matsuura et al., 1984). Of these three residues, the roles of Glu230 and Asp206 have been generally accepted as working for acid (proton donor) and base (nucleophile) catalyst, respectively (Janecek, 1997; Svensson, 1994). The catalytic mechanism has been discussed mostly on the basis of these two residues. However, the critical role of the third residue Asp297 seems to be still undefined and under dispute, except the facts that it plays an important role in the distortion of the substrate (Uitdehaag et al., 1999).

The generally accepted catalytic mechanism of the α-amylase family is that of the α-retaining double displacement (Koshland, 1953). The mechanism involves two catalytic residues in the active site; a glutamic acid as acid/base catalyst and an aspartate as the nucleophile. It involves five steps: (i) after the substrate has bound in the active site, the glutamic acid in the acid form donates a proton to the glycosidic bond oxygen, i.e. the oxygen between two glucose molecules at the subsites −1 and +1 and the nucleophilic aspartate attacks the C1 of glucose at subsite −1; (ii) an oxocarbonium ion-like transition state is formed followed by the formation of a covalent intermediate; (iii) the protonated glucose molecule at subsite +1 leaves the active site while a water molecule or a new glucose molecule moves into the active site and attacks the covalent bond between the glucose molecule at subsite −1 and the aspartate; (iv) an oxocarbonium ion-like transition state is formed again; (v) the base catalyst glutamate accepts a hydrogen from an incoming water or the newly entered glucose molecule at subsite +1, the oxygen of the incoming water or the newly entered glucose molecule at subsite +1 replaces the oxocarbonium bond between the glucose molecule at subsite −1 and the aspartate forming a new hydroxyl group at the C1 position of the glucose at subsite −1 (hydrolysis) or a new glycosidic bond between the glucose at subsite −1 and +1 (transglycosylation). Studies with cyclodextrin glycosyltransferase have shown that the intermediate indeed has a covalently linked bond with the enzyme (Uitdehaag et al., 1999).

Other conserved amino acid residues e.g. histidine, arginine, and tyrosine play a role in positioning the substrate into the correct orientation into the active site, proper orientation of the nucleophile, transition state stabilization, and polarization of the electronic structure of the substrate (Lawson et al., 1994; Strokopytov et al., 1996; Uitdehaag et al., 1999).

5. Biotechnological application

Nowadays, α-amylases represent one of the most important enzyme groups within the field of biotechnology. These enzymes are present in numerous biotechnological and industrial applications such as in food, detergents and textiles as well as in paper industry, for the hydrolysis of starch. They can also be of potential use in the pharmaceutical and fine chemical industries.

5.1. Industrial production of glucose and fructose from starch

The acid hydrolysis method for glucose production has been replaced recently by enzymatic treatment, with three or four different enzymes, in which α-amylase is the first (Crabb & Shetty, 1999).

For the complete conversion into high glucose syrup, the first step is the liquefaction into soluble, short-chain dextrins. Dry solids starch slurry (30–35%) of pH 6 is mixed with α-amylase and passed through a jet cooker after which the temperature is kept at 95–105 °C for 90 min to assure the removal of lipid–starch complexes. The dextrose equivalent value of starch hydrolysate syrup depends on the time of incubation and the amount of added enzyme. The drawback of the currently used α-amylases is that they are not active at a pH below 5.9 at the high used temperatures. Therefore, the pH has to be adjusted from the natural pH 4.5 of the starch slurry to pH 6. Also Ca^{2+} needs to be added because of the Ca^{2+}-dependency of these enzymes (Tonkova, 2006). The next step is the saccharification of the starch hydrolysate syrup to high concentration glucose syrup, with more than 95% glucose. This is done by using an exo-acting glucoamylase. The final step is the conversion of high glucose syrup into high fructose syrup by using glucose isomerase (Bhosale et al., 1996).

5.2. Bakery and anti-staling

The baking industry is a large consumer of starch and starch-modifying enzymes. Amylases can be added to degrade the damaged starch in the flour into smaller dextrins, which are subsequently fermented by the yeast. Upon storage, all undesirable changes together are called staling. Retrogradation of the starch fraction in bread is considered very important in staling (Kulp & Ponte, 1981). Staling is of considerable economic importance for the baking industry since it limits the shelf life of baked products. Several additives may be used in bread baking (Spendler & Jórgensen, 1997) to delay staling and improve texture, volume and flavor of bakery products. Enzymes active on starch have been suggested to act as anti-staling agents, especially α-amylases (Sahlstrom & Brathen, 1997).

α-amylase supplementation in flour not only enhances the rate of fermentation and reduces the viscosity of dough resulting in improvements in the volume and texture of the product (De Stefanis & Turner, 1981) but also it generates additional sugar in the dough, which improves the taste, crust colour and toasting qualities of the bread (Van Dam & Hille, 1992). A recent trend is to use intermediate temperature stable (ITS) α-amylases (Ahuja et al., 1998, as cited in Gupta et al., 2003) since they become inactive much before the completion of the baking process which avoid sickliness in bread.

5.3. Cyclodextrin formation

Cyclodextrins are cyclic α,1-4 linked oligosaccharides mainly consisting of 6, 7, or 8 glucose residues. The glucose residues in the rings are arranged in such a manner that the inside is hydrophobic while the outside is hydrophilic. This enables cyclodextrins to form inclusion complexes with a variety of hydrophobic guest molecules. The formation of inclusion complexes leads to changes in the chemical and physical properties of the guest molecules. These altered characteristics of the encapsulated compounds have led to various applications of cyclodextrins in analytical chemistry, agriculture, pharmacy, food and cosmetics. For the industrial production of cyclodextrins, starch is first liquefied by a heat-stable α-amylase and then the cyclization occurs with a cyclodextrin glycosyltransferase (Riisgaard, 1990).

5.4. Detergent industries

A growing area of application of α-amylases is in the fields of laundry, dish-washing detergents and spot removers (Borchet et al., 1995; Kennedy et al., 1988). Amylases have the function of facilitating the removal of starchy stains by means of catalytic hydrolysis of the starch polysaccharide, and have been used for this purpose for a fairly long time in dishwashing detergents and textile laundering (Speckmann et al., 2001).

Traditional detergent enzymes are functional under alkaline conditions, thermostable and resistant to various detergent ingredients, such as surfactants, chelating and oxidative agents. On the other hand, a modern trend among consumers is to use colder temperatures for doing the laundry or dishwashing. At these lower temperatures, the removal of starch from cloth and porcelain becomes more problematic. Detergents with α-amylases optimally working at moderate temperatures and alkaline pH can help solve this problem (Marc et al., 2002).

5.5. Ethanol production

For large-scale processing, the bioconversion of starchy materials to ethanol is very useful because it can be used as a biofuel and as the starting material for various chemicals. The production of ethanol from starchy biomass commonly involves three-step processes: liquefaction of starch by an endoamylase such as α-amylase to reduce the viscosity of the gelatinized starch produced after the cooking of the grains, enzymatic saccharification of the low-molecular-weight liquefaction products to produce glucose, and fermentation of glucose.

However, the present process for ethanol production from starchy materials via fermentation requires improvement of cost production. Although noncooking and low-temperature-cooking fermentation systems (Matsumoto et al., 1982) have succeeded in reducing energy consumption by approximately 50% (Matsumoto et al., 1982), it is still necessary to add large amounts of amylolytic enzymes to hydrolyze the starchy materials. Many researchers have reported attempts to resolve this problem by using recombinant glucoamylase-expressing yeasts with the ability to ferment starch to ethanol directly (Kondo et al., 2002). Also, a noncooking fermentation system using a cell surface-engineered yeast strain promises to be very effective in reducing the production costs of ethanol (Shigechi et al., 2004). On the other hand, fermentation of starch to ethanol in one step using co-cultures of two different strains has been suggested and has potential application for the direct bioconversion of starch into ethanol (Zeikus, 1979).

5.6. Miscellaneous applications

Besides amylases' use in the saccharification or liquefaction of starch, these are also used for the clarification of formed haze in fruit juices, the pretreatment of animal feed to improve the digestibility (Marc et al., 2002). α-amylase is used for the production of low viscosity, high molecular weight starch for coating of paper (Bruinenberg et al., 2004). Starch is a good sizing agent for the finishing of paper. It is added to the paper in the size press and paper picks up the starch by passing through two rollers that transfer the starch slurry. The temperature of this process lies in the range of 45–60 °C. A constant viscosity of the starch is required for reproducible results at this stage. The mill also has the flexibility of varying the starch viscosity for different paper grades. The viscosity of the natural starch is too high for paper sizing and is adjusted by partially degrading the polymer with α-amylases in a batch or continuous processes. Also, good desizing of starch sized textiles is achieved by the application of α-amylases, which selectively remove the size and do not attack the fibers. It also randomly cleaves the starch into dextrins that are water soluble and can be removed by washing.

Furthermore, high molecular weights amylases were found in culture supernatants of an environmentally derived microbial mixed culture selected for its ability to utilize starch-containing plastic films as sole carbon sources (Burgess-Cassler et al., 1991). This suggests a new application for amylases in biodegradation. With the advent of new frontiers in biotechnology, the spectrum of amylase applications has expanded into many other fields, such as clinical, medicinal and analytical chemistry (Becks et al., 1995).

A modern trend is to use starch for production of a more efficient and specific degradation products through a particular combination of activities. Amylase from *Aspergillus niger*, a saccharifying enzyme which produces maltose, maltotriose and some glucose, is capable of alcoholysis for the synthesis of methyl-glucosides from starch in the presence of methanol. As these products are a series of methyloligosaccharides, from methyl-glucoside to methyl-hexomaltoside, the biotechnological applications of using starch as substrate for the production of alkyl-glucosides is analyzed (Santamaria et al., 1999).

Moreover, it becomes possible to produce lactic acid directly from starch by an efficient simultaneous saccharification and fermentation from soluble starch by recombinant *Lactobacillus* strains (Okano et al., 2009). Finally, α-amylase is suggested as an enzyme that contributes to the reduction of $AuCl_4^-$ to gold nanoparticles (Au-NPs) which makes it ideal for the production of Au-NPs (Kalishwaralal et al., 2010).

6. Conclusion

Despite the fact that several different α-amylase preparations are available with various enzyme manufacturers for specific use in varied industries, amylase biotechnology demands extension in terms of both quality and quantity. Qualitative improvements in amylase gene and its protein can be achieved by recombinant DNA technology and protein engineering. Quantitative enhancement needs strain improvement through site directed mutagenesis and/or standardizing the nutrient medium for the overproduction of active α-amylases. Another approach is to screen for novel microbial strains from extreme environments.

All the above-mentioned approaches are aimed to increase stability, improve product specificity, alter pH optimum, improve thermostability, achieve free Ca^{+2} requirement; by using the currently available insights into the structure–function relationships of the amylase family enzymes.

Author details

Amira El-Fallal, Mohammed Abou Dobara, Ahmed El-Sayed and Noha Omar
Faculty of Science, Damietta Branch, Egypt

Acknowledgement

The authors are greatly acknowledging Damietta Branch, Mansoura University for supporting this work.

7. References

Abou Dobara, M.I., El-Sayed, A.K., El-Fallal, A.A & Omar, N.F. (2011). Production and partial characterization of high molecular weight extracellular α-amylase from *Thermoactinomyces vulgaris* isolated from Egyptian soil. *Polish Journal of Microbiology*, Vol. 60, No. 1, (January 2011), pp. 65- 71, ISSN 1733-1331

Aiyer, P. (2004). Effect of C: N ratio on alpha amylase production by *Bacillus licheniformis* SPT 27. *African Journal of Biotechnology*, Vol. 3, No. 10, (October 2004), pp. 519-522, ISSN 1684-5315

Aleshin, A., Firsov, L. & Honzatko, R. (1994). Refined structure for the complex of acarbose with glucoamylase from *Aspergillus awamori* Var. X100 to 2.4-A resolution. *J. Biol. Chem.*, Vol. 269, No. 22, (June 1994), pp. 15631-15639, ISSN 1083-351X

Aleshin, A., Firsov, L. & Honzatko, R. (1994). Refined structure for the complex of acarbose with glucoamylase from *Aspergillus awamori* Var. X100 to 2.4-A resolution. *J. Biol. Chem.*, Vol. 269, No. 22, (June 1994), pp. 15631-15639, ISSN 1083-351X

Andrews, L. & Ward, J. (1987). Extracellular amylases from *Streptomyces albus*. *J. Biochem. Soc. Trans.*, Vol. 15, No.3, (June 1987), pp. 522-523, ISSN 1470-8752

Aqeel, B. & Umar, D. (2010). Effect of alternative carbon and nitrogen sources on production of alpha-amylase by Bacillus megaterium. *World Applied Sciences Journal*, Vol. 8, Special Issue of Biotechnology & Genetic Engineering, pp. 85-90, ISSN 0264-8725

Arnesen, S., Eriksen, S.H., Olsen, J. & Jensen, B. (1998). Increased production of alpha amylase from *Thermomyces lanuginosus* by the addition of Tween-80. *Enzyme Microb. Technol.*, Vol. 23, No. 3-4, (November 1998), pp. 249–252, ISSN 0141-0229

Asoodeh, A., Chamanic, J. & Lagzian, M. (2010). A novel thermostable, acidophilic α-amylase from a new thermophilic "*Bacillus* sp. Ferdowsicous" isolated from Ferdows hot mineral spring in Iran: Purification and biochemical characterization. *Int. J. Biol. Macromol.*, Vol. 46, No. 3, (April 2010), pp. 289-297, ISSN 0141-8130

Aunstrup, K. (1979). Production, isolation and economics of extracellular enzymes, In: *Applied Biochemistry and Bioengineering*, J. Wingard, L. Katchalski – Katzir & L. Golstein (Ed.), pp. (27-69), Academic Press, ISBN 0120411024, New York, USA

Azhari, R. & Lotan, N. (1991). Enzymic hydrolysis of biopolymers via single-scission attack pathways: a unified kinetic model. *J. Mater. Sci. Mater. Med.*, Vol. 2, No. 1, (January 1991), pp. 9–18, ISSN 1573-4838

Bahrami, A., Shojaosadati, S., Khalilzadeh, R., Mohammadian, J., Farahani, E. & Masoumian, M. (2009). Prevention of human granulocyte colony-stimulating factor protein aggregation in recombinant Pichia pastoris fed-batch fermentation using additives. *Biotechnol. Appl. Biochem.*, Vol. 52, No. 2, (February 2009), pp. 141–148, ISSN 1470-8744

Bahri, S. & Ward, J. (1990). Cloning and expression of an α-amylase gene from *Streptomyces thermoviolaceus* CUB74 in *Escherichia coli* JM107 and *S. lividans* TK24. *J. Gen. Microbiol.*, Vol. 136, No. 5, (May 1990), pp. 811-818, ISSN 1465-2080

Bajpai, P. & Bajpai, P.K. (1989). High-temperature alkaline α-amylase from *Bacillus licheniformis* TCRDC-B13. *Biotechnol. Bioeng.*, Vol. 33, No. 1, (January 1989), pp. 72–78, ISSN 1097-0290

Barros, M., Silva, R., Ramada, M., Galdino, A., Moraes, L., Torres, F. & Ulhoa, C. (2009). The influence of N-glycosylation on biochemical properties of Amy1, an α-amylase from the yeast *Cryptococcus flavus*. *Carbohydrate Research*, Vol. 344, No. 13, (September 2009), pp. 1682–1686, ISSN 0008-6215

Bayer, E., Morag, E., Wilcheck, M., Lamed, R., Yaron, S. & Shoham, Y. (1995). Cellulosome domains for novel biotechnological application. In: *Carbohydrate bioengineering*, B. Petersen, B. Srensson & S. Pedersen (Ed.), pp. (251-259), Progress in biotechnology, Elsevier, ISBN 0-444-82223-2, The Netherlands.

Becks, S., Bielawaski, C., Henton, D., Padala, R., Burrows, K. & Slaby, R. (1995). Application of a liquid stable amylase reagent on the ciba corning express clinical chemistry *system*. *Clin. Chem.*, Vol. 41, No. 2, (February 1995), pp. 186, ISSN 1530-8561

Bertoft, E. (2007). Composition of clusters and their arrangement in potato amylopectin. *Carbohydrate Polymers*, Vol. 68, No. 3, (April 2007), pp. 433-446, ISSN 0144-8617

Bhella, R.S. & Altosaar, I. (1988). Role of CAMP in the mediation of glucose catabolite repression of glucoamylase synthesis in *Aspergillus awamori*. *Curr. Genet*, Vol. 14, No. 3, (September 2004), pp. 247-252, ISSN 1432-0983

Bhosale, S., Rao, M. & Deshpande, V. (1996). Molecular and industrial aspects of glucose isomerase. *Microbiol. Rev.*, Vol. 60, No. 2, (June 1996), pp. 280–300, ISSN 1098-5557

Bijttebier A., Goesaert H. & Delcour J.A. (2007). Temperature impacts the multiple attack action of amylases. *Biomacromolecules*, Vol. 8, No. 3, (March 2007), pp. 765–772, ISSN 1526-4602

Bijttebier, A., Goesaert, H. & Delcour, J. (2008). Amylase action pattern on starch polymers. *Biologia*, Vol. 63, No. 6, (December 2008), pp. 989-999, ISSN 1336-9563

Borchet, T., Lassen, S., Svendsen, A. & Frantzen, H. (1995). Oxidation stable amylases for detergents. In: *Carbohydrate bioengineering: Progress in biotechnology*. Vol. 10, S. Petersen,

B. Svensson & S. Pedersen (Ed.), pp. (175-179), Elsevier, ISBN 0444822232, The Netherlands.

Bozic, N., Ruizb, J., López-Santínb, J. & Vujci´c, Z. (2011). Production and properties of the highly efficient raw starch digesting α-amylase from a *Bacillus licheniformis* ATCC 9945a. *Biochemical Engineering Journal*, Vol. 53, No. 2, (January 2011), pp. 203–209, ISSN 1369-703X

Brooker, J.D. & McCarthy, J.M. (1997). Gene knockout of the intracellular amylase gene by homologous recombination in *Streptococcus bovis*. *Curr. Microbiol.*, Vol. 35, No. 3, (September 1997), pp. 133–138, ISSN 1432-0991

Bruinenberg, P.M., Hulst, A.C., Faber, A. & Voogd, RH. (2004). A process for surface sizing or coating of paper, In: *European Patent Application* no. EP0690170, April 2012, Available from :< http://www.freepatentsonline.com/EP0690170.html>

Burgess-Cassler, A., Imam, S. & Gould, J. (1991). High-Molecular-Weight Amylase Activities from Bacteria Degrading Starch-Plastic Films. *Appl. Environ. Microbiol.*, Vol. 57, No. 2, (February 1991), pp. 612–614, ISSN 098-5336

Busch, J.E. & Stutzenberger, F.J. (1997). Amylolytic activity of *Thermomonospora fusca*. *World J. Microbiol. Biotechnol.* Vol. 13, No. 6, (November 1997), pp. 637-642, ISSN 1573-0972

Calamai, M., Canale, C., Relini, A., Stefani, M., Chiti, F. & Dobson, C.M. (2005). Reversal of protein aggregation provides evidence for multiple aggregated states. *J. Mol. Biol.* Vol. 346, No. 2, (February 2005), pp. 603–616, ISSN 0022-2836

Cesk, M., Brown, B. & Birath, K. (1969). Ranges of α-amylase activities in human serum and urine and correlations with some other α-amylase methods. *Clin. Chem. Acta.*, Vol. 26, No. 3, (December 1969), pp. 445-453, ISSN 0009-8981

Collins, B., Kelly, C., Fogarty, W. & Doyle, E. (1993). The high maltose-producing alpha-amylase of the thermophilic actinomycete *Thermomonospora curvata*. *Appl. Microbiol. Biotechnol.*, Vol. 39, No.1, (April 1993), pp. 31–35, ISSN 1432-0614

Coutinho, P. & Henrissat, B. (1999). Carbohydrate active enzymes. In: *Recent Advances in Carbohydrate bioengineering*, H. Gilbert, G. Davies, B. Svensson & B. Henrissat (Ed.), pp. (3-12), Royal Society of Chemistry, ISBN 0854047743, Cambridge

Crabb, W. & Shetty, J. (1999). Commodity scale production of sugars from starches. *Curr. Opin. Microbiol.*, Vol. 2, No. 3, (June 1999), pp. 252– 256, ISSN 1369-5274

De Mot, R., Andries, K. & Verachtert, H. (1984a). Comparative study of starch degradation and amylase production by ascomycetous yeast species. *Syst. Appl. Microbiol.* Vol. 5, No. 1, (April 1984), pp. 106-118, ISSN 0723-2020

De Mot, R., Demeersman, M. & Verachtert. H. (1984b). Comparative study of starch degradation and amylase production by non-ascomycetous yeast species. *Syst. Appl. Microbiol.* Vol. 5, No. 3, (October 1984), pp. 421-432, ISSN 0723-2020

De Mot, R., Van Oudendijck, E., Hougaerts, S. & Verachtert. H. (1984c). Effect of medium composition on amylase production by some starch-degrading yeasts. *FEMS Microbiol. Lett.*, Vol. 25, No. 2, (December 1984), pp. 169-173, ISSN 1574-6968

De Stefanis, V. & Turner, E. (1981). Modified enzyme system to inhibit bread firming method for preparing same and use of same in bread and other bakery products, In:

United states Patent Application no. US4299848, April 2012, Available from: <http://patft.uspto.gov/netacgi/nphParser?Sect2=PTO1&Sect2=HITOFF&p=1&u=/netaht ml/PTO/searchbool.html&r=1&f=G&l=50&d=PALL&RefSrch=yes&Query=PN/4299848 >

Dhawale, M., Wilson, J., Khachatourians, G. & Mike, W. (1982). Improved method for detection of starch hydrolysis. *Appl. Eenviron. Microbiol.*, Vol. 43, No. 4, (April 1982), pp. 747-750, ISSN 1098-5336

Domsch, K.H., Gams, W. & Anderson, T.H. (1995). *Compendium of soil fungi*, p. 859, IHW-Verlag, ISBN 3980308383, Alemanha

Dona, A., Pages, G., Gilbert, R. & Kuchel, P. (2010). Digestion of starch: In vivo and in vitro kinetic models used to characterise oligosaccharide or glucose release. *Carbohydrate Polymers.*, Vol. 80, No. 3, (May 2010), pp. 599-617 , ISSN 0144-8617

Doyle, E.M., Noone, A.M., Kelly, C.T., Quigley, T.A. & Fogarty, W.M. (1998). Mechanisms of action of the maltogenic α-amylase of *Byssochlamys fulva*. *Enzyme Microb. Technol*. Vol. 22, No. 7, (May 1998), pp. 612–616, ISSN 0141-0229

EL-Banna, T.E., Abd-Aziz, A.A., Abou-Dobara, M.I. & Ibrahim, R.I. (2007). Production and immobilization of α-amylase from *Bacillus subtilis*. *Pakistan J. of Biological Sciences*, Vol. 10, No. 12, (June 2007), pp. 2039-2047, ISSN 1812-5735

EL-Banna, T.E., Abd-Aziz, A.A., Abou-Dobara, M.I. & Ibrahim, R.I. (2008). Optimization and immobilization of α-amylase from *Bacillus licheniformis*, *Proceedings of the second international conference on the role of genetics and biotechnology in conservation of natural resources*, Ismailia, Egypt, July 2007.

Eriksen, S., Jensen, B. & Olsen, J. (1998). Effect of N-linked glycosylation on secretion, activity and stability of α-amylase from *Aspergillus oryzae*. *Curr. Microbiol.*, Vol. 37, No. 2, (August 1998), pp. 117-122, ISSN 1432-0991

Espargaro, A., Castillo, V., de Groot, N. & Ventura, S. (2008). The in vivo and in vitro aggregation properties of globular proteins correlate with their conformational stability: the SH3 case. *J. Mol. Biol.*, Vol. 378, No. 5, (May 2008), pp. 1116–1131, ISSN 0022-2836

Farber, G. & Petsko, G. (1990). The evolution of α/β–barrel enzymes. *TIBS*, Vol. 15, No. 6, (June 1990), pp. 228-234, ISSN 0968-0004

French, D. (1984). Organization of starch granules, In: *starch: chemistry and technology*, R.L. Whistler, J.N. Bemiller & E.F. Paschall (Ed.), pp. (183-207), Academic press, ISBN 78-0-12-746275-2, London.

Friedrich, J., Cimerman, A. & Steiner, W. (1989). Submerged production of pectinolytic enzymes by *Aspergillus niger*: effect of different aeration/agitation regimes. *Appl. Microbiol. Biotechnol.* , Vol. 31, No. 5-6, (Octobar 1989), pp. 490-494, ISSN 1432-0614

Fukumoto, J. & Okada, S. (1963). Studies on bacterial amylase, Amylase types of *Bacillus subtilis* species. *J. Ferment. Technol.*, Vol. 41, No. 1, pp. 427-434, ISSN 0385-6380

Gerard, C., Planchot, V., Colonna, P. & Bertoft, E. (2000). Relationship between branching density and crystalline structure of A- and B-type maize mutant starches. *Carbohydr. Res.* Vol. 326, No. 2, (June 2000), pp. 130–144, ISSN 0008-6215

Gigras, P., Sahai, V. & Gupta, R. (2002). Statistical media optimization and production of its alpha amylase from *Aspergillus oryzae* in a bioreactor. *Curr. Microbiol.*, Vol. 45, No. 3, (September 2002), pp. 203–208, ISSN 1432-0991

Giraud, E., Brauman, A., Keleke, S., Lelong, B. & Raimbault, M. (1991). Isolation and physiological study of an amylolytic strain of *Lactobacillus plantarum*. *Appl. Microbiol. Biotechnol.*, Vol. 36, No. 3, (December 1991), pp. 379-383, ISSN 1432-0614

Gupta R., Paresh, G., Mohapatra, H., Goswami, V., & Chauhan, B. (2003) Microbial α-amylases: a biotechnological perspective. *Process Biochem.*, Vol. 38, No. 12, (July 2003), pp. 1599–1616, ISSN 1359-5113

Hanashiro, I., Abe, J. & Hizukuri, S. (1996). A periodic distribution of the chain length of amylopectin as revealed by high-performance anion-exchange chromatography. *Carbohydr. Res.* Vol. 283, No. 1, (March 1996), pp. 151-159, ISSN 0008-6215

Hang, M., Furuyoshi, S., Yagi, T. & Yamamoto, S. (1996). Purification and characterization of raw starch digesting α-amylase from *Streptomyces thermocyaneoviolaceus* IFO14271. *J. Appl. Glycosci.*, Vol. 43, pp. 487-497, ISSN 1880-7291

Hao, Y., Chu, J., Wang, Y., Zhuang, Y. & Zhang, S. (2007). The inhibition of aggregation of recombinant human consensus interferon-alpha mutant during *Pichia pastoris* fermentation. *Appl. Microbiol. and Biotechnol.*, Vol. 74, No. 3, (March 2007), pp. 578–584, ISSN 1432-0614

Hayashi, T., Akiba, T. & Horikoshi, K. (1988). Production and purification of new maltohexaose-forming amylases from alkalophilic *Bacillus* sp. H-167. *Agric. Biol. Chem.*, Vol. 52, No. 2, (February 1988). pp. 443–448, ISSN 1881-1280

Henrissat, B. & Bairoch, A. (1993). New families in the classification of glycosyl hydrolases based on amino acid sequence similarities. *Biochem. J.*, Vol. 293, No. 3, (August 1993), pp. 781-788, ISSN 0264-6021

Henrissat, B. (1991). A classification of glycosyl hydrolases based on amino acid sequence similarities. *Biochem.J.*, Vol. 280, No. 2, (December 1991), pp. 309–316, ISSN 0264-6021

Hofemeister, B., Konig, S., Hoang, V., Engel, J., Mayer, G., Hansen, G. & Hofemeister, J. (1994). The gene *amyE* (TV1) codes for a nonglucogenic α-amylase from *Thermoactinomyces vulgaris* 94-2A in *Bacillus subtilis*. *Appl. Environ. Microbiol.*, Vol. 60, No. 9, (September 1994), pp. 3381-3389, ISSN 1098-5336

Hyun, H.H. & Zeikus, J.G. (1985a). General biochemical characterization of thermostable extracellular β-amylase from *Clostridium thermosulfurogenes*. *Appl. Environ. Microbiol.*, Vol. 49, No. 5, (May 1985), pp. 1162-1167, ISSN 1098-5336

Imberty, A., Buleon, A., Tran, V. & Perez, S. (1991). Recent advances in knowledge of starch structure. *Starch/Starke*, Vol. 43, No. 10, (October 1991), pp. 375-384, ISSN 1521-379X

Janecek, S. (1997). α-Amylase family: molecular biology and evolution. *Prog. Biophys. Mol. Biol.* Vol. 67, No. 1, (January 1997), pp. 67-97, ISSN 0079-6107

Jang, S., Cheong, T., Shim, W., Wan Kim, J. & Park, K. (1994). Purification of *Bacillus licheniformis* thermostable α-amylase by immunoaffinity chromatography. *Korean Biochem. J.*, Vol. 27, No. 1, (January 1994), pp. 38-41, ISSN 0368-4881

Kadziola, A., Abe, J., Svensson, B. & Haser, R. (1994). Crystal and molecular structure of barley α-amylase. *J.Mol.Biol.*, Vol. 239, No. 1, (may 1994), pp. 104-121, ISSN 0022-2836

Kalishwaralal, K., Gopalram, S., Vaidyanathan, R., Deepak, V., Pandian, S. & Gurunathan, S. (2010). Optimization of α-amylase production for the green synthesis of gold nanoparticles. *Colloids and Surfaces B: Biointerfaces*, Vol. 77, No. 2, (June 2010), pp. 174–180, ISSN 0927-7765

Kaneko, T., Ohno, T. & Ohisa, N. (2005). Purification and characterization of a thermostable raw starch digesting amylase from a *Streptomyces* sp. isolated in a milling factory. *Biosci. Biotechnol. Biochem.*, Vol. 69, No. 6, (June 2005), pp. 1073-1081, ISSN 1347-6947

Kelly, R., Dijkhuizen, L. & Leemhuis, H. (2009). Starch and glucan acting enzymes, modulating their properties by directed evolution. *Biotechnology*, Vol. 140, No. 3-4, (March 2009), pp. 184–193, ISSN 0168-1656

Kennedy, J., Cabalda, V. & White, C. (1988). Enzymic starch utilization and genetic engineering. *Tibtech.*, Vol. 6, pp. 184-189, ISSN 0167-7799

Kim, T.U., Gu, B.G., Jeong, J.Y., Byun, S.M. & Shin, Y.C. (1995). Purification and characterization of maltotetraose-forming alkaline α-amylase from an alkalophilic *Bacillus* strain, GM8901. *Appl. Environ. Microbiol.*, Vol. 61, No. 8, (August 1995), pp. 3105-3112, ISSN 1098-5336

Kimura, K., Tsukamoto, A., Ishii, Y., Takano, T. & Yamane, K. (1988). Cloning of a gene for maltohexaose producing amylase of an alkalophilic *Bacillus* and hyper-production of the enzyme in *Bacillus subtilis* cells. *Appl. Microbiol. Biotechnol.* Vol. 27, No. 4, (January 1988), pp. 372–377, ISSN 1432-0614

Klein, B., Foreman, J. & Searcy, R. (1970). New chromogenic substrate for determination of serum amylase activity. *Clin. Chem.*, Vol. 16, No. 1, (January 1970), pp. 32-38, ISSN 1530-8561

Kondo, A., Shigechi, H., Abe, M., Uyama, K., Matsumoto, T., Takahashi, S., Ueda, M., Tanaka, A., Kishimoto, M. & Fukuda, H. (2002). High-level ethanol production from starch by a flocculent *Saccharomyces cerevisiae* strain displaying cell-surface glucoamylase. *Appl. Microbiol. Biotechnol.*, Vol. 58, No. 3, (March 2002), pp. 291–296, ISSN 1432-0614

Koshland, D.E. (1953). Stereochemistry and the mechanism of enzymatic reactions. *Biol. Rev.*, Vol. 28, No. 4, (November 1953), pp. 416-436, ISSN 1469-185X

Kramhøft B., Bak-Jensen K., Mori H., Juge N., Nohr J. & Svensson B. (2005). Involvement of individual subsites and secondary substrate binding sites in multiple attack on amylase by barley α-amylase. *Biochemistry*, Vol. 44, No. 6, (February 2005), pp. 1824–1832, ISSN 1520-4995

Kulp, K. & Ponte, J.G. (1981). Staling white pan bread: fundamental causes. *Crit. Rev. Food Sci. Nutr.*, Vol. 15, No. 1, (January 1981), pp. 1–48, ISSN 1549-7852

Kunst A., Drager, B. & Ziegenhorn, J. (1984). Colorimetric methods with glucose oxidase and peroxidase. In: *Methods of enzymatic analysis*, H. Bermeyer (Ed.), pp. (178-185), Weiheim, ISBN 3527255982, Verleg Chemie

Kuo, M. & Hartman, P. (1966). Isolation of amylolytic strains of *Thermoactinomyces vulgaris* and production of thermophilic actinomycete amylases. *J. Bacteriol.*, Vol. 92, No. 3, (September 1966), pp. 723-726, ISSN 1098-5530

Lawson, C.L., van Montfort, R., Strokopytov, B., Rozeboom, H.J., Kalk, K.H., de Vries, G.E., Penninga, D., Dijkhuizen, L. & Dijkstra, B.W. (1994). Nucleotide sequence and X-ray structure of cyclodextrin glycosyltransferase from *Bacillus circulans* strain 251 in a maltose-dependent crystal form. *J. Mol. Biol.* Vol. 236, No. 2, (February 1994), pp. 590–600, ISSN 0022-2836

Leuschner, C. & Antranikian, G. (1995). Heat-stable enzymes from extremely thermophilic and hyperthermophilic microorganisms. *World J Microb Biot.*, Vol. 11, No. 1, (January 1995), pp. 95–114, ISSN 1573-0972

Liebl, W., Stemplinger, I. & Ruile, P. (1997). Properties and gene structure of the *Thermotoga maritima* alpha-amylase AmyA, a putative lipoprotein of a hyperthermophilic bacterium. *J. Bacteriol.* Vol. 179, No.3, (February 1997), pp. 941-948, ISSN 1098-5530

Lim, W., Park, S., An, C., Lee, J., Hong, S., Shin, E., Kim, E., Kim, J., Kim, H.& Yun, H. (2003). Cloning and characterization of a thermostable intracellular alpha-amylase gene from the hyperthermophilic bacterium *Thermotoga maritima* MSB8. *Res. Microbiol.* Vol. 154, No.10, (December 2003), pp. 681–687, ISSN 0923-2508

Linden, A., Mayans, O., Meyer-Claucke, W., Antranikian, G. & Wilmanns, M. (2003). Differential regulation of a hyperthermophilic α-amylase with a novel (Ca, Zn) two-metal center by zinc. *J. Biological Chemistry*, Vol. 278, No. 11, (March 2003), pp. 9875-9884, ISSN 1083-351X

Lonsane, B. & Ramesh M. (1990). Production of bacterial thermostable [alpha]-Amylase by solid-state fermentation: a potential tool for achieving economy in enzyme production and starch hydrolysis. In: *Advances in Applied Microbiol.*, Vol. 35, S. Neidleman & A. Laskin (Ed.), pp. (1-56), Elsevier, ISBN 978-0-12-002635-7, The Netherlands.

Machius, M., Wiegand, G. & Huber, R. (1995). Crystal structure of calcium depleted *Bacillus licheniformis* α-amylase at 2.2 Å resolution. *J.Mol.Biol.*, Vol. 246, No. 4, (March 1995), pp. 545-559, ISSN 0022-2836

Malhotra, R., Noorvez, S. & Satyanarayana, T. (2000). Production and partial characterization of thermostable and calcium-independent α-amylase of an extreme thermophile *Bacillus thermooleovorans* NP54. *Letters in Applied Microbiology*, Vol. 31, No. 5, (November 2000), pp. 378-384, ISSN 1472-765X

Marc, J.E., van der Maarel, C., Joost, B.V., Uitdehaag, C.M., Leemhuis, H. & Dijkhuizen, L. (2002). Properties and applications of starch-converting enzymes of the α-amylase family. *J. Biotechnology*, Vol. 94, No. 2, (March 2002), pp. 137-155, ISSN 0168-1656

Marciniak, G. & Kula, M. (1982). Vergleichende untersuchung der methoden zur bestimmung der aktivität bakterieller alpha amylasen. *Starch/starke*, Vol. 34, No. 12, (December 1982), pp. 422-430, ISSN 1521-379X

Matsumoto, N., Fukushi, O., Miyanaga, M., Kakihara, K., Nakajima, E. & Yoshizumi, H. (1982). Industrialization of a noncooking system for alcoholic fermentation from grains. *Agric. Biol. Chem.*, Vol. 46, No. 6, (June 1982), pp. 1549–1558, ISSN 1881-1280

Matsuura, Y., Kusunoki, M., Harada, W. & Kakudo, M. (1984). Structure and possible catalytic residues of Taka amylase A. J. Biochem., Vol. 95, No. 3, (March 1984), pp. 697-702, ISSN 1756-2651

Matsuzaki, H., Yamane, K., Yamaguchi, K., Nagata, Y. & Maruo, B. (1974). Hybrid α-amylase produced by transformants of *Bacillus subtilis* I. Immunological and chemical properties of α-amylases produced by the parental strain and the transformants. *Biochim Biophys Acta.*, Vol. 367, No. 2, (October 1974), pp. 235-247, ISSN 0005-2736

McCann, A.K. & Barnett. J.A. (1984). Starch utilization by yeasts: mutants resistant of carbon catabolite repression. *Curr. Genet.* Vol. 8, No. 7, (September 1984), pp. 525-530, ISSN 1432-0983

Miller, G.L. (1959). Use of dinitrosalisylic acid reagent for determination of reducing sugar. *Anal. Chem.*, Vol. 31, No. 3, (March 1959), pp. 426-428, ISSN 1520-6882

Mishra, R.S. & Maheshwari, R. (1996). Amylases of the thermophilic fungus *Thermomyces lanuginosus*: their purification, properties, action on starch and response to heat. *J Biosci.*, Vol. 21, No. 5, (September 1996), pp. 653-672, ISSN 0973-7138

Morkeberg, R., Carlsen, M. & Nielsen, J. (1995). Induction and repression of α-amylase production in batch and continuous cultures of *Aspergillus oryzae*. *Microbiol.*, Vol. 141, No. 10, (October 1995), pp. 2449-2454, ISSN 1465-2080

Murthy, P., Naidu, M. & Srinivas, P. (2009). Production of α-amylase under solid-state fermentation utilizing coffee waste. *J. Chem. Technol. Biotechnol.*, Vol. 84, No. 8, (August 2009), pp. 1246–1249, ISSN 1097-4660

Narang, S. & Satyanarayana, T. (2001). Thermostable a-amylase production by an extreme thermophile *Bacillus thermooleovorans*. *Letters in Applied Microbiology*, Vol. 32, No. 1, (January 2001), pp. 31–35, ISSN 1472-765X

Obi, S. & Odibo, F. (1984). Some properties of a highly thermostable α-Amylase from a *Thermoactinomyces* sp. *Can. J. Microbiol.*, Vol. 30, No. 6, (June 1984), pp. 780-785, ISSN 1480-3275

Okano, K., Zhang, O., Shinkawa, S., Yoshida, S., Tanaka, T., Fukuda, H. & Kondo, A. (2009). Efficient Production of Optically Pure D-Lactic Acid from Raw Corn Starch by Using a Genetically Modified L-Lactate Dehydrogenase Gene-Deficient and á-Amylase-Secreting *Lactobacillus plantarum* Strain. *Appl. Environ. Microbiol.*, Vol. 75, No. 2, (January 2009), pp. 462-467, ISSN 098-5336

Olesen, T. (1991). Antistaling process and agent. Patent application, In: *European Patent Application* no. EP19900915100, April 2012, Available from: <http://www.freepatentsonline.com/EP0494233.html >

Omar, N., Abou-Dobara, M. & El-Sayed, A. (2011). *Studies on amylase produced by some actinomycetes*, Lambert, Academic publishing ISBN 9783847334361, Germany

Pandey, A., Nigam, P., Soccol, C., Soccol, V., Singh, D. & Mohan, R. (2000). Advances in microbial amylases. *Biotechnol. Appl. Biochem.*, Vol. 31, No. 2, (April 2000), pp. 135–152, ISSN 1470-8744

Petruccioli, M. & Federici, R.G. (1992). A note on the production of extracellular hydrolytic enzymes by yeast-like fungi and related microorgnisms. *Ann. Microbiol. Enzimol.*, Vol. 42, No.1 , pp. 81-86, ISSN 0003-4649

Prakash, O. & Jaiswal, N. (2010). α-Amylase: An Ideal Representative of Thermostable Enzymes. *Appl Biochem Biotechnol.*, Vol. 160, No. 8, (April 2010), pp. 2401-2414, ISSN 1559-0291

Primarini, D. & Ohta, Y. (2000). Some enzyme properties of raw starch digesting amylases from *Streptomyces* sp. no. 4. *Starch/Starke*, Vol. 52, No. 1, (January 2000), pp. 28–32, ISSN 1521-379X

Ratanakhanokchai, K., Kaneko, J., Kamio, Y. & Izaki, K. (1992). Purification and properties of a maltotetraose and maltotriose producing amylase from *Chloroflexus aurantiacus*. *Appl. Environ. Microbiol.* Vol. 58, No. 8, (August 1992), pp. 2490-2494, ISSN 1098-5336

Rauscher, E. (1984). Determination of the degradation products maltose and glucose. In: *Methods of enzymatic analysis*, H. Bermeyer (Ed.), pp. (890-894), Weiheim, ISBN 3527255982, Verleg Chemie

Rhodes, C., Strasser, J. & Friedberg, F. (1987). Sequence of an active fragment of *Bacillus polymyxa* β-amylase. *Nucleic Acids Res.*, Vol. 15, No. 9, (May 1987), pp. 3934, ISSN 1362-4962

Riisgaard, S. (1990). The enzyme industry and modern biotechnology. In: *Proceedings of the Fifth European Congress on Biotechnology*, Vol. 1, C. Christiansen, L. Munck, J. Villadsen (Ed.), pp. (31-40), ISBN 8716106172, Munksgaard, Copenhagen.

Robyt, J.F. & Whelan, W.J. (1972). Reducing value methods for maltodextrins: chain length dependence of alkaline 3, 5-dinitrosalisylate and chain length independence of alkaline copper. *Anal. Biochem.*, Vol. 45, No. 2, (February 1972), pp. 510-516, ISSN 0003-2697

Robyt, J.F. (1998). *Essentials of Carbohydrate Chemistry*, Springer, ISBN 0-387-94951-8, New York, USA.

Russell, J.B. (1984). Factors influencing competition and composition of the rumen bacterial flora. In: *Herbivore nutrition in the subtropics and tropics*. F.M.C. Gilchrist and R.I. Mackie (Ed.), pp. (313-345), The Science Press, ISBN 0907997031, Craighall, South Africa

Ryan, S.M., Fitzgerald, G.F. & Van Sinderen, D. (2006). Screening for and identification of starch-, amylopectin-, and pullulan-degrading activities in *Bifidobacterial* strains. *Appl. Environ. Microbiol.*, Vol. 72, No. 8, (August 2006), pp. 5289-5296, ISSN 1098-5336

Sahlstrom, S. & Brathen, E. (1997). Effects of enzyme preparations for baking, mixing time and resting time on bread quality and bread staling. *Food Chem.*, Vol. 58, No. 1, (January 1997), pp. 75–80, ISSN 0308-8146

Sambrook, J., Fritsch, E.F. & Maniatis, T. (1989). *Molecular cloning: a laboratory manual*, 2nd ed. N. Harbor (Ed.), Cold Spring Harbor Laboratory Press, ISBN 0879693096, New York

Santamaria, R.I., Del Rio, G., Saab, G., Rodriguez, M.E., Soberón, X. and Loöpez-Munguia, A. (1999). Alcoholysis reactions from starch with α-amylases. *FEBS Letters*, Vol. 452, No. 3, (June 1999), pp. 346-350, ISSN 0014-5793

Satoh, E., Uchimura, T., Kudo, T. & Komagata, K. (1997). Purification, characterization, and nucleotide sequence of an intracellular maltotrioseproducing α-amylase from

Streptococcus bovis 148. *Appl. Environ. Microbiol.*, Vol. 63, No. 12, (December 1997), pp. 4941-4944, ISSN 1098-5336

Saxena, R.K., Dutt, K., Agarwal, L. & Nayyar, P. (2007). A highly thermostable and alkaline amylase from a *Bacillus* sp. PNS. *Biores. Technol.*, Vol. 98, No. 2, (January 2007), pp. 260-265, ISSN 0960-8524

Schäfer, T., Duffner, F. and Borchert, T.V. (2000). Extremophilic enzymes in industry: screening, protein engineering and application, *Proceedings of Third International Congress on Extremophiles*, Hamburg, Germany, 3-7 September 2000.

Shental-Bechor, D. & Levy, Y. (2009). Folding of glycoproteins: toward understanding the biophysics of the glycosylation code. *Current Opinion in Structural Biology*, Vol. 19, No. 5, (October 2009), pp. 524–533, ISSN 0959-440X

Shigechi, H., Koh, J., Fujita, Y., Matsumoto, T., Bito, Y., Ueda, M., Satoh, E., Fukuda, H. & Kondo, A. (2004). Direct Production of Ethanol from Raw Corn Starch via Fermentation by Use of a Novel Surface-Engineered Yeast Strain Codisplaying Glucoamylase and α-Amylase. *Appl. Environ. Microbiol.*, Vol. 70, No. 8, (August 2004), pp. 5037-5040, ISSN 098-5336

Shimizu, M., Kanno, M., Tamura, M. & Suckane, M. (1978). purification and some properties of a novel α-Amylase produced by a strain of *Thermoactinomyces vulgaris. Agric. Biol. Chem.*, Vol. 42, No. 9, (September 1978), pp. 1681-1688, ISSN 1881-1280

Shinke, R., Nishira, H. & Mugibayashi, N. (1974). Isolation of, B-amylase producing microorganisms. *Agric. Biol. Chem.*, Vol. 38, No. 3, (March 1974), pp. 665- 666, ISSN 1881-1280

Siggens, K. (1987). Molecular cloning and characterization of the B-amylase gene from *Bacillus circulans. Mol. Microbiol.* Vol. 1, No.3, (November 1987), pp. 86-91, ISSN 1365-2958

Sills, A., Sauder, M. & Stewart, G. (1984). Isolation and characterization of the amylolytic system of *Schwanniomyces castelli . J Inst Brew.*, Vol. 90,pp. 311-314, ISSN 0046-9750

Smith, A.M. (1999). Making starch. *Curr. Opin. Plant Biol.*, Vol. 2, No. 6, (December 1999), pp. 223–229, ISSN 1369-5266

Somers, W., Lojenja, A.K., Bonte, A., Rozie, H., Visser, J., Rombouts, F. & Riet, V. (1995). Isolation of α-amylase on crosslinked starch. *Enzyme and Microb. Technol.*, Vol. 17, No. 1, (January 1995), pp. 56-62, ISSN 0141-0229

Speckmann, H., Kottwitz, B., Nitsch, C. & Maurer, K.H. (2001). Detergents containing amylase and protease, In: *Patent Application* no. US6380147B1, April 2012, Available from:
<http://www.google.com.eg/patents?hl=ar&lr=&vid=USPAT6380147&id=jd8KAAAAEB AJ&oi=fnd&dq=Detergents+containing+amylase+and+protease&printsec=abstract#v=on epage&q=Detergents%20containing%20amylase%20and%20protease&f=false>

Spendler, T. & Jørgensen, O. (1997). Use of a branching enzyme in baking, In: *Patent Application* no. WO97/41736, April 2012, Available from <http://www.wipo.int/patentscope/search/en/detail.jsf?docId=WO1997041736&recNum =1&maxRec=&office=&prevFilter=&sortOption=&queryString=&tab=PCT+Biblio>

Srivastava, R. (1984). Studies on extracellular and intracellular purified amylases from a thermophilic *Bacillus stearothermophilus*. *Enzyme Microb Technol.*, Vol. 6, No. 9, (September 1984), pp. 426-422, ISSN 0141-0229

Strokopytov, B., Knegtel, R.M.A., Penninga, D., Roozeboom, H.J., Kalk, K.H., Dijkhuizen, L. and Dijkstra, B.W. (1996). Structure of cyclodextrin glycosyltransferase complexed with a maltononaose inhibitor at 2.6 A° resolution. Implications for product specificity. *Biochemistry*. Vol. 35, No. 13, (April 1996), pp. 4241-4249, ISSN 1520-4995

Sun H., Zhao, P., Ge, X., Xia, Y., Hao, Z., Liu, J. & Peng, M. (2010). Recent Advances in Microbial Raw Starch Degrading Enzymes. *Appl. Biochem. Biotechnol.*, Vol. 160, No. 4, (February 2010), pp. 988–1003, ISSN 1559-0291

Sunna, A., Moracci, M., Rossi, M. & Antranikian, G. (1997). Glycosyl hydrolases from hyperthermophiles. *Extremophiles*, Vol. 1, No. 1, (February 1997), pp. 2–13, ISSN 1433-4909

Svensson B., Jensen M., Mori H., Bak-Jensen K., Bonsager B., Nielsen P., Kramhøft B., Praetorius-Ibba M., Nohr J., Juge N., Greffe L., Williamson G. & Driguez H. (2002). Fascinating facets of function and structure of amylolytic enzymes of glycoside hydrolase family 13. *Biologia Bratislava*, Vol. 57, Suppl. 11,pp. 5–19, ISSN 0006-3088

Svensson, B. & Sogaard, M. (1993). Mutational analysis of glycosylase function. *J. Biotechnol.* Vol. 29, No. 1, (May 1994), pp. 1-37, ISSN 0168-1656

Svensson, B. (1994). Protein engineering in the α-amylase family: catalytic mechanism, substrate specificity, and stability. *Plant Mol. Biol.*, Vol. 25, No. 2, (May 1994), pp. 141-157, ISSN 1573-5028

Takasaki, Y. (1976). Purification and enzymatic properties of β- amylase and pullulanase from *Bacillus cereus* var. mycoides. *Agric. Biol. Chem.*, Vol. 40, No. 8, (August 1974), pp. 1523-1530, ISSN 1881-1280

Thompson, D.B. (2000). On the non-random nature of amylopectin branching. *Carbohydrate Polymers*, Vol. 43, No. 3, (November 2000), pp. 223–239, ISSN 0144-8617

Tonkova, A., (2006). Microbial starch converting enzymes of the α-Amylase family. In: *Microbial Biotechnology in Horticulture*, R.C. Ray & O.P. wards (Ed.), pp. (421-472), , Science Publishers, Enfield, ISBN 9781578084173, New Hampshire, USA

Tuohy, M. & Coughlan, M. (1992). Production of thermostable xylan degrading enzymes by *Talaromyces emersonii*. *Bioresource Technol.*, Vol. 39, No. 2, (February 1992), pp. 131-137, ISSN 0960-8524

Uguru, G., Akinyanju, J. & Sani. A. (1997). The use of sorghum for thermostable amylase production from *Thermoactinomyces thalpophilus*. *Letters in Applied Microbiology.*, Vol. 25, No. 1, (July 1997), pp. 13–16, ISSN 1472-765X

Uitdehaag, J., Mosi, R., Kalk, K., van der Veen, B., Dijkhuizen, L., Withers, S. & Dijkstra, B. (1999). X-ray structures along the reaction pathway of cyclodextrin glycosyltransferase elucidate catalysis in the α-amylase family. *Nat. Struct. Biol.*, Vol. 6, No. 5, (May 1999), pp. 432-436, ISSN 1545-9985

Van Dam, H. & Hille, J. (1992). Yeast and enzymes in bread making. *Cereal Foods World*, Vol. 37, No. 5, (September 1992), pp. 245–252, ISSN 0146-6283

Vigal, T., Gil, J., Daza, A., Garcia-Gonzales, M. & Martin, J.F. (1991). Cloning, characterization and of an α-amylase gene from *Streptomyces griseus* IMRU3570. *Mol. Gen. Genet.*, Vol. 225, No. 2, (February 1991), pp. 278-288, ISSN 1432-1874

Vihinen, M. & Mantsala, P. (1989). Microbial amylolytic enzymes. *Crit Rev Biochem Mol Biol.*, Vol. 24, No. 4, (April 1989), pp. 329-418, ISSN 1549-7798

Virolle, M., Long, C., Chang, S. & Bibb, M. (1988). Cloning, characterization and regulation of an α-amylase gene from *Streptomyces venezuelae*. *Gene*, Vol. 74, No. 2, (December 1988), pp. 321-334, ISSN 0378-1119

Wang, J.P., Mason, J.C. & Broda, P. (1993). Xylanases from *Streptomyces cyaneus*: their production, purification and characterization. *J. Gen. Micrbiol.*, Vol. 139, No. 9, (September 1993), pp. 1987-1993, ISSN 0022-1287

Wang, J.P., Zeng, A.W., Liu, Z. & Yuan, X.G. (2006). Kinetics of glucoamylase hydrolysis of corn starch. *J. Chem. Techno. and Biotechno.*, Vol. 81, No. 5, (May 2006), pp. 727–729, ISSN 1097-4660

Wang, Y.W. & Wang, Y. (1990) Study on nutrient physiology of some species of *Ganoderma*. *Edible Fungi of China*, Vol. 9, No. 5, (September 1990), pp. 7–10, ISSN 1003-8310

Wang, Y.W., Nema, S. & Teagarden, D. (2010). Protein aggregation Pathways and influencing factors. *Int. J. Pharm.*, Vol. 390, No. 2, (May 2010), pp. 89-99, ISSN 0378-5173

Wardi, A. & Michos, G. (1972). Alcian blue staining of glycoproteins in acrylamide disc electrophoresis. *Anal. Biochem.*, Vol. 49, No. 2, (October 1972), pp. 607-609, ISSN 0003-2697

Whistler, R.L. & Daniel, J.R. (1985). Carbohydrates, In: *Food chemistry*, O.R. Fennema, (Ed.), pp. (69-137), Marcel Dekker, ISBN 0-87055-504-9, New York, USA

Whitehead, T.R. & Cotta, M.A. (1995). Identification of intracellular amylase activity in *Streptococcus bovis* and *Streptococcus salivarius*. *Curr. Microbiol.*, Vol. 30, No. 3, (March 1995), pp. 143–148, ISSN 1432-0991

Wolfenden, R., Lu, X. & Young, G. (1998). Spontaneous hydrolysis of glycosides. *J. Am. Chem. Soc.*, Vol. 120, No. 27, (July 1998), pp. 6814–6815, ISSN 1520-5126

Yamamoto, T. (1988). Bacterial a-amylase (liquefying- and saccharifying types) of *Bacillus subtilis* and related bacteria. In: *Handbook of amylases and related enzymes*. The amylase research society of Japan (Ed.), pp. (40-45), Pergamon Press, ISBN 0080361412, Oxford

Yamane, K., Yamaguchi, K & Maruo, B. (1973). Purification and properties of a cross-reacting material related to α-amylase and biochemical comparison with parental α-amylase. *Biochim Biophys Acta.*, Vol. 298, No. 2, (March 1973), pp. 295-323, ISSN 0005-2736

Zare-Maivan, H. & Shearer, C.A. (1988). Extracellular enzyme production and cell wall degradation by freshwater lignicolous fungi. *Mycology*, Vol. 80, No. 3, (May 1988), pp. 365-375, ISSN 00275514

Zeikus, J. (1979). Thermophilic bacteria: ecology, physiology, and technology. *Enzyme Microb. Technol.*, Vol. 1, No. 4, (October 1979), pp. 243-251, ISSN 0141-0229

Zhu, F., Corke, H. & Bertoft, E. (2011).Amylopectin internal molecular structure in relation to physical properties of sweetpotato starch. *Carbohydrate Polymers*, Vol. 84, No. 3, (March 2011), pp. 907-918, ISSN 0144-8617

Zobel, H.F. (1992). Starch granule structure, In: *Developments in Carbohydrate chemistry*, R. J. Alexander & H.F. Zobel (Ed.), pp. (1-36), American association of cereal chemistry, ISBN 0913250767, Minnesota, USA.

Chapter 11

Biological Applications
of Plants and Algae Lectins: An Overview

Edson Holanda Teixeira, Francisco Vassiliepe Sousa Arruda,
Kyria Santiago do Nascimento, Victor Alves Carneiro, Celso Shiniti Nagano,
Bruno Rocha da Silva, Alexandre Holanda Sampaio and Benildo Sousa Cavada

Additional information is available at the end of the chapter

1. Introduction

More than 120 years ago, Peter Hermann Stillmark in his doctoral thesis presented in 1888 to the University of Dorpat, gave the earliest step in the study of proteins that have a very interesting feature: the ability to agglutinate erythrocytes. These proteins were initially referred as to hemagglutinins or phytoagglutinins, since they were originally isolated from extracts of plants [1]. The first hemagglutinin isolated by Stillmark was extracted from seeds of the castor tree (*Riccinus communis*) and was named ricin [2]. This hemagglutinin was strongly used by Paul Ehrlich as model antigens for immunological studies [2,3].

Thirty-one years after Stillmark, James B. Sumner, isolated from jack bean (*Canavalia ensiformis*) a protein that he called concanavalin A (ConA). For the first time a pure hemagglutinin had been obtained [4]. However, the report that ConA agglutinates cells such as erythrocytes and yeasts and also precipitates glycogen from solution was given by Summer and Howell nearly two decades after its isolation. In addition, the findings of these researchers showed that the hemagglutination by ConA was inhibited by sucrose, demonstrating for the first time the sugar specificity of lectins. Thus, they suggested that the hemagglutination induced by ConA might be due to the reaction of the plant protein with carbohydrates expressed on the surface of the erythrocytes [1,4].

In 1907, Landsteiner and Raubitschek analyzed the hemagglutination of red blood cells from different animals by various seeds extracts. They found that the relative hemagglutinating activity was quite different for each extract tested [1]. However, it was only in the 1940s that Willian Boyd and Karl Renkonen, working independently, made the important discovery of human blood groups specificity for hemagglutinins. They found that crude extracts of two leguminous plants, *Phaseolus limensis* and *Vicia cracca*, agglutinated blood type A erythrocytes but not blood type B or O cells, whereas the extract of *Lotus tetragonolobus* agglutinated only blood type O erythrocytes [1,5].

The specific interaction between lectins and carbohydrates of erythrocytes played a crucial role in the investigations of the antigens associated with the ABO blood group system. In the subsequent decade, Morgan and Watkins found that the agglutination of type A erythrocytes by extracts of *Phaseolus limensis* was best inhibited by α-linked N-acetyl-D-galactosamine, while the agglutination of O cells by the extract of *L. tetragonolobus* was best inhibited by α-linked L-fucose [6].

Around thirty years after Boyd, the research on lectins reached the molecular level studies. It was clear the need to a better understanding on the structural aspects of lectins. Then, in 1972 Edelman and colleagues established the primary sequence of ConA [6]. In the same year, Edelman's group and independently Karl Hardman with Clinton Ainsworth, solved the 3D structure of ConA by X-ray crystallography [7,8].

2. What exactly is a lectin?

In 1954 Boyd and Shapleigh proposed the term lectin, from the Latin verb *legere* (which means "to select"). This term was based on the fact that these proteins have the ability to distinguish between erythrocytes of different blood types [9].

Lectins were early defined as carbohydrate-binding proteins of nonimmune origin that agglutinate cells or as carbohydrate-binding proteins other than antibodies or enzymes. However, these definitions were updated, since some plant enzymes are fusion proteins composed of a carbohydrate-binding and a catalytic domain, for instance, type 2 RIPs, such as ricin and abrin, are fusion products of a catalytically active A-chain (which has the N-glycosidase activity) and a carbohydrate-binding B-chain, both linked by a disulfide bond [10]. Furthermore, there is in nature carbohydrate-binding proteins possessing only one binding site and, therefore, are not capable of precipitating glycoconjugates or agglutinating cells [11].

Thus, in 1995 Peumans and Van Damme proposed the most suitable definition for lectins. According to the "new" definition, all plant proteins that possess at least one noncatalytic domain that binds reversibly to a specific mono- or oligosaccharide are considered as lectins [12,13].

2.1. Plant lectins

Lectins are proteins widely distributed in nature such in microorganisms, plants, animals and humans, acting as mediators of a wide range of biological events that involve the crucial step of protein-carbohydrate recognition, such as cell communication, host defense, fertilization, cell development, parasitic infection, tumor metastasis, inflammation, etc [14-15].

Peumans and Van Damme classified the plant lectins according to their overall structure. *Merolectins* consist exclusively of a single carbohydrate-binding domain (e.g. hevein, a chitin-binding latex protein isolated from the rubber tree *Hevea brasiliensis*). Since merolectins have a unique carbohydrate-binding site, they are incapable of precipitating glycoconjugates or agglutinating cells. *Hololectins* are also built of carbohydrate-binding domains. However, they contain at least two such domains that are identical or very similar. Because these lectins are di- or multivalent they can agglutinate cells and/or precipitate glycoconjugates. Most plant lectins are hololectins. *Superlectins* are built of at least two carbohydrate-binding domains. Unlike hololectins, these domains are not identical or similar. Thus, superlectins recognize structurally different sugar (e.g. TxLCI, a tulip bulb lectin that recognizes mannose and N-acetyl-galactosamine). *Chimerolectins* are fusion proteins that consist of two different chains, one of them with a remarkable catalytic activity (or another biological activity). RIPs type 2 are examples of chimerolectins [11-12].

The most thoroughly investigated lectins have been isolated from plants, particularly that extracted from members of the Leguminosae family. Legume lectins are a large group of proteins that share a high degree of structural similarity with distinct carbohydrate specificities. The subtribe Diocleinae (Leguminosae) comprises 13 genera, mostly of them from the New World. However, only 3 of these genera (i.e. *Canavalia, Cratylia* and *Dioclea*) are considered as the main sources for protein purification [16].

Concerning the biological activity, legume lectins are considered as enigmatic proteins. Despite the philogenetic proximity as well the high degree of similarity shared between them, they possess different biological activities such as histamine release from rat peritoneal mast cells, lymphocyte proliferation and interferon-γ production, peritoneal macrophage stimulation and inflammatory reaction as well as induction of paw edema and peritoneal cell immigration in rats [16].

2.2. Plant lectins as biotechnological tools

Significant progress has been reached in last years in understanding the crucial roles of lectins in several biological processes [17]. The importance of lectins as biotechnological tools has been established early in the studies involving its biological application. In 1960 a major step in immunology was given in order to determine the role of these proteins on the lymphocytes cell division. It was found that the lectin of the red kidney bean (*Phaseolus vulgaris*), known as phytohemagglutinin (PHA), possesses the ability to stimulate lympho-cytes to undergo mitosis [18]. After these findings, many studies have been performed to evaluate the role of lectins on different models involving the immune response and its products, for instance the stimulation of citokine secretion [19], functional activation of monocytic and macrophage-like cells [20] and ROS production by spleen cells [21].

In addition to immunological studies, recent works have been investigated the influence of lectins in the field of microbiology, since lectins can be considered as valuable tools to verify the role of interaction between the pathogen and carbohydrates present in host cells and its importance to disease development. For instance, it has been proposed that the pathogen *Helicobcter pylori* infects human cells through an interaction involving a lectin [22].

3. Carbohydrates and the neoplasic process

Currently, malignancies are considered a major problem in the public health, especially given their increasing incidence and prevalence rates observed in recent decades. In this context, the malignant tumors, or cancers, account for approximately 7.8 million deaths per year, thus becoming the second greatest cause of death worldwide, only behind the cardiovascular disease [23].

The cancer can be defined as a set of more than 100 diseases that have in common the uncontrolled growth of cells, which invade tissues and organs and can spread to other body regions [23]. Thus, both the processes of cellular mutation that affects the neoplastic cell and metastasis, involve a series of genetic changes that culminate in modifications in the pattern of several receptors and signaling molecules present on the cell surface [24].

Carbohydrates are biomolecules that have enormous potential for encoding biological information. These combined-molecules (Glycoproteins and Glycolipids) are responsible for different biological interactions between the cell and the extracellular environment [26]. Regarding the neoplastic cells, the glycosylation of these proteins and lipids is changed, which generates membrane signaling molecules capable of inducing several processes directly related to tumor progression such as cell adhesion, angiogenesis, cellular mitosis and metastasis, in addition, in some cases, be responsible for inhibition of apoptosis induction triggered by the cells of the immune system [26].

Certain changes in glycans occur frequently in neoplastic cells and may be considered "tumor-specific", establishing a correlation between the stage of disease progression and prognosis of the same [25]. Some classic examples of these changes are the antigens of the ABH and Lewis system. ABH antigens are not expressed in cells of healthy human colon but significantly expressed on tumor cells [27]. Since the antigen Lewis^y can be observed in several carcinomas and has been correlated with poor prognosis in breast tumors [28]. Another example is the glycosylated antigens sialyl-Lewis^A and sialyl-Lewis^x, which are significantly up-regulated in carcinomas of the colon and appear to be related to tumor progression [29].

Thus, due to the intrinsic role of carbohydrates in the tumorigenesis, the glycosylation process as well as the identification of glycosylated antigens have been intensively focused, given the fact that glycosylation of antigens can vary extensively depending on the stage of the disease, which can provide, when properly identified, a better possibility of correct diagnosis and treatment.

4. Application of plant lectins in the diagnosis of malignant tumors

Because the peculiar characteristic of specific binding to carbohydrates, lectins have been used as tools to identify aberrant glycans expressed by neoplastic cells. Such methods have been essential to obtain a more precise diagnosis that allows a more accurate prognosis.

Several methods regarding the use of lectins as tools for recognizing aberrant glycans have been proposed in recent days [30,31,32]. The technique most common and widespread is the use of lectins in immunohistochemical assays.

In this context, the study conducted by [30], showed that leguminous lectins from *Canavalia ensiformis* (ConA) and *Ulex europaeus* (UEA-1) were used as histochemical markers of parotid gland mucoepidermoid carcinoma with low, intermediate, and high grade dysplasia. The authors stated that ConA localization in the cytoplasm and/or plasma membrane was significantly associated with neoplastic cells from the three grades of severity, whereas UEA-1 was associated with low and intermediate grade dysplasia. The authors obtained similar results after the analysis of other cell regions.

Another recent methodology was addressed by [33]. This methodology exploits the fact that glycoproteins produced by cancer cells have altered glycan structures, although the proteins themselves are common, ubiquitous, abundant, and familiar. However, as cancer tissue at the early stage probably constitutes less than 1% of the normal tissue in the relevant organ, only 1% of the relevant glycoproteins in the serum should have altered glycan structures [34]. With that in mind, the strategy to approach the detection of these low-level glycoproteins is based in: (a) a quantitative real-time PCR array for glycogenes to predict the glycan structures of secreted glycoproteins; (b) analysis by lectin microarray to select lectins that distinguish cancer-related glycan structures on secreted glycoproteins; and (c) an isotope-coded glycosylation site-specific tagging high-throughput method to identify carrier proteins with the specific lectin epitope [33].

Therefore, further analyses of lectins as biomarkers have been undertaken to improve our understanding of the processes involved in malignant tumor formation. As well as enable us to acquire new methods of identification of neoplastic cells at an early stage, enabling a better prognosis with appropriate treatment and low cost.

5. Application of plant lectins in the treatment of malignant tumors

Apoptosis is a mechanism by which cells undergo death to control cell proliferation or in response to irreparable DNA damage. It is featured by unique morphological and biochemical changes, such as nucleus condensation and margination, membrane blebbing, and internucleosomal DNA cleavage [35]. As the type I programmed cell death (PCD), apoptosis occurs through two major pathways, the extrinsic pathway triggered by the Fas death receptors, and the mitochondria-dependent pathway that brings about the release of Cytocrome *c* (Cyto *c*) and activation of the death signals under stimulus. In both ways the caspases, which belong to a family of cysteine proteases, have been well established as major players in apoptosis-causing mechanisms [36].

Several studies have demonstrated that lectins can induce apoptotic cell death. In the mitochondrial-dependent pathway, ConA treatment results in a decrease of mitochondrial membrane potential, and thus collapsing mitochondrial transmembrane potential. Cyto *c* is subsequently released, making up apoptosome with Apaf-1 and procaspase-9. After conjugating apoptosome, procaspase-3 turns into active caspase-3 that eventually triggers apoptosis [37].

In [38], it was evaluated the pro-apoptotic activity of a lectin isolated from *Artocarpus incisa* (frutalin) on HeLa cells derived from human cervical cancer. In this study, frutalin possessed a remarkable antiproliferative effect on HeLa cells. This effect was irreversible as well as time and dose dependent. When the lectins were added, serious visible cellular morphology changes were observed, possibly as a result of cell stress.

An interesting study conducted by [39] showed the pro-apoptotic caspase-dependent activity of the lectin isolated from *Astragalus membranaceus* in leukemia cell line (CML K562). These results showed a close relationship with the low expression of BCL-2 (anti-apoptotic protein) indicating that the lectin is active through the mitochondrial apoptotic pathway [40]. Furthermore, structural changes in cell membrane and different levels of Caspase-3 contributed to support their hypothesis.

Despite the apoptotic activity of several lectins have been demonstrated in different studies, the precise mechanism of how this process is triggered as well the mode of internalization is unknown until the present date.

6. Carbohydrates and the inflammatory process

The inflammation is a nonspecific event of immune response that occurs in reaction to any type of tissue injury. This process is capable of triggering a series of physiological changes such as increased blood flow, elevated cellular metabolism, vasodilation, release of soluble mediators, extravasation of fluids and cellular influx [41].

The continuity of cell recruitment and tissue damage in addition to chemical mediators released by the injured tissue as well as resident cells on site activate various mechanisms, in turn, induce the migration of more immune cells as well as increasing local tissue perfusion [42].

Both, acute and chronic inflammations have specific characteristics and the innate immune system plays a central role, since it mediates the initial response. Infiltration of innate immune cells, specifically neutrophils and macrophages, characterizes the acute inflammation, while infiltration of T lymphocytes and plasma cells are features of chronic inflammation [41,42].

As discussed previously, carbohydrates can act as the intermediates of communication in biological processes such as differentiation, proliferation and certain cell–cell interactions that are crucially important in both physiological and pathological phenomena [43,44]. The information contained in the enormous variety of oligosaccharide structures normally conjugated to proteins or lipids on cell surfaces (glycocodes) is recognized and deciphered by a specialized group of structurally diverse proteins, the lectins [44].

Galectins (formerly "S-type lectins"), an evolutionarily conserved family of endogenous animal lectins, share unique features, including their highly conserved structure, exquisite carbohydrate specificity, and ability to differentially regulate a myriad of biological responses [45].

Although galectins have been implicated in many biological activities, most of the functional studies reported to date link galectins to early developmental processes, such as neovascularization, regulation of immune cell homeostasis and inflammation [44,46,47]. Through deciphering glycan-containing information about host immune cells or microbial structures, galectins can modulate a diversity of signaling events that lead to cellular proliferation, survival, chemotaxis, trafficking, cytokine secretion and cell–cell communication [46,47].

These findings are extremely important because they demonstrate the importance of the glicocodes in the process of cell recognition and inflammation. In this context, plant lectins have been widely used to understand the pro-inflammatory mechanisms, as well as the design of new compounds with pro-healing effect.

7. Plant lectins and its immunomodulatory activity

An immune system is a system of biological structures and processes within an organism that protects against disease. In order to function properly, an immune system must detect a wide variety of agents, from viruses to cancer cells, and distinguish them from the organism's own healthy tissue [41]. The immune system is composed of many cells and molecules that act in a complex and harmonious way with the ultimate goal of annihilating the aggressive factor [42].

In the immune system, two phases of activity can be clearly established: the innate immune response and the adaptive immune response. In the innate immune response, there is the activity of cells and cytokines in a nonspecific way with the main purpose to annihilate quickly the local damaging agents. At this stage, we highlight the neutrophils, eosinophils, basophils and macrophages, cells with well-established activities but with the common function of production and release of cytokines. These cytokines are molecules with various functions in the inflammatory process, such as chemotaxis, activation of certain cell groups and increased tissue perfusion [48].

On the other hand, the adaptive immune response is composed by another set of cells that acts in a more specific way, the lymphocytes. Such cells are responsible for producing antibodies specific for certain invading microorganisms and the activation of mechanisms of apoptosis in abnormal cells [49].

Thus, the use of molecules capable of inducing cell recruitment as well as cytokine production and lymphocytes proliferation is of special scientific interest.

Korean mistletoe (*Viscum album* L. var. *coloratum*) is traditionally used as a sedative, analgesic, anti-spasmolytic, cardiotonic and anticancer agent, in Korea. An important lectin has been isolated from this plant and its immunomodulatory activity was analyzed [50]. It was shown that KML differentially modulated macrophage-mediated immune responses. It also enhanced the expression of various cytokines (IL-3, IL-23 and TNF-a), ROS generation, phagocytic uptake and surface levels of some glycoproteins (co-stimulatory molecules, PRRs and adhesion molecules). Nevertheless, the functional activation of adhesion molecules assessed by cell–cell or fibronectin adhesion events was up-regulated by KML treatment.

A recent study [51] demonstrated the immunomodulatory activity of ConBr, a lectin isolated from *Canavalia brasiliensis* seeds. The assays showed that ConBr was able to induce *in vitro* proliferation of splenocytes with minimal damage to the cellular structure. Furthermore, ConBr increased in the production of cytokines such as IL-2, IL-6 and IFN-γ production and decreased IL-10. These findings indicate the potential immunomodulatory effect of this lectin in conjunction with the intrinsic role of carbohydrates in intercellular communication related to the inflammatory process.

Regarding the activity of lectins on lymphocytes, a recent study [52] evaluated the effect of lectin extracted from seeds of *Cratylia mollis* Mart. (Cramoll 1,4) on experimental cultures of mice lymphocytes. In this study, aspects directly related to inflammation as cytokine production, cytotoxicity and cellular production of nitric oxide (NO) were evaluated. Cramoll 1,4 did not show cytotoxicity at the concentrations tested, in addition, was able to induce IFN-γ and showed an anti-inflammatory activity through the supression of NO production.

The biological function of carbohydrates in inflammation events is well-defined. In this context, proteins that bind specifically to such glycans are of great interest because of their possible functions and applications in biotechnological studies.

8. Plant lectins and its pro-healing activity

Recently, researches have undertaken efforts at the possible pro-healing activity of some lectins [53,54]. This goal is supported by the fact that such molecules may interfere with the inflammatory process. This effect is not yet fully elucidated, however peculiar and interesting results can be observed.

Experiment conducted in a murine model, employing the lectin isolated from *Bauhinia variegata* seeds (BVL) topically on surgically induced wounds, revealed the pro-healing potential of such molecule. Although not yet elucidated, it is suggested that this lectin appears to stimulate the mitogenic activity of resident cells, turning them into potent chemotactic agents for the recruitment of neutrophils through the release of cytokines [54]. Furthermore, it is suggested that the BVL stimulates the differentiation of fibroblasts into myofibroblast, which is an extremely important event in the remodeling of connective tissue [55].

Although promising, this issue requires further studies to better characterize the mechanisms involved in pro-healing role played by lectins.

9. Marine algae lectins

To date, there are fewer than 100 publications describing the presence of lectins in marine red, green and brown macroalgae. Moreover, and in marked contrast to higher land plant lectins, marine algal lectins have been isolated and characterized at a much lower pace since the first report of haemagglutinating activity in these organisms appeared more than 46 years ago [56]. Thereafter, other studies describing the presence and/or purification of algal lectins were reported by groups from England [57], Japan [58], Spain [59], United States [60] and Brazil [61].

Currently, the presence of lectins was analyzed at about 800 algae species. However, this number is still small, considering that there are thousands of species of marine algae. Together, the research shows that approximately 60% of the analyzed species show hemagglutinating activity. The number of positive species could be higher since in the first screenings the authors used a limited number of red blood cells and without enzyme treated erythrocytes.

The improvement in the methodologies of both, extraction and hemagglutination assays could increase the number of positive species. In fact, there appears to be coincidence that the rabbit erythrocytes treated with papain are most suited for the hemagglutinating activity detection in marine macroalgae [62,63].

Although marine algal lectins show proteinaceous content similar to lectins from terrestrial plants, they differ in some aspects. Early publications on this issue, reported that in general, lectins from algae have low molecular masses, no affinity for monosaccharides, strong specificity for complex oligosaccharides and/or glycoproteins. Moreover, they appear to have no requirement for metal ions, showing high content of acidic residues and even in high concentrations tend to stay in the monomeric form [64,65,66]. However there are a few reports showing that some of these molecules may be inhibited by simple sugars and are cation dependent as showed for the lectins from the green marine alga genus, *Codium* [67] and red marine alga genus, *Ptilota* [68,69,70].

The classical methods used to purify marine algae lectins include methods such as protein precipitation (using salt or ethanol), liquid chromatography (especially affinity) and electrophoresis [69,71]. Ion exchange chromatography has been effectively used in the isolation of lectins from seaweed, mainly in initial stages in purification. In this step, the lectins were separated from pigments present in the extracts [66,72,73]. In the protein extracts, phycobilins are co-extracted with lectins, becoming an undesirable contaminant in the purification process [72].

Lectins from marine algae *Cystoclonium purpureum* [74], *Gracilaria verrucosa* [75], *Palmaria palmata* [76], *Solieria robusta* [62], *Gracilaria tikvahiae* [60], *Bryothamnion seaforthii* and *B.triquetrum* [66], *Solieria filiformis* [77], *Enantiocladia duperreyi* [78], *Amansia multifida* [73], *Hypnea musciformis* [79], *Gracilaria ornata* [80], *Hypnea cervicornis* [81] and *Georgiela confluens* [82] were isolated by exchange chromatography, usually on DEAE cellulose. In contrast, due usually algal lectins have binding specificity of complex sugars, affinity chromatography has been used a few times, such as the lectins of green algae of the genus *Codium* [83,84,85], lectins from *Ulva lactuca* [86], *Caulerpa cupressoides* [87], *Enteromorpha prolifera* [88], *Ulva pertusa* [89], *Bryopsis plumosa* [90,91,92], *Bryopsis hypnoides* [93] and lectins from red marine algae of the genus *Ptilota* [68,69,70].

To date, only 31 lectins from Rhodophyceae and 17 lectins from Chlorophyceae were isolated and characterized. The virtual absence of lectins isolated from brown algae (Phaeophyta) is mainly due to the amount of polyphenols present in plants. It is known that polyphenols are released in extraction and that these compounds and their oxidation products, quinones, bind tightly to proteins [94] causing a false hemagglutination [83,95].

Even with the increase in the publications related to marine algae lectins, biochemical and structural information on algal lectins is scarce and from only a few species, and hence the functional and phylogenetic classification of these lectins remains unclear. The available structural information indicates the existence of different carbohydrate-binding proteins in the marine algae investigated. Moreover, the complete amino acid sequences of only 14 algal lectins have been determined. In red marine algae, *Bryothamnion triquetrum* lectin (BTL) was the first lectin to be determined its primary structure [96].

In same year, [97] reported the primary structure of *Hypnea japonica* agglutinin (HJA) that shares sequence similarity with BTL and with lectin from *Bryothamnion seaforthii* [98] and these lectins constitute the first marine red alga lectin family. Based on identity between HML and HCA and in the differences in amino acid sequences compared with BTL/HJA, [99] suggest that HCA/HML constitute another algal lectin family.

On the other hand, the lectins isolated from *Eucheuma serra, E. amakusaensis, E. cottonii* [100] have masses around 28 kDa, presents a monomer, share N-terminal sequence similarity with the complete amino acid sequence of isolectin 2 from *Eucheuma serra* (ESA-2) [101]. Also, the primary structure of ESA-2 contains repeated domains in their primary structure. These data suggest that lectins from genus *Eucheuma* can be grouped in a thirty family of red marine alga haemagglutinins. The amino acid sequence of lectin from red marine alga *Griffthisia sp.* [102] displays sequence similarity with lectin from jack fruit (*Artocarpus integrifolia*). The common methodology employed to determine the primary structures from red marine alga lectins was a combination of Edman degradation of sets of overlapping peptides and mass spectrometry. From green marine algae, the first primary structure determined was the lectin from *Ulva pertusa* (UPL-1) [89]. [103] described a 19 kDa protein expressed in strictly freshwater conditions in species of *Ulva limnetica Ichihara et Shimada* (freshwater alga), and this protein (named ULL) was identified and sequenced by cDNA cloning. The protein encoded by the cDNA showed 30% identity to UPL-1. However, the ULL should be considered a lectin-like since its haemaglutination activity was not yet characterized. UPL-1 and ULL did not show amino acid sequence similarity with known plant and animal lectins.

Recently [90] reported that the aggregation of cell organelles of *Bryopsis plumosa* in seawater was mediated by a lectin–carbohydrate complementary system and the purified lectin (named bryohealin) is involved in protoplast regeneration. The primary structure of bryohealin and of lectin from *Bryopsis hypnoides* [104] had few similarity with any known plant lectin, but rather resembled animal lectins with fucolectin domains. In addition, *Bryopsis plumosa*, has other two lectins described.

The lectin BPL-2 is a 17 kDa protein specific to D-mannose (ref). The authors found no similarity with others proteins in specific databases. The BPL-3 [92] possesses specificity to N-acetyl-D-galactosamine/N-acetyl-D-glucosamine and share the same sugar specificity with bryohealin. However, the primary structures of the two lectins were completely different. The homology sequence analysis of BPL-3 showed that it might belong to H lectin group from Roman snails (*Helix pomatia*). The latest primary structure published was of lectin from *Boodlea coacta* (BCA) [105]. BCA consisted of 3 internal tandem-repeated domains, each one containing the sequence motif similar to the carbohydrate-binding site of land plant lectin from *Galanthus nivalis*. It should be noted that the primary structures from green marine algal lectins were mainly determined with combination of Edman degradation and cDNA cloning.

Another observation is that a large number of sequenced algal lectins have the presence of cysteinyl residues or the duplication of internal domain. Still a lot of work needs to be done on the structure of algal lectins, since a few amount of primary structures has been determined. Further structural studies will contribute to understanding the differences in their biochemical characteristics as well as to the evolutionary aspects upon lectin presence in land plants and marine algae.

10. Marine algae lectins and its biotechnological role

There are few studies in the literature about the biotechnological applicability of lectins from marine algae. Probably due to low rentability of lectins obtained through the purification processes. It is noteworthy that the majority of lectins isolated so far were extracted from red algae, which are rich in carbohydrates and not in proteins. Moreover, during the extraction of algae proteins, phycobiliproteins are extracted simultaneously, and therefore the addition of steps to remove these phycobiliproteins causes losses of other proteins, among these are the lectins.

However, several studies on biological applications of algae lectins demonstrate that these molecules have an additional benefit; they are molecules with low molecular weight and may be less antigenic when used in biological models.

In cancerology, it was demonstrated that a lectin from the red marine algal *Eucheuma serra* (ESA) induced cell death in human cancer cells through the induction caspase-3 activity and DNA fragmentation in human colon adenocarcinoma (Colo201) cells [106]. ESA also induced cell death in a dose-dependent manner via apoptosis pathway in *in vitro* studies with Colon adenocarcinoma (Colon26) Cells derived from BALB/c mice [107].

In current studies, Span 80 vesicles (a potential type of nonionic vesicular drug delivery system) with ESA and PEGylated (EPV) lipids immobilized, showed hemagluttinating activity similar to free ESA and decreased the viability of Colo201 cancer cells *in vitro* and not affected the growth of normal cells. EPV caused the anti-tumor effect *in vivo* by inducing apoptosis in tumor cells [108].

Bryothamnion seaforthii lectin (BSL) and *Bryothamnion triquetrum* lectin (BTL) were able to differentiate human colon carcinoma cell variants. Differentiation was, probably, in function to cell membrane glycoreceptors and could be exploited to investigate structural modification of cell membrane glycoconjugates in cancer cell systems. In addition, it has been shown that the binding of both lectins to the carcinoma cells results in their internalization, which is a very interesting property that could be used in future applications, such as drug delivery [109].

The lectins isolated from the red marine algae *Hypnea cervicornis* (HCA), *Pterocladiela capillacea* (PcL) and *Caulerpa cupressoides* (CcL) have been tested as anti-inflamatory and antinociceptive agents. The data indicated that HCA, PcL and CcL have actions anti-inflammatory and antinociceptive (in formalin and acetic acid models). However, these lectins did not present significant antinociceptive effects in the hot plate test [110,111,112].

Concerning the mitogenic activity, the lectin from the red marine alga *Carpopeltis flabellate* (Carnin) was the first lectin that showed mitogenic activity for T lymphocytes from mouse spleen at a concentration of 10,5 µg/ml. Carnin inhibited the normal embryonic development of the sea urchin *Hemicentrotus pulcherrimus* at the stage of blastula and also inhibited the gasturulation induced in starfish *Asterina pectinifera*, at concentrations of 10 and 5 µg/mL, respectively. In addition, it was showed an interaction between a lectin of macroalgae with a microorganism of the marine ecosystem [113].

The antibacterial effect was too evaluated. The lectins from the red marine algal *Eucheuma serra* (ESA) and *Galaxaura marginata* (GMA) have an antibiotic activity. ESA and GMA strongly inhibited *Vibrio vulnificus*, a fish pathogen, but not were able to inactive *V. neresis* and *V. pelagius* [114]. BSL and BTL also were able to avoid the bacterial adhesion of streptococci strains in enamel pellicles. BTL was more efficient in avoiding the adherence of *Streptoccocus sobrinus* and *S. mitis* and BSL was able to reduce adherence of *S. mutans*. This fact can contribute with preventing caries at early stages [115].

Red marine algae lectins from *Amansia multifida* (AML), *Bryothamnion seaforthii* (BSL), *Bryothamnion triquetrum* (BTL) and *Gracilaria caudata* (GCL) induced neutrophil migration *in vitro* and *in vivo* in the peritoneal cavity or dorsal air pouch of rats or mice. The results showed that BT had the most potent effect in neutrophil migration when tested on rats. In mice, BS required four times higher than the dose of BT to induce neutrophil migration. However, when the algae lectins were assayed in mice, AML was the most potent [116].

Concerning the antiviral effects, a lectin named KAA-2 from the red marine algal *Kappaphycus alvarezii* (specific to high mannose type *N*-glycans) showed antiviral action against H1N1 virus [117]. The red alga lectin from *Griffithsia* sp. called griffithsin (GRFT) is a small mannose specific lectin that binds carbohydrates on HIV envelope glycoproteins and block HIV entry into target cells. GRFT is a candidate for development of anti-HIV microbicides [118,119]. GRFT was also able to inhibit the action of the pathogenic agent responsible for respiratory syndrome (SARS), the SARS coronavirus (SARS-CoV) [120].

11. Lectins as biotechnological tools to study the microbial biofilms

Biofilms are microbial complex communities established in a wide variety of surfaces and are generally associated with an extracellular matrix composed by several types of polymers [121]. This type of microbial association can develop on biotic and abiotic surfaces, including living tissues, medical devices and/or industrial water piping systems and marine environments [122,123,124].

Bacterial infections involving biofilm formation are usually chronic and often present a arduous treatment [125,126,127]. The growth and proliferation of microorganisms inside the biofilm provides reduction or prevents the penetration of various antimicrobial agents [128,129] and thus become extremely difficult or impossible to eradicate them [130,131,132]. For some antibiotics, the concentration required to eliminate the biofilm can be up to a thousand times higher than required to planktonic form of the same specie [127].

The ability of microorganisms to form pathogenic cell aggregates is a worldwide concern. In attempt to remedy this problem, pharmaceutical companies associated to research groups work avidly to the development of new options for the treatment of infections caused by bacteria organized in biofilms.

Biofilm formation is a process in which bacteria has a change in lifestyle, it goes from a state unicellular in suspension to a multicellular sessile, where the growth and cell differentiation results in structured communities. The biofilm begins with the setting free of microorganisms in a given area. The first microorganisms adhere initially by weak interactions, mainly of the van der Waals forces [133]. If the colonies are not immediately removed from the surface, they can anchor by cell adhesion molecules existing in the pili and / or flagella [134]. The first colonies facilitate the arrival of other cells through adhesion sites and begin to build the matrix that will form the biofilm. Only a few species are able to adhere to a surface *per se*. Others may anchor to the matrix or directly on existing colonies. Since the colonization has started, the biofilm grows through cell division and combination of the recruitment of other cells [135].

According to [136], the biofilm development occurs through three events. At first, there is a distribution of fixed cells in surface through cell motility. Then, occurs the proliferation of fixed cells by division, expanding to upward and sides forming agglomerated of cells, similar to the formation of colonies on agar plates [137]. Finally, clusters of cells attached to the biofilm are recruited by the action of the environment itself to the development of other biofilms, reaching a climax community [136]. These general stages provide guidance to the study of biofilms by bacteria furniture, although many details of regulation of this process may vary between species.

One of the biofilm-forming components of great importance in the maintenance of cell clusters is the matrix polymeric called EPS (*Extracellular Polymeric Substances*) [138].

Consisting of proteins, polysaccharides and environmental waste results in a solid structure highly hydrated with small channels between the microcolonies [139]. This matrix holds the biofilm together, one of those factors responsible for providing an increased resistance to antibiotics, disinfectants, and ultraviolet radiation [140]. In most biofilms, microorganisms that make up these agglomerates constitute less than 10% dry weight, while the extracellular matrix may contribute up to 90% [141].

The first step to biofilm formation, the early adhesion, is considered essential for colonization and infection by pathogenic bacteria. Macromolecules surfaces are directly involved in this stage [142]. Proteins known as adhesins are able to recognize specific polysaccharide substrate present on the surface to be colonized, for example, the presence of carbohydrates existing in the film of saliva that covers the teeth in the oral environment [143]. Glycidic epitopes present on surfaces of microorganisms (early colonizers) can also be recognized by these proteins to mediate an event known as coaggregation, which will start the formation of a multi-species community [144].

One etiological factor for the development of teeth biofilms is the adhesion of pathogenic bacteria in the dental enamel [145]. However, the microorganisms are not deposited directly on the tooth surface, but bind to a thin acellular layer composed of salivary proteins and other macromolecules that cover the tooth surfaces called acquired pellicle [146,147].

The acquired pellicle is formed by glycoproteins and carbohydrates that serve as receptors for bacteria containing proteins with glucan-binding domains [148]. Bacteria interact with the film by several specific mechanisms, including the interaction lectin-like involving the bacterial adhesins and receptors existing in the pellicle [148,149]. Next, other bacteria can adhere to the film as well as the bacteria pre-existing in the biofilm [150].

The event coaggregation is a phenomenon widely observed in diverse microbial communities [151,152,153]. Cells can interact in suspension, forming cell aggregates, as well as connect directly on biofilm in the process of formation. In the first case, plancktonic cells recognize specifically species genetically distinct creating the coaggregates. In other situations, coaggregates in the form of secondary colonizers can adhere on biofilm in development, a process known as coadhesion. Both cases have an important role in the integration and establishment of a mature biofilm [154].

Thus, carbohydrate residues have an important role in formation and maintainability of microbial biofilms. They act as mediators of the binding between bacteria and the surface that will serve as substrate for biofilm formation [155], as well as site of interaction between microorganisms to form cell aggregates [156,157]. Furthermore, through EPS matrix, maintains the biofilm attached in the surface, conferring a greater resistance to antimicrobial agents in general [158,159].

Molecules able to bind specifically and selectively to carbohydrates have a key importance in the development of research related to microbial biofilms. Thus, lectins have been shown as powerful tools for analyze the glycidic structures of those aggregates from microbial origin [160,161].

Studies of microbial biofilms through the interaction with lectins have two main objectives: visualization and characterization of polymeric matrix (EPS) formed by different species of microorganisms [162,163] and inhibition of oral biofilm formation by blocking the bacterial binding sites present in the pellicle of saliva in the form of glycoproteins and / or carbohydrates [164].

The application of lectins in the characterization of EPS is already widely exploited by many research groups. Lectin of wheat germ (WGA) was used as a marker of *Staphylococcus epidermidis* microcolonies, mainly in the study of the mechanisms involved in bacterial organization during the formation of the cell aggregates [165]. In a study published by the same group, WGA was used to quantify the production of GlcNAcβ-1,4n, a sugar component of the extracellular matrix involved in biofilm formation [166].

In 1980s was developed a system called ELLA (*Enzyme-linked Lectin Assay*) that uses enzyme-lectin conjugates to detect specific carbohydrate units on the surface of cells. This assay allows better detection and quantitation of the sugars by standard immunofluorescence with fluorescein-conjugated lectins [167]. The lectins concanavalin A (ConA) and WGA peroxidase-labeled were used to detect D-glucose or D-mannose and N-acetyl-D-glucosamine or N-acetyl-neuraminic acid, respectively, on the biofilm of several species [168].

Recent studies demonstrated the characteristics and distribution of glycoconjugates in cyanobacteria biofilm using lectins with different specificities. In this study the authors stated that the distribution of carbohydrates in the matrix is very variable. Based on lectin specificity, glycoconjugates produced by cyanobacteria biofilm contained mainly fucose, N-acetylglucosamine or -galactosamine and sialic acid [169].

Lectins may be a suitable antiadhesion agent for *streptococci*, since it has been reported that a lectin-dependent mechanism is involved in its adhesion [170]. However, the application of lectins as tools to interfere with biofilm formation is still poorly explored [115,171]. As shown by [172], plant lectins appear as a new strategy to reduce the development of dental caries by inhibiting the early adherence and subsequent formation of of *Streptococcus mutans* biofilm.

Author details

Edson Holanda Teixeira, Francisco Vassiliepe Sousa Arruda and Bruno Rocha da Silva
LIBS, Integrated Laboratory of Biomolecules, Faculty of Medicine of Sobral, Federal University of Ceará, Fortaleza, CE, Brazil

Kyria Santiago do Nascimento, Victor Alves Carneiro and Benildo Sousa Cavada
BioMol, Laboratory of Biologically Active Molecules, Department of Biochemistry and Molecular Biology, Federal University of Ceará, Fortaleza, CE, Brazil

Celso Shiniti Nagano and Alexandre Holanda Sampaio
BioMar, Laboratory of Marine Biotechnology, Department of Fishing Engineering, Federal University of Ceará, Fortaleza, CE, Brazil

Acknowledgement

We would like to thank CNPq, CAPES, FUNCAP and UFC for financial support. AHS, BSC, CSN, EHT, KSN are senior investigators of CNPq.

12. References

[1] Sharon N, Lis H (2004) History of lectins: from hemagglutinins to biological recognition molecules. Glycobiology. 14: 53R-62R.

[2] Franz H (1988) The ricin story. Adv Lectin Res. 1: 10-25.

[3] Olsnes S (2004) The history of ricin, abrin and related toxins. Toxicon. 44: 361-70.

[4] Sumner JB, Howell SF (1936) The identification of the hemagglutinin of the jack bean with concanavalin A. J Bacteriol. 32: 227-237.

[5] Boyd WC, Shapleigh E (1954) Specific precipitation activity of plant agglutinins (lectins). Science. 119: 419.

[6] Morgan WT, Watkins WM (2000) Unraveling the biochemical basis of blood group ABO and Lewis antigenic specificity. Glycoconj J. 17: 501-530.

[7] Edelman GM, Cunningham BA, Reeke GN, Becker JW, Waxdal MJ, Wang JL (1972) The covalent and threedimensional structure of concanavalin A. Proc Natl Acad Sci USA. 69: 2580-2584.

[8] Hardman KD, Ainsworth CF (1972) Structure of concanavalin A at 2.4-A° resolution. Biochemistry. 11: 4910-4919.

[9] Hou FJ, Xu H, Liu WY (2003) Simultaneous existence of cinnamomin (a type II RIP) and small amount of its free A- and B-chain in mature seeds of camphor tree. Int J Biochem Cell Biol. 35: 455-64.

[10] Van Damme EJM, Balzarini J, Smeets K, Van Leuven F, Peuinans WJ (1994) The monomeric and dimeric mannose binding proteins from the Orchidaceae species *Listeru ovata* and *Epipactis helleborine*: sequence homologies and differences in biological activities. Glycoconjugate J. 11: 321-332.

[11] Peumans WJ, Van Damme EJ (1995) Lectins as plant defense proteins. Plant Physiol. 109: 347-52.

[12] Van Damme EJ, Peumans WJ, Barre A, Rougé P (1998) Plant lectins: a composite of several distinct families of structurally and evolutionary related proteins with diverse biological roles. Crit Rev Plant Sci. 17: 645-662.

[13] Sharon N, Lis H (1989) Lectins as cell recognition molecules. Science. 246: 227-234.

[14] Sharon N (2007) Carbohydrate-specific reagents and biological recognition molecules. J Biol Chem. 282: 2753-2764.

[15] Cavada BS, Barbosa T, Arruda S, Grangeiro TB, Barral-Netto M (2001) Revisiting *proteus*: do minor changes in lectin structure matter in biological activity? Lessons from and potential biotechnological uses of the Diocleinae subtribe lectins. Curr Protein Pept Sci. 2: 123-135.

[16] Wu AM, Lisowska E, Duk M, Yang Z (2009) Lectins as tools in glycoconjugate research. Glycoconj J. 26: 899-913.

[17] Nowell PC (1960) Phytohemagglutinin: an initiator of mitosis in culture of animal and human leukocytes. Cancer Res. 20: 462-466.

[18] de Oliveira Silva F, das Neves Santos P, de Melo CM, Teixeira EH, de Sousa Cavada B, Arruda FV, Cajazeiras JB, Almeida AC, Pereira VA, Porto AL (2011) Immunostimulatory activity of ConBr: a focus on splenocyte proliferation and proliferative cytokine secretion. Cell Tissue Res. 346: 237-44.

[19] Lee JY, Kim JY, Lee YG, Byeon SE, Kim BH, Rhee MH, Lee A, Kwon M, Hong S, Cho JY (2007) In vitro immunoregulatory effects of Korean Mistletoe lectin on functional activation of monocytic and macrophage-like cells. Biol Pharm Bull 30: 2043-2051.

[20] Melo CML, Paim BA, Zecchin KG, Morari J, Chiarrati MR, CorreiaMTS, Coelho LCBB, Paiva PMG (2010) Cramoll 1,4 lectin increases ROS production, calcium levels and cytokine expression in treated spleen cells of rats. Mol Cell Biochem. 342: 163-169.

[21] Ringnér M, Valkonen KH, Wadström T (1994) Binding of vitronectin and plasminogen to Helicobacter pylori. FEMS Immunol Med Microbiol. 9: 29-34.

[22] Bennett HJ, Roberts IS (2005) Identification of a new sialic acid-binding protein in Helicobacter pylori. FEMS Immunol Med Microbiol. 44: 163-9.

[23] Curado M, Edwards B, Shin H, Storm HH, Ferlay J, Heanue M, Boyle P (2008) Cancer incidence in five continents. Lyon: International Agency for Research on Cancer. 837 p.

[24] Saeland E, van Kooyk Y (2011) Highly glycosylated tumour antigens: interactions with the immune system. Biochem Soc Trans. 39: 388-92.

[25] Nangia-Makker P, Conklin J, Hogan V, Raz A (2002) Carbohydrate-binding proteins in cancer, and their ligands as therapeutic agents. Trends Mol Med. 8: 187-92.

[26] Taniguchi N, Korekane H (2011) Branched N-glycans and their implications for cell adhesion, signaling and clinical applications for cancer biomarkers and in therapeutics. BMB Rep. 44: 772-81.

[27] Cooper HS, Haesler WE Jr (1978) Blood group substances as tumor antigens in the distal colon. Am J Clin Pathol. 69: 594-8.

[28] Madjd Z, Parsons T, Watson NF, Spendlove I, Ellis I, Durrant LG (2005) High expression of Lewis y/b antigens is associated with decreased survival in lymph node negative breast carcinomas. Breast Cancer Res. 7: 80-7.

[29] St Hill CA, Farooqui M, Mitcheltree G, Gulbahce HE, Jessurun J, Cao Q, Walcheck B (2009) The high affinity selectin glycan ligand C2-O-sLex and mRNA transcripts of the core 2 beta-1,6-N-acetylglucosaminyltransferase (C2GnT1) gene are highly expressed in human colorectal adenocarcinomas. BMC Cancer. 9: 79.

[30] Sobral AP, Rego MJ, Cavalacanti CL, Carvalho LB Jr, Beltrão EI (2010) ConA and UEA-I lectin histochemistry of parotid gland mucoepidermoid carcinoma. J Oral Sci. 52: 49-54.

[31] de Lima AL, Cavalcanti CC, Silva MC, Paiva PM, Coelho LC, Beltrão EI, dos S Correia MT (2010) Histochemical evaluation of human prostatic tissues with Cratylia mollis seed lectin. J Biomed Biotechnol. 2010: 179817

[32] Beltrão EI, Medeiros PL, Rodrigues OG, Figueredo-Silva J, Valença MM, Coelho LC, Carvalho LB Jr (2003) Parkia pendula lectin as histochemistry marker for meningothelial tumour. Eur J Histochem. 47: 139-42.

[33] Narimatsu H, Sawaki H, Kuno A, Kaji H, Ito H, Ikehara Y (2010) A strategy for discovery of cancer glyco-biomarkers in serum using newly developed technologies for glycoproteomics. FEBS J. 277: 95-105.

[34] Rüegg C (2006) Leukocytes, inflammation, and angiogenesis in cancer: fatal attractions. J Leukoc Biol. 80: 682–684.

[35] Ghobrial IM, Witzig TE, Adjei AA (2005) Targeting apoptosis pathways in cancer therapy. CA Cancer J Clin. 55: 178-94.

[36] Michael OH (2000) The biochemistry of apoptosis. Nature. 407: 770–7.

[37] Li CY, Xu HL, Liu B, Bao JK (2010) Concanavalin A, from an old protein to novel candidate anti-neoplastic drug. Curr Mol Pharmacol. 3: 123-8.

[38] Oliveira C, Nicolau A, Teixeira JA, Domingues L (2011) Cytotoxic effects of native and recombinant frutalin, a plant galactose-binding lectin, on HeLa cervical cancer cells. J Biomed Biotechnol. 2011: 568932.

[39] Huang LH, Yan QJ, Kopparapu NK, Jiang ZQ, Sun Y (2011) *Astragalus membranaceus* lectin (AML) induces caspase-dependent apoptosis in human leukemia cells. Cell Prolif. 45:15-21.

[40] Lavrik IN, Golks A, Krammer PH (2005) Caspases pharmacological manipulation. J Clin Invest. 115: 2665–72.

[41] Ferrero-Miliani L, Nielsen OH, Andersen PS, Girardin SE (2007) Chronic inflammation: importance of NOD2 and NALP3 in interleukin-1beta generation. Clin Exp Immunol. 147: 227-35.

[42] Margetic S (2012) Inflammation and haemostasis. Biochem Med (Zagreb). 22: 49-62.

[43] Hurwitz ZM, Ignotz R, Lalikos JF, Galili U (2012) Accelerated porcine wound healing after treatment with α-gal nanoparticles. Plast Reconstr Surg. 129: 242e-251e.

[44] Di Lella S, Sundblad V, Cerliani JP, Guardia CM, Estrin DA, Vasta GR, Rabinovich GA (2011) When galectins recognize glycans: from biochemistry to physiology and back again. Biochemistry. 50: 7842-57.

[45] Rabinovich GA, Toscano M, Jackson DA, Vasta G (2007) Functions of cell surface galectin-glycoprotein lattices. Curr Opin Struct Biol. 17: 513–520.

[46] Rabinovich GA, Ilarregui JM (2009) Conveying glycan information into T-cell homeostatic programs: A challenging role for galectin-1 in inflammatory and tumor microenvironments. Immunol Rev. 230: 144–159.

[47] Rabinovich GA, Toscano M (2009) Turning "sweet" on immunity: Galectin-glycan interactions in immune tolerance and inflammation. Nat Rev Immunol. 9: 338–352.

[48] Rodríguez RM, López-Vázquez A, López-Larrea C (2012) Immune systems evolution. Adv Exp Med Biol. 739: 237-51.

[49] Fiocchi C (2011) Early versus late immune mediated inflammatory diseases. Acta Gastroenterol Belg. 74: 548-52.

[50] Lee JY, Kim JY, Lee YG, Byeon SE, Kim BH, Rhee MH, Lee A, Kwon M, Hong S, Cho JY (2007) *In vitro* immunoregulatory effects of Korean mistletoe lectin on functional activation of monocytic and macrophage-like cells. Biol Pharm Bull. 30: 2043-51.

[51] de Oliveira Silva F, das Neves Santos P, de Melo CM, Teixeira EH, de Sousa Cavada B, Arruda FV, Cajazeiras JB, Almeida AC, Pereira VA, Porto AL (2011)

Immunostimulatory activity of ConBr: a focus on splenocyte proliferation and proliferative cytokine secretion. Cell Tissue Res. 346: 237-44.

[52] de Melo CM, de Castro MC, de Oliveira AP, Gomes FO, Pereira VR, Correia MT, Coelho LC, Paiva PM (2010) Immunomodulatory response of Cramoll 1,4 lectin on experimental lymphocytes. Phytother Res. 24: 1631-6.

[53] Brustein VP, Souza-Araújo FV, Vaz AF, Araújo RV, Paiva PM, Coelho LC, Carneiro-Leão AM, Teixeira JA, Carneiro-da-Cunha MG, Correia MT (2012) A novel antimicrobial lectin from *Eugenia malaccensis* that stimulates cutaneous healing in mice model. Inflammopharmacology. 20.

[54] Neto LG, Pinto Lda S, Bastos RM, Evaristo FF, Vasconcelos MA, Carneiro VA, Arruda FV, Porto AL, Leal RB, Júnior VA, Cavada BS, Teixeira EH (2011) Effect of the lectin of *Bauhinia variegata* and its recombinant isoform on surgically induced skin wounds in a murine model. Molecules. 16: 9298-315.

[55] Li B, Wang JH (2011) Fibroblasts and myofibroblasts in wound healing: force generation and measurement. J Tissue Viability. 20: 108-20.

[56] Boyd WC, Almodovar LR, Boyd LG (1966) Agglutinins in marine algae for human erythrocytes. Transfusion 6: 82-83.

[57] Blunden G, Rogers DJ, Farnham WF (1975) Survey of British seaweeds for hemagglutinins. Lloydia 38: 162-168.

[58] Kamiya H, Ogata K, Hori K (1982) Isolation and characterization of a new lectin in the red alga *Palmaria palmata* (L.) O. Kuntze. Bot Marina. 15: 537-540.

[59] Fabregas J, Munoz A, Llovo J, Abalde J (1984) Agglutinins in marine red algae. IRCS Medical Science 12: 298-299.

[60] Chiles TC, Bird KT (1989) A comparative study of animal erythrocyte agglutinins from marine algae. Comp Biochem Physiol 94: 107-111.

[61] Ainouz IL, Sampaio AH (1991) Screening of Brazilian marine algae for hemaglutinins. Bot Marina 34: 211-214.

[62] Hori K, Ikegami S, Miyazawa K, Ito K (1988) Mitogenic and antineoplastic isoagglutinins from red alga *Solieria robusta.* Phytochemistry. 27: 2063-2067.

[63] Ainouz IL, Sampaio AH, Benevides NMB, Freitas ALP, Costa FHF, Carvalho MR, Pinheiro-Joventino F (1992) Agglutination of enzyme treated erythrocytes by Brazilian marine algal extracts. Bot Marina 35: 475-479.

[64] Rogers DJ, Hori K (1993) Marine algal lectins: new developments. Hydrobiologia, 260/261: 589-593.

[65] Hori K, Miyazawa K, Ito K (1990) Some common properties of lectins from marine algae. Hydrobiologia. 205: 561-566.

[66] Ainouz IL, Sampaio AH, Freitas ALP, Benevides NMB, Mapurunga S (1995) Comparative study on hemagglutinins from the red marine algae *Bryothamnion triquetrum* and *B. Seaforthii.* Rev Bras Fisiol Vegetal. 7: 15-19.

[67] Rogers DJ, Swain L, Carpenter BG, Critchley AT (1994) Binding of N-acetyl-D-galactosamine by lectins from species of green marine alga genus, *Codium.* Clin Biochem. 10: 162-165.

[68] Sampaio AH, Rogers DJ, Barwell CJ (1998) A galactose specific lectin from the red marine alga *Ptilota filicina*. Phytochemistry 48: 765-769.

[69] Sampaio AH, Rogers DJ, Barwell CJ, Saker-Sampaio S, Costa FHF, Ramos MV (1999) A new isolation and further characterization of the lectin from the red marine alga *Ptilota serrata*. J Appl Phycol 10: 539-546.

[70] Sampaio AH, Rogers DJ, Barwell CJ, Saker-Sampaio S, Nascimento KS, Nagano CS, Farias WRL (2002) New affinity procedure for the isolation and further characterization of the blood group B specific lectin from the red marine alga *Ptilota plumosa*. J Appl Phycol. 14: 489-496.

[71] Sharon N, Lis H (1990) Legume lectins – a large family of homologous proteins. FASEB J. 4: 3198-3208.

[72] Rogers DJ, Fish B, Barwell CJ, Loveless RW (1988) Lectins from marine algae associated with photosynthetic accessory proteins. In.INTERNATIONAL LECTIN MEETING, Berlin/New York. Proceedings of the 9th Lectin Meeting. Berlin/New York: Walter de Gruyter. 6: 373-376.

[73] Costa FHF, Sampaio AH, Neves SA, Rocha MLA, Benevides NMB, Freitas ALP (1999) Purification and partial characterization of a lectin from the red marine alga *Amansia multifida*. Physiol Mol Biol Plants. 5:53-61.

[74] Kamiya H, Shiomi K, Shimizu Y (1980) Marine biopolymers with cell specificity III agglutinins in the red alga *Cystoclonium purpureum*: isolation and characterization. J Natur Products. 43: 136-139.

[75] Shiomi K, Yamanaka H, Kikuchi T (1981) Purification and physicochemical properties of a hemagglutinin (GVA-1) in the red alga *Gracilaria verrucosa*. Bull Jap Soc Sci Fisheries. 47: 1079-1084.

[76] Kamiya H, Ogata K, Hori K (1982) Isolation and characterization of a: new agglutinin in the red alga *Palmaria palmata* (L.) O. Kuntze. Bot Marina. 25: 537-540.

[77] Benevides NMB, Leite AM, Freitas ALP (1996) Atividade hemaglutinante na alga vermelha *Solieria filiformis*. R Bras Fisiol Vegetal 8: 117-122.

[78] Benevides NMB, Holanda ML, Melo FR, Freitas ALP, Sampaio AH (1998) Purification and partial characterization of the lectin from the red marine alga *Enantiocladia duperreyi* (C. Agardh) Faalkenberg. Bot Marina 41: 521-525.

[79] Nagano CS, Moreno FB, Bloch Jr C, Prates MV, Calvete JJ, Saker-Sampaio S, Farias WR, Tavares TD, Nascimento KS, Grangeiro TB, Cavada BS, Sampaio AH (2002) Purification and characterization of a new lectin from the red marine alga *Hypnea musciformis*. Protein Pept Lett. 9: 159-66.

[80] Leite YFMM, Silva LMCM, Amorim RCN, Freire EA, Jorge DMM, Grangeiro TB, Benevides NMB (2005) Purification of a lectin from the marine red alga *Gracilaria ornata* and its effect on the development of the cowpea weevil *Callosobruchus maculatus* (Coleoptera: Bruchidae). Bioch et Bioph Acta. 1724: 137-145.

[81] Nascimento KS, Nagano CS, Nunes EV, Rodrigues RF, Goersch GV, Cavada BS, Calvete JJ, Saker-Sampaio S, Farias WRL, Sampaio AH (2006) Isolation and characterization of a new agglutinin from the red marine alga *Hypnea cervicornis* J. Agardh. Biochem Cel Biol. 84: 49-54.

[82] Souza BWS, Andrade FK, Teixeira DIA, Mansilla A, Freitas ALP (2010) Haemagglutinin of the antarctic seaweed *Georgiella confluens* (Reinsch) Kylin: isolation and partial characterization. Polar Biol. 1: 1-8.

[83] Rogers DJ, Loveless RW, Balding P (1986) Isolation and characterization of the lectins from sub-species of *Codium fragile*. In. INTERNATIONAL LECTIN MEETING, Berlin/New York. Proceedings of the 7th Lectin Meeting. Berlin/New York: Walter de Gruyter. 5: 155-160.

[84] Fabregas J, Muñoz A, Llovo J, Carracedo A (1988) Purification and partial purification of tomentine. An N-acetylglucosamine-specific lectin from green alga *Codium tomentosum* (huds) Stackh. J Exp Mar Biol Ecology. 124: 21-30.

[85] Rogers DJ, Flangu H (1991) Lectins from *Codium* species. Br Phycol J. 26: 95-96.

[86] Sampaio AH, Rogers DJ, Barwell CJ (1998) Isolation and characterization of the lectin from the green marine alga *Ulva lactuca*. Bot Marina 41: 765-769.

[87] Benevides NMB, Holanda ML, Melo FR, Pereira MG, Monteiro ACO, Freitas ALP (2001) Purification and partial characterization of the lectin from the marine green alga Caulerpa cupressoides (Vahl) C. Agardh. Bot Marina 44: 17-22.

[88] Ambrosio A, Sanz L, Sanchez EI, Wolfenstein-Todel C, Calvete JJ (2003) Isolation of two novel mannan- and L-fucose-binding lectins from the green alga *Enteromorpha prolifera*: biochemical characterization of EPL-2. Archives of Biochemistry and Biophysics 415: 245-250.

[89] Wang S, Zhong FD, Zhang YJ, Wu ZJ, Lin QY, Xie LH (2004) Molecular characterization of a new lectin from the marine alga *Ulva pertusa*. Acta Biochim Biophys Sin. 36: 111-117.

[90] Kim GH, Klochkova TA (2005) Purification and Characterization of a Lectin, Bryohealin, Involved in the Protoplast Formation of a Marine Green Alga *Bryopsis Plumosa* (Chlorophyta). J Phycol. 42: 86-95.

[91] Han JW, Jung MG, Kim MJ, Yoon KS, Lee KP, Kim GH (2010) Purification and characterization of a D-mannose specific lectin from the green marine alga, *Bryopsis plumose*. Phycol Res. 58: 143-150.

[92] Han JW, Yoon KS; Klochkova TA, Hwang M-S, Kim GH (2011) Purification and characterization of a lectin, BPL-3, from the marine green alga *Bryopsis plumosa*. J Appl Phycol. 23: 745-753.

[93] Niu J, Wang G, Lü F, Zhou B, Peng G (2009) Characterization of a new lectin involved in the protoplast regeneration of *Bryopsis hypnoides*. Chin J Oceanol Limnol. 27: 502-512.

[94] Loomis WD (1974) Overcoming problems of phenolics and quinines in the isolation of plant enzymes and organelles. Methods Enzymol. 16: 528-544.

[95] Blunden G, Roger DJ, Loveless RW, Patel A.V (1986) Haemagglutinins in marine algae: Lectins or Phenols? In: "Lectins: Biology, Biochemistry, Clin Biochem. 5: 139-145.

[96] Calvete JJ, Costa FHF, Saker-Sampaio S, Moreno-Murciano MP, Nagano CS, Cavada BS, Grangeiro TB, Ramos MV, Bloch Jr C, Silveira SB, Freitas BT, Sampaio AH (2000) The amino acid sequence of the agglutinin isolated from the red marine alga *Bryothamnion triquetrum* defines a novel lectin structure. Cell Mol Lif Sci 57: 343-350.

[97] Hori K, Matsubara K, Miyazawa K (2000) Primary structures of two hemagglutinins from the red alga *Hypnea japonica*. Bioch Biophys Acta. 1474: 226-236.

[98] Medina-Ramirez G, Gibbs RV, Calvete JJ (2006) Micro-heterogeneity and molecular assembly of the haemagglutinins from the red algae *Bryothamnion seaforthii* and *B. triquetrum* from the Caribbean Sea. Eur J Phycol. 42: 105-112.

[99] Nagano CS, Debray H, Nascimento KS, Pinto VPT, Cavada BS, Saker-Sampaio S, Farias WRL, Sampaio AH, Calvete JJ (2005) HCA and HML isolated from the red marine algae *Hypnea cervicornis* and *Hypnea musciformis* define a novel lectin family. Protein Sci. 14: 2167-2176.

[100] Kawabuko A, Makino H, Ohnishi J, Hirohara H, Hori K (1999) Occurrence of highly yielded lectins homologous within the genus *Eucheuma*. J Appl Phycol. 11: 149-156.

[101] Hori K, Sato Y, Ito K, Fujiwara Y, Iwamoto Y, Makino H, Kawakubo A (2007) Strict specificity for high-mannose type N-glycans and primary structure of a red alga *Eucheuma serra* lectin. Glycobiology, 17: 479-9.

[102] Mori T, O'Keefe BR, Sowder II RC, Bringans S, Gardella R, Berg S, Cochran P, Turpin JA, Buckheit Jr RW, MacMahon JB, Boyd MR (2005) Isolation and characterization of griffithsin, a novel HIV-inactivating protein, from the red alga *Griffithsia* sp. J Biol Chem, 280: 9345-9353.

[103] Ishihara K, Arai S, Shimada S (2009) cDNA cloning of a lectin-like gene preferentially expressed in freshwater from macroalga *Ulva limnetica* (ulcales, Chlorophyta). Phycol Res 57: 104-110.

[104] Yoon KS, Lee KP, Klochkova TA, Kim GH (2008) Molecular characterization of the lectin, bryohealin, involved in protoplast regeneration of the marine alga *Bryopsis plumosa* (chlorophyta). J Phycol 44: 103-112.

[105] Sato Y, Hirayama M, Morimoto K, Yamamoto N, Okuyama S, Hori K (2011) High Mannose-binding Lectin with preference for the cluster of α1-2 Mannose from green Alga *Boodlea coacta* is potent inhibitor of HIV-1 and Influenza Viruses. J Biol Chem. 286: 19446-19458.

[106] Sugahara T, Ohama Y, Fukuda A, Hayashi M, Kawakubo A, Kato K (2001) The cytotoxic effect of *Eucheuma serra* agglutinin (ESA) on cancer cells and its application to molecular probe for drug delivery system using lipid vesicles. Cytotechnology. 36: 93-99.

[107] Fukuda Y, Sugahara T, Ueno M, Fukuta Y, Ochi Y, Akiyama K, Miyazaki T, Masuda S, Kawakubo A, Kato K (2006) The anti-tumor effect of Euchema serra agglutinin on colon cancer cells *in vitro* and *in vivo*. Anticancer Drugs. 17: 943-7.

[108] Omokawa Y, Miyazaki T, Walde P, Akiyama K, Sugahara T, Masuda S, Inada A, Ohnishi Y, Saeki T, Kato K (2010) In vitro and in vivo anti-tumor effects of novel Span 80 vesicles containing immobilized *Eucheuma serra* agglutinin. Int J Pharm. 15: 157-167.

[109] Pinto VP, Debray H, Dus D, Teixeira EH, de Oliveira TM, Carneiro VA, Teixeira AH, Filho GC, Nagano CS, Nascimento KS, Sampaio AH, Cavada BS (2009) Lectins from the Red Marine Algal Species *Bryothamnion seaforthii* and *Bryothamnion triquetrum* as Tools to Differentiate Human Colon Carcinoma Cells. Adv Pharmacol Sci. 2009: 862162.

[110] Bitencourt FS, Figueiredo JG, Mota MR, Bezerra CC, Silvestre PP, Vale MR, Nascimento KS, Sampaio AH, Nagano CS, Saker-Sampaio S, Farias WR, Cavada BS, Assreuy AM, de Alencar NM (2008) Antinociceptive and anti-inflammatory effects of a mucin-binding agglutinin isolated from the red marine alga *Hypnea cervicornis*. Naunyn Schmiedebergs Arch Pharmacol. 377: 139-48.

[111] Silva LM, Lima V, Holanda ML, Pinheiro PG, Rodrigues JA, Lima ME, Benevides NM (2010) Antinociceptive and anti-inflammatory activities of lectin from marine red alga *Pterocladiella capillacea*. Biol Pharm Bull. 33: 830-5.

[112] Vanderlei ES, Patoilo KK, Lima NA, Lima AP, Rodrigues JA, Silva LM, Lima ME, Lima V, Benevides NM (2010) Antinociceptive and anti-inflammatory activities of lectin from the marine green alga *Caulerpa cupressoides*. Int Immunopharmacol. 10: 1113-8.

[113] Hori K, Matsuda H, Miyazawa K, Ito K (1987) A mitogenic agglutinin from the red alga *Carpopeltis flabellate*. Phytochemistry. 26: 1335-1338.

[114] Liao WR, Lin JY, Shieh WY, Jeng WL, Huang R (2003) Antibiotic activity of lectins from marine algae against marine vibrios. J Ind Microbiol Biotechnol. 30: 433-439.

[115] Teixeira EH, Napimoga MH, Carneiro VA, de Oliveira TM, Nascimento KS, Nagano CS, Souza JB, Havt A, Pinto VP, Gonçalves RB, Farias WR, Saker-Sampaio S, Sampaio AH, Cavada BS (2007) *In vitro* inhibition of oral streptococci binding to the acquired pellicle by algal lectins. J Appl Microbiol. 103: 1001-6.

[116] Neves SA, Dias-Baruff M, Freitas AL, Roque-Barreira MC (2001) Neutrophil migration induced in vivo and in vitro by marine algal lectins. Inflamm Res. 50: 486-90.

[117] Sato Y, Morimoto K, Hirayama M, Hori K (2011) High mannose-specific lectin (KAA-2) from the red alga Kappaphycus alvarezii potently inhibits influenza virus infection in a strain-independent manner. Biochem Biophys Res Commun. 405: 291-6.

[118] Mori T, O'Keefe BR, Sowder RC 2nd, Bringans S, Gardella R, Berg S, Cochran P, Turpin JA, Buckheit RW Jr, McMahon JB, Boyd MR. Isolation and characterization of griffithsin, a novel HIV-inactivating protein, from the red alga *Griffithsia sp*. J Biol Chem. 280: 9345-53.

[119] Emau P, Tian B, O'keefe BR, Mori T, McMahon JB, Palmer KE, Jiang Y, Bekele G, Tsai CC (2007) Griffithsin, a potent HIV entry inhibitor, is an excellent candidate for anti-HIV microbicide. J Med Primatol. 36: 244-53.

[120] O'Keefe BR, Giomarelli B, Barnard DL, Shenoy SR, Chan PK, McMahon JB, Palmer KE, Barnett BW, Meyerholz DK, Wohlford-Lenane CL, McCray PB Jr (2010) Broad-spectrum *in vitro* activity and *in vivo* efficacy of the antiviral protein griffithsin against emerging viruses of the family *Coronaviridae*. J Virol. 84: 2511-21.

[121] Abee T, Kovács AT, Kuipers OP, Van Der Veen S (2010) Biofilm formation and dispersal in Gram-positive bacteria. Curr Opin Biotechnol. 22: 1-8.

[122] Merritt J, Anderson MH, Park NH, Shi W (2001) Bacterial biofilm and dentistry. J Calif Dent Assoc. 29: 355-60.

[123] Donlan RM, Costerton JW (2002) Biofilms: survival mechanisms of clinically relevant microorganisms. Clin Microbiol Rev. 15: 167-193.

[124] Onurdağ FK, Ozkan S, Ozgen S, Olmuş H, Abbasoğlu U (2010) *Candida albicans* and *Pseudomonas aeruginosa* adhesion on soft contact lenses. Graefes Arch Clin Exp Ophthalmol. Dez. 2010.

[125] Prosser BT, Taylor PA, Cix PA, Cluland R (1987) Method of evaluating effects of antibiotics on bacterial biofilms. Antimicrob. Agents Chemother. 31: 1502-1506.

[126] Costerton JW, Lewandowski Z, Caldwell DE, Korber DR, Lappin-Scott HM (1995) Microbial biofilms. Ann. Rev. Microbiol. 49: 711-745.

[127] Mah TF, O'Toole GA (2001) Mechanisms of biofilm resistance to antimicrobial agents. Trends Microbiol. 9(1): 34-39.

[128] Costerton JW, Stewart PS, Greenberg EP (1999) Bacterial biofilms: A common cause of persistent infections. Scien. 284: 1318-1322.

[129] Costerton W, Veeh R, Shirtliff M, Pasmore M, Post C, Ehrlich G (2003) The application of biofilm science to the study and control of chronic bacterial infections. J. Clin. Invest. 112: 146-1477.

[130] Fux CA, Stoodley P, Hall-Stoodley, Costerton JW (2003) Bacterial biofilms: a diagnostic and therapeutic challenge. Exp Rev Anti-infect Ther. 1(4): 667-683.

[131] Alhede M, Bjarnsholt T, Jensen PO (2009) *Pseudomonas aeruginosa* recognizes and responds aggressively to the presence of polymorphonuclear leukocytes. Microbiol. 155: 3500-3508.

[132] Van Gennip M, Christensen LD, Alhede M (2009) Inactivation of the rhlA gene in *Pseudomonas aeruginosa* prevents rhamnolipid production, disabling the protection against polymorphonuclear leukocytes. APMIS. 117: 537-546.

[133] Busscher HJ, Bos R, van der Mei HC (1995) Initial microbial adhesion is a determinant for the strength of biofilm adhesion. FEMS Microbiol. Lett. 128: 229–234.

[134] Wu C, Mishra A, Reardon ME, Huang IH, Counts SC, Das A, Ton-That H (2012) Structural determinants of *Actinomyces sortase* SrtC2 required for membrane localization and assembly of type 2 fimbriae for interbacterial coaggregation and oral biofilm formation. J Bacteriol. Mar 23.

[135] Periasamy S, Kolenbrander PE (2010) Central role of the early colonizer *Veillonella sp.* in establishing multispecies biofilm communities with initial, middle, and late colonizers of enamel. J Bacteriol. 192(12): 2965-72.

[136] Stoodley P, Sauer K, Davies DG, Costerton JW (2002) Biofilms as complex differentiated communities. Annu Rev Microbiol. 56: 187-209.

[137] Heydorn A, Ersboll BK, Hentzer M, Parsek MR, Givskov M, Molin S (2000) Experimental reproducibility in flow-chamber biofilms. Microbiol. 146: 2395-2407.

[138] Xiao J, Klein MI, Falsetta ML, Lu B, Delahunty CM, Yates JR 3rd, Heydorn A, Koo H (2012) The Exopolysaccharide matrix modulates the interaction between 3D architecture and virulence of a mixed-species oral biofilm. PLoS Pathog. 8(4): e1002623.

[139] López D, Vlamakis H, Kolter R (2010) Biofilms. Cold Spr Harb Perspect Biol. 2(7): a000398.

[140] Flemming HC, Neu TR, Wozniak DJ (2007) The EPS matrix: the "house of biofilm cells". J Bacteriol. 189(22): 7945-7.

[141] Flemming HC, Wingender J (2010) The biofilm matrix. Nat Rev Microbiol. 8(9): 623-33.

[142] Fronzes R, Remaut H, Waksman G (2008) Architectures and biogenesis of non-flagellar protein appendages in Gram-negative bacteria. EMBO J. 27: 2271-80.

[143] Dorkhan M, Chávez de Paz LE, Skepö M, Svensäter G, Davies J (2012) Effects of saliva or serum coating on adherence of *Streptococcus oralis* strains to titanium. Microbiol. 158(2): 390-7.

[144] Mukherjee J, Karunakaran E, Biggs CA (2012) Using a multi-faceted approach to determine the changes in bacterial cell surface properties influenced by a biofilm lifestyle. Biofouling. 28(1): 1-14.

[145] Marsh PD (1994) Microbial ecology of dental plaque and its significance in health and disease. Adv. Dent. Res. 8: 263-271.

[146] Nyvad B (1993) Microbial colonization of human tooth surfaces. APMIS Suppl. 32: 1-45.

[147] Thylstrup A, Fejerskov O (1995) Cariologia clínica 2 ed. São Paulo: Santos, Cap. 3, p. 45-49.

[148] Ruhl S, Sandberg AL, Cisar JO (2004) Salivary receptors for the proline-rich protein-binding and lectin-like adhesins of oral actinomyces and streptococci. J Dent Res. 83(6): 505-10.

[149] Jenkinson HF, Lamont RJ (1997) Streptococcal adhesion and colonization. Crit Rev Oral Biol Med. 8(2): 175-200.

[150] Korea GK, Ghigo JM, Beloin C (2011) The sweet connection: Solving the riddle of multiple sugar-binding fimbrial adhesins in *Escherichia coli*: Multiple *E. coli* fimbriae form a versatile arsenal of sugar-binding lectins potentially involved in surface-colonisation and tissue tropism. Bioessays 33: 300-311.

[151] Kolenbrander PE, Andersen RN, Blehert DS, Egland PG, Foster JS, Palmer Jr RJ (2002) Communication among oral bacteria. Microbiol Mol Biol Rev. 66: 486–505.

[152] Ferreira CL, Grześkowiak L, Collado MC, Salminen S (2011) In vitro evaluation of *Lactobacillus gasseri* strains of infant origin on adhesion and aggregation of specific pathogens. J Food Prot. 74(9): 1482-7.

[153] Okuda T, Okuda K, Kokubu E, Kawana T, Saito A, Ishihara K (2012) Synergistic effect on biofilm formation between *Fusobacterium nucleatum* and *Capnocytophaga ochracea*. Anaerobe. 18(1): 157-61.

[154] Handley PS, Rickard AH, High NJ, Leach SA (2001) Coaggregation - is it a universal phenomenon? In: Biofilm community interactions: Chance or Necessity (Gilbert, P., Allison, D., Verran, J., Brading, M. and Walker, J., Eds.), pp. 1–10.

[155] Rodrigues DF, Elimelech M (2009) Role of type 1 fimbriae and mannose in the development of *Escherichia coli* K12 biofilm: from initial cell adhesion to biofilm formation. Biofoul. 25(5): 401-11.

[156] Holmes AR, Gopal PK, Jenkinson HF (1995) Adherence of *Candida albicans* to a cell surface polysaccharide receptor on *Streptococcus gordonii*. Infect Immun. 63(5): 1827-34.

[157] Rosen G, Sela MN (2006) Coaggregation of *Porphyromonas gingivalis* and *Fusobacterium nucleatum* PK 1594 is mediated by capsular polysaccharide and lipopolysaccharide. FEMS Microbiol Lett. 256(2): 304-10.

[158] Ryu JH, Beuchat LR (2005) Biofilm formation by *Escherichia coli* O157:H7 on stainless steel: effect of exopolysaccharide and curli production on its resistance to chlorine. Appl Environ Microbiol. 71(1): 247-54.

[159] Simões LC, Lemos M, Pereira AM, Abreu AC, Saavedra MJ, Simões M (2011) Persister cells in a biofilm treated with a biocide. Biofoul. 27(4): 403-11.

[160] Becer CR (2012) The glycopolymer code: Synthesis of glycopolymers and multivalent carbohydrate-lectin interactions. Macromol Rapid Commun. Apr 16.

[161] Kurz K, Garimorth K, Joannidis M, Fuchs D, Petzer A, Weiss G (2012) Altered immune responses during septicaemia in patients suffering from haematological malignancies. Int J Immunopathol Pharmacol. 25(1): 147-56.

[162] Strathmann, M, Wingender J, Flemming H (2002) Application of fluorescently labelled lectins for the visualization and biochemical characterization of polysaccharides in biofilms of *Pseudomonas aeruginosa*. J. Microbiol. Methods 50: 237-248.

[163] Wawrzynczyk J, Szewczyk E, Norrlöw O, Dey E (2007) Application of enzymes, sodium tripolyphosphate and cation exchange resin for the release of extracellular polymeric substances from sewage sludge. Characterization of the extracted polysaccharides/glycoconjugates by a panel of lectins. J. Biotechnol. 130: 274-281.

[164] Cavalcante TT, Rocha BAM, Carneiro VA, Arruda FVS, AS FN, Sá NC, Nascimento KS, Cavada BS, Teixeira EH (2011) Effect of lectins from Diocleinae subtribe against oral Streptococci. Molecul. 16(5): 3530-43.

[165] Sanford BA; De Feijter AW, Wade MH, Thomas VL (1996) A dual fluorescence technique for visualization of *Staphylococcus epidermidis* biofilm using scanning confocal laser microscopy. J Ind Microbiol. 16(1): 48-56.

[166] Thomas VL, Sanford BA, Moreno R, Ramsay MA (1997) Enzyme-linked lectin sorbent assay measures N-acetyl-D-glucosamine in matrix of biofilm produced by *Staphylococcus epidermidis*. Curr Microbiol. 35(4): 249-54.

[167] Mccoy Jr JP, Varani J, Goldstein IJ (1984) Enzyme-linked lectin assay (ELLA): Detection of carbohydrate groups on the surface of unfixed cells. Exper Cell Res. 1(51): 96-103.

[168] Leriche V, Sibille P, Carpentier B (2000) Use of an enzyme-linked lectin sorbent assay to monitor the shift in polysaccharide composition in bacterial biofilms. Appl Environ Microbiol. 66(5): 1851-6.

[169] Zippel B, Neu TR (2011) Characterization of glycoconjugates of extracellular polymeric substances in tufa-associated biofilms by using fluorescence lectin-binding analysis. Appl Environm Microbiol. 77: 505-516.

[170] Rebiere-Huët J, Di Martino P, Hulen C. (2004) Inhibition of *Pseudomonas aeruginosa* adhesion to fibronectin by PA-IL and monosaccharides: involvement of a lectin-like process. Can J Microbiol. 50(5): 303-12.

[171] Teixeira EH, Napimoga MH, Carneiro VA, de Oliveira TM, Cunha RM, Havt A, Martins JL, Pinto VP, Gonçalves RB, Cavada BS (2006) In vitro inhibition of Streptococci binding to enamel acquired pellicle by plant lectins. J Appl Microbiol. 101(1): 111-6.

[172] Islam B, Khan SN, Naeem A, Sharma V, Khan AU (2009) Novel effect of plant lectins on the inhibition of *Streptococcus mutans* biofilm formation on saliva-coated surface. J Appl Microbiol. 106(5): 1682-9.

Chapter 12

Algal Polysaccharides, Novel Applications and Outlook

Stefan Kraan

Additional information is available at the end of the chapter

1. Introduction

Marine algae contain large amounts of polysaccharides, notably cell wall structural, but also mycopolysaccharides and storage polysaccharides (Kumar et al. 2008b; Murata and Nakazoe 2001). Polysaccharides are polymers of simple sugars (monosaccharides) linked together by glycosidic bonds, and they have numerous commercial applications in products such as stabilisers, thickeners, emulsifiers, food, feed, beverages etc. (McHugh 1987; Tseng 2001; Bixler and Porse, 2010). The total polysaccharide concentrations in the seaweed species of interest range from 4-76 % of the dry weight (Table 1). The highest contents are found in species such as *Ascophyllum*, *Porphyra* and *Palmaria*, however, green seaweed species such as *Ulva* also have a high content, up to 65 % of dry weight.

Seaweeds are low in calories from a nutritional perspective. The lipid content is low and even though the carbohydrate content is high, most of this is dietary fibres and not taken up by the human body. However, dietary fibres are good for human health as they make an excellent intestinal environment (Holt and Kraan, 2011).

The cell-wall polysaccharides mainly consist of cellulose and hemicelluloses, neutral polysaccharides, and they are thought to physically support the thallus in water. The building blocks needed to support the thalli of seaweed in water are less rigid/strong compared to terrestial plants and trees. The cellulose and hemicellulose content of the seaweed species of interest in this review is 2-10 % and 9 % dry weight respectively. Lignin is only found in *Ulva* sp. at concentrations of 3 % dry weight (Table 2).

The cell wall and storage polysaccharides are species-specific and examples of concentrations are given in Table 3. Green algae contain sulphuric acid polysaccharides, sulphated galactans and xylans, brown algae alginic acid, fucoidan (sulphated fucose), laminarin (β-1, 3 glucan) and sargassan and red algae agars, carrageenans, xylans, floridean starch (amylopectin like glucan), water-soluble sulphated galactan , as well as porphyran as

Carbohydrates: Integrated Research on Glycobiology and Glycotechnology

| | Brown Algae | | | | | Green Algae | Red Algae | | | |
	Laminaria & Saccharina	Fucus	Ascophyllum	Undaria	Sargassum	Ulva	Chondrus	Porphyra	Gracilaria	Palmaria
Polysaccharides Total	38 % [b] 48 % [s] 58 % [a] 61 % [s]	62 % [s] 66 % [a]	42-64 % [s] 44 % [c] 70 % [a]	35-45 % [d,e]	4 % [f] 68 % [f]	15 %-65 % [g-i,q] 18 % [p] 42-46 % [s]	55-66 % [s]	40 % [e] 41 % [r] 50-76 % [s] 54 % [b]	36 % [f] 62 % [b] 63 % [f]	38-74 % [j] 50 % [s] 66 % [k]
Structural and dietary fibres Total	36 % [l]			35-46 % [l,m]	49-62 % [l,m]	38 % [l,m]		35-49 % [l,m]		
Soluble			38 % [c]	30 % [l,m]	33 % [l,m]	21 % [l,m]		18 % [l,m]		
Lignin						3 % [g]				
Cellulose	10 % in stipe [n] 2-4.5 % [n] 4.5-9 % [o]		2 % [o] 3.5-4.6 % [n]			9 % [g]				
Hemicelluloses						9 % [g]				

a= (Rioux et al. 2007), b= (Wen et al. 2006), c= (Tseng 2001), d= (Je et al. 2009), e= (Murata and Nakazoe 2001), f= (Marinho-Soriano et al. 2006), g= (Ventura and Castañón 1998), h= (Ortiz et al. 2006), i= (Sathivel et al. 2008), j= (Heo and Jeon 2008), k= (Mishra et al. 1993), l= (Dawczynski et al. 2007), m= (Lahaye 1991), n= (Horn 2000), o= (Black 1950), p= (Foster and Hodgson 1998), q= (Wong and Cheung 2000), r= (Arasaki and Arasaki 1983), s= (Morrissey et al. 2001)

Table 1. Content of total polysaccharides (% of dry weight) and structural and dietary fibres (% of dry weight) in seaweed species of interest in North-west Europe.

Class	Genus	Uses
Chlorophyta		
	Monostroma	edible, human food
	Enteromorpha	edible, human food
Phaeophyta		
	Laminaria	alginates, edible, human food
	Undaria	edible, human food
	Cladosiphon	edible, human food
Rhodophyta		
	Asparagopsis	medical applications
	Gelidiella	agar, food and medical
	Gelidiopsis	agar, food and medical
	Gelidium	agar, food and medical
	Gracilaria	agar, food and medical
	Pterocladia	agar, food and medical
	Chondrus	carrageenan, human food
	Eucheuma	carrageenan, human food
	Kappaphycus	carrageenan, human food
	Gigartina	carrageenan, human food
	Hypnea	carrageenan, human food
	Iridaea	carrageenan, human food
	Palmaria	human and animal feed
	Porphyra	human food

Table 2. Most common genera and uses of seaweeds produced in aquaculture.

Product	Global Production	Retail Price	Approximate Gross Market Value	Amount Used	
	(ton/year)	(US$/kg)	(US$million/year)	Food (%)	Pharmacy(%)
Agars	9,600	18	173	80	ca 15
Alginates	26,500	12	318	30	5
Carrageenans	50,000	10.5	525	80	10

Table 3. Phycocolloids and their global production, retail price, gross value and percentage used in food and pharmacy(Data; McHugh, 2003; Bixler and Porse, 2010)

mucopolysaccharides located in the intercellular spaces (Table 3; Kumar et al. 2008b; Murata and Nakazoe 2001). Contents of both total and species-specific polysaccharides show seasonal variations. The mannitol content varied markedly in the fronds of *Saccharina* and *Laminaria* species with maximum amounts found during summer and autumn, from June to November. The laminaran showed extreme variations during the year with very small amounts or none at all in February to June and maximum in September to November (Haug and Jensen 1954; Jensen and Haug 1956). The maximum content of alginic acid in the fronds of *Saccharina* and *Laminaria* species was generally found from March to June and the minimum from September to October (Haug and Jensen 1954). However, highest contents of alginic acid were found during winter in other seasonal studies on *Laminaria* species from the same areas in Norway (Jensen and Haug 1956).

Further investigations on the hydrolysates of some brown algae showed complex mixtures of monosaccharides. The components of galactose, glucose, mannose, fructose, xylose, fucose and arabinose were found in the total sugars in the hydrolysates. The glucose content was 65 %, 30 % and 20 % of the total sugars in an autumn sample of 50 individual plants of *Saccharina, Fucus (serratus and spiralis)* and *Ascophyllum*, respectively (Jensen 1956).

Several other polysaccharides are present in and utilised from seaweed e.g. furcellaran, funoran, ascophyllan and sargassan, however these are not described in this chapter.

Seaweed polysaccharides are separated into dietary fibres, hydrocolloids etc. in the following paragraphs, even though the polysaccharides belong to more than just one of the functional groups.

2. Seaweed production and extraction

Harvesting or aquaculture of marine algae or seaweeds is an extensive global industry.

This seaweed industry provides a wide variety of products that have an estimated total annual production value of about US$ 6 billion. Food products for human consumption contribute about US$ 5 billion to this figure. Substances that are extracted from seaweeds – hydrocolloids – account for a large part of the remaining billion dollars, while smaller, miscellaneous uses, such as fertilizers and animal feed additives, make up the rest. The industry uses almost 20 million ton of wet seaweed annually (FAO, 2012 http://www.fao.org/fishery/statistics/global-aquaculture-production/en), harvested either from naturally growing (wild) seaweed or from cultivated (farmed) crops. The farming of seaweed has expanded rapidly as demand has outstripped the supply available from natural resources. Commercial harvesting occurs in about 35 countries, spread between the northern and southern hemispheres, in waters ranging from cold, through temperate, to tropical (Mc Hugh, 2003).

In Asia, seaweed cultivation is by far more important in terms of output and value than any other form of aquaculture. Looking at a global scale cultivated, seaweeds account for 87% of all seaweed harvested and processed of which the bulk is derived from aquaculture in Asia. The most valued of the cultivated seaweeds is the red alga Porphyra, or Nori. It is a major source of food for humans throughout the world, although it is almost exclusively cultivated in Japan, South Korea and China. Worldwide production has an annual value of over € 1.5 billion. In addition to Porphyra, other edible seaweeds include Gracilaria, Undaria, Laminaria and Caulerpa with their collective value exceeding € 3.0 billion. New applications of seaweeds and specific seaweed compounds in different sectors, such as food supplements, cosmetics, biomedicine and biotechnology are developed (Chritchley et al., 2006)

Seaweeds are also the industrial sources of polysaccharides such as carrageenans (Chondrus, Eucheuma and Kappaphycus), alginates (Ascophyllum, Laminaria, and

Macrocystis) and agars (Gelidium and Gracilaria; Table 2) and have a global value of approximately $ US 1 billion (Table 3). These important polysaccharides are used in the food, textile, paint, biotechnological and biomedical industries and have recently come under the spotlight as functional food ingredients. (Critchley et al., 2006).The majority of these species are used in some form for food or, in a few cases, for chemical extracts. The costs of production of the biomass tend to exceed the value of the biomass as a raw material for phycocolloid extraction, although it is known that some Chinese kelps produced by aquaculture are used for the production of salts of alginic acid, and applying low-technology extensive forms of aquaculture are used to produce Gracilaria for agar extraction (table 2). The value of red seaweed produced by aquaculture showed a declining trend to US$1.3 - 1.4 billion over the 1997 - 2000 period. This is probably due to the high volumes of carrageenophytes Kappaphycus and Eucheuma produced by cultivation in south-east Asia and, to some extent, Gracilaria cultivated in South America, and as such, showing trends towards commoditization.

3. Phycolloid industry

The term phycocolloid is used to refer to three main products (alginate, carrageenan and agar) which are extracted from Brown and red seaweeds respectively. The estimated world market value for phycocolloids is $ US 1 Billion (Bixler and Porse, 2010). European output of phycocolloids is estimated to have an annual wholesale value of around €130 million which is 97.5% of the total for all algal products in Europe (Earons, 1994).

3.1. Brown seaweeds and alginates

Global alginate production is ca. 26,500 tons and valued at US$318 million annually, and is extracted from brown seaweeds, most of which are harvested from the wild. The more useful brown seaweeds grow in cold waters, thriving best in waters up to about 20 °C. The main commercial sources of phaeophytes for alginates are *Ascophyllum, Laminaria,* and *Mycrocystis*. Other minor sources include *Sargassum, Durvillea, Eklonia, Lessonia,* and *Turbinaria* (Bixler and Porse, 2010)..

Brown seaweeds are also found in warmer waters, but these are less suitable for alginate production and are rarely used as food due to non-desirable chemical compounds such as terpenes. A wide variety of species are used, harvested in both the northern and southern hemispheres. Countries producing alginate include Argentina, Australia, Canada, Chile, Japan, China, Mexico, Norway, South Africa, the United Kingdom (mainly Scotland) and the United States. Most species are harvested from natural resources; cultivated raw material is normally too expensive for alginate production with the exception of China. While much of the *Laminaria* cultivated in China is used for food, when there is surplus production this can also be used in the alginate industry.

Seaweed hydrocolloids, such as alginate, agar and carrageenan, compete with plant gums (such as guar and locust bean) and cellulose derivatives (such as carboxy methyl cellulose (CMC) and methyl cellulose) that are often cheaper.

Alginates are used in the manufacture of pharmaceutical, cosmetic creams, paper and cardboard, and processed foods (Chapman, 1970). Alginate represents the most important seaweed colloid in term of volume. Grades of alginate are available for specific applications. Sodium alginate, pharmaceutical grade (US$ 13-15.5 per kg), food grade (US$ 6.5-11.0 per kg). In Japan and Korea, high demands for *Laminaria japonica* (trade name kombu) have resulted in high prices and necessitated the import of supplies for alginate extraction (Critchley, 1998). Annual growth rate for alginates is ca. 2-3 percent, with textile printing applications accounting for about half of the global market. Pharmaceutical and medical uses are about 20 percent by value of the market and have stayed buoyant, with 2-4 percent annual growth rates, driven by ongoing developments in controlled release technologies and the use of alginates in wound care applications. Food applications are worth about 20 percent of the market. That sector has been growing only slowly, and recently has grown at only 1-2 percent annually. The paper industry takes about 5 percent and the sector is very competitive, not increasing but just holding its own. The alginate industry faces strong competition from Chinese producers, whose prices do not reflect the real expense of cultivating Laminaria japonica, even in China, yet they do not appear to import sufficient wild seaweeds to offset those costs. The result is low profitability for most of the industry, with the best opportunities lying in the high end of the market, such as pharmaceutical and medical applications.

3.2. Red seaweeds and Carrageenans

Carrageenans are a group of biomolecules composed of linear polysaccharide chains with sulphate half-esters attached to the sugar unit. These properties allow carrageenans to dissolve in water, form highly viscous solutions, and remain stable over a wide pH range. There are three general forms (kappa, lambda and iota), each with their own gelling property. Kappa carrageenan today is almost exclusively obtained from farmed *Kappaphycus alvarezii* and iota from farmed *Eucheuma denticulatum* (Rasmussen and Morrissey 2007).

The Carrageenan market is worth US$ 527 m with most Carrageenan used as human food-grade semi-refined carrageenan (90%) and the rest going into pet food as semi-refined carrageenan. *Chondrus crispus* and *Kappaphycus* sp. are species containing up to 71 % and 88 % of carrageenan, respectively (Chopin et al. 1999; Rodrigueza and Montaño 2007). Food applications for carrageenans (E 407) are many, including canned foods, desert mousses, salad dressings, bakery fillings, ice cream, instant deserts and canned pet-foods. In the 1970s, an energy efficient process was developed in the Philippines to make a lower cost, strong-gelling kappa carrageenan and a weakly gelling iota. These semi-refined products gradually replaced the use of refined carrageenan as the gelling agent in canned meat pet foods. The process required lower capital investment than standard carrageenan refineries and the semi-refined extracts could be profitably sold for about two-thirds the price of conventionally refined carrageenan (Bixler and Porse, 2010). Industrial applications for purified extracts of carrageenans are equally diverse. They are used in the brewing industry for clarifying beer, wines and honeys, although less commonly than previously. There has been a fairly significant increase in production capacity for gel-press-refined carrageenan in recent years, particularly in Asia-Pacific.

3.3. Red seaweeds and agar

Agar is a mixture of polysaccharides, which can be composed of agarose and agropectin, with similar structural and functional properties as carrageenans. It is extracted from red seaweed such as *Gelidium* spp. and *Gracilaria* spp. (FAO 2008; Rasmussen and Morrissey 2007; Jeon et al. 2005). A total of 9600 mt were sold in 2009 with a value of US$ 173 m. Agar–agar is a typical and traditional food material in Japan and it is used as a material for cooking and Japanese-style confectionary. In addition, agar-agar is used in the manufacture of capsules for medical applications and as a medium for cell cultures, etc. Food applications continue to grow as shown by an increase of 2100 mt in over the last decade, and have been driven by the growth of processed foods in developing countries (Bixler and Porse, 2010).

Agar melts and gels at higher temperatures than carrageenan so it finds uses in pastry fillings and glazes that can be applied before the pastry is baked without melting in the pastry oven. In processed meats, carrageenan is the favored water binder or texturing agent, but agar hold on to the gelatine replacement market in canned meats and aspics. The texture of agar in fruit jellies also compete with kappa carrageenan jellies, but the agar texture is preferred in parts of Asia, particularly in Japan. Although significantly smaller, the markets for the specialty grades are quite attractive because of better profit margins. Agarose, a sulfate-free very pure form of agar finds widespread use today in gel electrophoretic analysis of the molecules of biotechnology.

4. Bioactive algal polysaccharides and functional properties

While food has long been used to improve health, our knowledge of the relationship between food components and health is now being used to improve food. Strictly speaking, all food is functional, in that it provides energy and nutrients necessary for survival. But the term "functional food" in use today conveys health benefits that extend far beyond mere survival. Food and nutrition science has moved from identifying and correcting nutritional deficiencies to designing foods that promote optimal health and reduce the risk of disease.

The combination of consumer desires, advances in food technology, and new evidence-based science linking diet to disease and disease prevention has created an unprecedented opportunity to address public health issues through diet and lifestyle. Widespread interest in select foods that might promote health has resulted in the use of the term "functional foods." Although most foods can be considered "functional," in the context of this Chapter the term is reserved for algal polysaccharides that have been demonstrated to provide specific health benefits beyond basic nutrition, such as, alginates, agars, carrageenans, fucoidan, mannitol, laminarin, and ulvan.

4.1. Dietary fibres

The dietary fibres are very diverse in composition and chemical structure as well as in their physicochemical properties, their ability to be fermented by the colonic flora, and their biological effects on animal and human cells (Lahaye and Kaeffer 1997). Edible seaweed contain 33-50 % total fibres on a dry weight basis, which is higher than the levels found in higher plants, and these fibres are rich in soluble fractions (Lahaye 1991). The dietary fibres included in marine algae are classified into two types, i.e. insoluble such as cellulose, mannans and xylan, and water soluble dietary fibres such as agars, alginic acid, furonan, laminaran and porphyran. The total content of dietary fibres is 58 % dry weight for *Undaria*, 50 % for *Fucus*, 30 % for *Porphyra* and 29 % for *Saccharina* (Murata and Nakazoe 2001). *Fucus* and *Laminaria* have the highest content of insoluble dietary fibres (40 % and 27 % respectively) and *Undaria pinnatifida* (wakame), *Chondrus* and *Porphyra* have the highest content of soluble dietary fibres (15 % – 22 %; Fleury and Lahaye 1991).

The undigested polysaccharides of seaweed can form important sources of dietary fibres, although they might modify digestibility of dietary protein and minerals. Apparent digestibility and retention coefficients of Ca, Mg, Fe, Na and K were lower in seaweed-fed rats (Urbano and Goñi 2002). The seaweed dietary fibres contain some valuable nutrients and substances, and there has been a deal of interest in seaweed meal, functional foods, and nutraceuticals for human consumption (McHugh 2003), because among other things polysaccharides show anti-tumour and anti-herpetitic bioactivity, they are potent as an anti-coagulant and decrease low density lipid (LDL)-cholesterols in rats (hypercholesterolemia), they prevent obesity, large intestine cancer and diabetes, and they have anti-viral activities (Lee et al. 2004; Murata and Nakazoe 2001; Amano et al. 2005; Athukorala et al. 2007; Ghosh et al. 2009; Murata and Nakazoe 2001; Ye et al. 2008). Moreover, glucose availability and absorption are delayed in the proximal small intestine after the addition of soluble fibres, thus reducing postprandial glucose levels (Jenkins et al. 1978). Water insoluble polysaccharides (celluloses) are mainly associated with a decrease in digestive tract transit time (see also digestibility of polysaccharides; Mabeau and Fleurence 1993).

4.2. Alginates

Alginates were discovered in the 1880s by a British pharmacist, E.C.C. Stanford; industrial production began in California in 1929. Algins/alginates are extracted from brown seaweed and are available in both acid and salt form. The acid form is a linear polyuronic acid and referred to as alginic acid, whereas the salt form is an important cell wall component in all brown seaweed, constituting up to 40 % - 47 % of the dry weight of algal biomass (Arasaki and Arasaki 1983; Rasmussen and Morrissey 2007). Alginates are anionic polysaccharides. They contain linear blocks of covalently (1–4)-linked β-D-mannuronate with the C5 epimer α-L-guluronate. The blocks may contain one or both of the monomers and the ratio of monomers A and B, as well as a number of units (m and n) in a block is species dependent.

It has been reported that alginic acid leads to a decrease in the concentration of cholesterol, it exerts an anti-hypertension effect, it can prevent absorption of toxic chemical substances, and it plays a major role as dietary fibre for the maintenance of animal and human health (Kim and Lee 2008; Murata and Nakazoe 2001; Nishide and Uchida 2003). These dietary polysaccharides are not found in any land plants. They help protect against potential carcinogens and they clear the digestive system and protect surface membranes of the stomach and intestine. They absorb substances like cholesterol, which are then eliminated from the digestive system (Burtin 2003; Ito and Tsuchida 1972) and result in hypocholesterolemic and hypolipidemic responses (Kiriyama et al. 1969; Lamela et al. 1989; Panlasigui et al. 2003). This is often coupled with an increase in the faecal cholesterol content and a hypoglycaemic response (Dumelod et al. 1999; Ito and Tsuchida 1972; Nishide et al. 1993).

Alginates, fucoidans and laminarin extracts were tested against nine bacteria, including *Escerichia coli, Staphycococcus, Salmonella* and *Listeria*. They appeared to be effective against *E. coli* and *Staphylococcus*. Sodium alginate seemed to demonstrate a strong anti-bacterial element. It not only binds but also kills the bacteria. Studies conducted on seaweed extracts found that fucoidan appeared to function as a good prebiotic (a substance that encourages the growth of beneficial bacteria in the intestines). An anti-inflammatory effect from some of the extracts has also been found, and so far no toxic effects have emerged in use for human health (Hennequart 2007). Furthermore, alginates with molecular weights greater than or equal to 50 kDa could prevent obesity, hypocholesterolemia and diabetes (Kimura et al. 1996). Clinical observations of volunteers who were 25 % - 30 % overweight showed that alginate, a drug containing alginic acid, significantly decreased body weight (Zee et al. 1991). In type II diabetes treatment, taking 5 g of sodium alginate every morning was found to prevent a postprandial increase of glucose, insulin, and C-peptide levels and slowed down gastric transit (Torsdottir et al. 1991). Meal supplemented with 5 % kelp alginates decreased glucose absorption balance over 8 hours in pigs. Similar studies have been done on rats and humans (Vaugelade et al. 2000). Another health effect is that the binding property of alginic acid to divalent metallic ions is correlated to the degree of the gelation or precipitation in the range of Ba<Pb<Cu<Sr<Cd<Ca<Zn<Ni<Co<Mn<Fe<Mg. No intestinal enzymes can digest alginic acid. This means that heavy metals taken into the human body are gelated or rendered insoluble by alginic acid in the intestines and cannot be absorbed into the body tissue (Arasaki and Arasaki 1983).In several countries such as the USA, Germany, Japan, Belgium and Canada, the use of alginic acid and its derivatives for the treatment of gastritis and gastroduodenal ulcers, as well as the use of alginates as anti-ulcer remedies, is protected by patents (Bogentoff 1981; Borgo 1984; Reckitt and Colman Products Ltd 1987; Sheth 1967). Several products of alginate containing drugs have been shown to effectively suppress postprandial (after eating) and acidic refluxes, binding of bile acids and duodenal ulcers in humans. Examples are "Gaviscon" (sodium alginate, sodium bicarbonate, and calcium carbonate), "Algitec" (sodium alginate and cimetidine, an H2 antagonist) and "Gastralgin" (alginic acid, sodium alginate, aluminium hydroxide, magnesium hydroxide and calcium carbonate) (Khotimchenko et al. 2001; Washington and Denton 1995). Clinical trials showed that sodium alginate promotes regeneration of the mucous membrane in the stomach, suppresses inflammation, eradicates colonies of *Helicobacter pylori* in the mucous membrane and normalizes non-specific resistance of the latter in 4 to 15-year-old children. It also promotes restoration of the intestinal biocenosis (Miroshnichenko et al. 1998). Other studies show positive dietry effects of alginates on faecal microbial fauna, changes in concentrations of compounds and acids, and prebiotic properties that can promote health (Terada et al. 1995; Wang et al. 2006).

Sodium alginate is often used as a powder, either pure or mixed with other drugs, on septic wounds. The polysaccharide base stimulates reparative processes, it prepares the wound for scarring, and it displays protective and coating effects, shielding mucous membranes and damaged skin against irritation from unfavourable environments. Calcium alginate promotes the proliferation of fibroblasts and inhibits the proliferation of microvascular endothelial cells and keratinocytes (Doyle et al. 1996; Glyantsev et al. 1993; Swinyard and Pathak 1985). Profound wound healing effects have also been reported for a gelatine-alginate sponge impregnated with anti-septics and anti-biotics (Choi et al. 1999).

Another use of alginates is the absorbing hemostatic effect exploited in surgery. Gauze dressings, cotton, swabs, and special materials impregnated with a solution of sodium alginate are produced and used for external use and for application onto bleeding points during abdominal operations on parenchymatous organs (Khotimchenko et al. 2001; Savitskaya 1986). Studies on the effect of alginate on prothrombotic blood coagulation and platelet activation have shown that the degree of these effects depends on the ratio between the mannuronic and guluronic chains in the molecule, as well as on the concentration of calcium. However, a zinc ion containing alginate was shown to have the most profound hemostatic effects (Segal et al. 1998). A "poraprezinc-sodium alginate suspension" has been suggested as a high performance mixture for the treatment of severe gingivostomatitis (cold sores) complicated by hemorrhagic erosions and ulcers (Katayama et al. 1998). When applied to the tooth surface, alginate fibres swell to form a gel like substance, a matrix for coagulation. Alginate dressings are used to pack sinuses, fistulas, and tooth cavities (Reynolds et al. 1982).

Furthermore, the algins have anti-cancer properties (Murata and Nakazoe 2001) and a bioactive food additive "Detoxal", containing calcium alginate, has anti-toxic effects on hepatitis. This drug decreases the content of lipid peroxidation products and normalizes the concentrations of lipids and glycogen in the liver (Khotimchenko et al. 2001).

4.3. Carrageenans

From a human health perspective it has been reported that carrageenans have anti-tumour and anti-viral properties (Skoler-Karpoff et al. 2008; Vlieghe et al. 2002; Yan et al. 2004; Zhou et al. 2006b). Furthermore, Irish Moss or Carrageen (*C. crispus* and *Mastocarpus stellatus*) has a large number of medical applications, some of which date from the 1830s. It is still used in Ireland to make traditional medicinal teas and cough medicines to combat colds, bronchitis, and chronic coughs. It is said to be particularly useful for dislodging mucus and has anti-viral properties. Carrageenans are also used as suspension agents and stabilisers in other drugs, lotions and medicinal creams. Other medical applications are as an anti-coagulant in blood products and for the treatment of bowel problems such as diarrhoea, constipation and dysentery. They are also used to make internal poultices to control stomach ulcers (Morrissey et al. 2001).

New research on the biocide properties shows that applications of carrageenan gels from *C. crispus* may block the transmission of the HIV virus as well as other STD viruses such as gonorrhoea, genital warts and the herpes simplex virus (HSV) (Caceres et al. 2000; Carlucci et al. 1997; Luescher-Mattli 2003; Shanmugam and Mody 2000; Witvrouw and DeClercq 1997). In addition, carrageenans are good candidates for use as vaginal microbicides because they do not exhibit significant levels of cytotoxicity or anti-coagulant activity (Buck et al. 2006; Zeitlin et al. 1997). Results of sexual lubricant gels raised the possibility that use of such lubricant products, or condoms lubricated with carrageenan-based gels, could block the sexual transmission of HPV (human papillomavirus) types that can cause cervical cancer and genital warts. However, carrageenan inhibition of herpes simplex virus and HIV-1 infectivity were demonstrated as about a thousand-fold higher than the IC50s observed for genital HPVs *in vitro* (Witvrouw and de Clerck 1997; Luescher-Mattli et al. 2003). A carrageenan-based vaginal microbicide called Carraguard has been shown to block HIV and other sexually transmitted diseases *in vitro*. Massive clinical trials by the Population Council Centre began in two severely affected African countries; Botswana and South Africa in 2002. Carraguard entered phase III clinical trials involving 6000 non-pregnant, HIV-negative women in South Africa and Botswana in 2003 (Spieler 2002).

Many reports exist of anti-coagulant activity and inhibited platelet aggregation of carrageenan (Hawkins et al. 1962; Hawkins and Leonard 1963; Kindness et al. 1979). Among the carrageenan types, λ carrageenan (primarely from *C. crispus*) has approximately twice the activity of unfractioned carrageenan and four times the activity of κ-carrageenan (*Eucheuma cottoni* and *E. spinosum*). The most active carrageenan has approximately one-fifteenth the activity of heparin (Hawkins et al. 1962), but the sulphated galactan from *Grateloupa indica* collected from Indian waters, exhibited anti-coagulant activity as potent as heparin (Sen et al. 1994). The principal basis of the anti-coagulant activity of carrageenan appeared to be an anti-thrombotic property. λ-carrageenan showed greater anti-thrombotic activity than κ-carrageenan, probably due to its higher sulphate content, whereas the activity of the unfractionated material remained between the two. It was found that toxicity of carrageenans depended on the molecular weight and not the sulphate content. Similar results were obtained with λ-carrageenan of *Phyllophora brodiaei* which gave the highest blood anti-coagulant activity (Sen et al. 1994). In addition, the hypoglycaemic effect of carrageenan may prove useful in the prevention and management of metabolic conditions such as diabetes (Dumelod et al. 1999).The use of carrageenan for food applications started almost 600 years ago. Due to its long and safe history of use, carrageenan is generally recognised as safe (GRAS) by experts from the US Food and Drug Administration (21 CFR 182.7255) and is approved as a food additive (21 CFR 172.620). The World Health Organisation (WHO) Joint Expert Committee of Food Additives has concluded that it is not necessary to specify an acceptable daily intake limit for carrageenans (van de Velde et al. 2002). Although carrageenans are used widely as a food ingredient, they are also used in experimental research in animals where they induce pleurisy and ulceration of the colon (Noa et al. 2000). Furthermore, carrageenans can cause reproducible inflammatory reaction and they remain a standard irritant for examining acute inflammation and anti-inflammatory drugs. Two test systems are used widely for the evaluation of non-steroidal anti-inflammatory drugs and cyclooxygenase activity: (1) The carrageenan air pouch model and (2) The carrageenan-induced rat paw edema assay (Dannhardt and Kiefer 2001). The role of carrageenans in promotion of colorectal ulceration formation is controversial and much seems to depend on the molecular weight of the carrageenan used. The international agency for research on cancer classified degraded carrageenan as a possible human carcinogen but native carrageenan remains unclassified in relation to a causative agent of human colon cancer and as mentioned it has GRAS status (Carthew 2002; Tobacman 2001).

4.4. Agar

The agar content in *Gracilaria* species can reach 31 % . It has been reported that agar-agar leads to decreases in the concentration of blood glucose and exerts an anti-aggregation effect on red blood cells. It has also been reported to affect absorption of ultraviolet rays (Murata and Nakazoe 2001). Anti-tumour activity was also found in an agar-type polysaccharide from cold water extraction of another *Gracilaria* species and hydrolysates of agar resulted in agaro-oligosaccharides with activity against α-glucosidase and antioxidant ability (Chen et al. 2005; Fernandez et al. 1989). Agarose can be separated from the agar with a yield of 42 %, and the agar content varied seasonally from 26 % - 42 % in *Gelidium* spp. in another experiment (Mouradi-Givernaud et al. 1992; Jeon et al. 2005). Agaro-oligosaccharides have also been shown to suppress the production of a pro-inflammatory cytokine and an enzyme associated with the production of nitric oxide (Enoki et al. 2003).

4.5. Fucoidan/fucans/fucanoids

Fucoidans are a group of polysaccharides (fucans) primarily composed of sulphated L-fucose with less than 10 % of other monosaccharides. They are widely found in the cell walls of brown seaweed, but not in other algae or higher plants (Berteau and Mulloy 2003).

Fucoidan is considered as a cell wall reinforcing molecule and seems to be associated with protection against the effects of desiccation when the seaweed is exposed at low-tide. Fucoidans were first isolated by Kylin almost one century ago and have interesting bioactivities (Kylin 1913). According to Table 4b the species *Fucus vesiculosus* contains the highest concentration of fucoidans (up to 20 % on a dry weight basis). Fucanoids can make up more than 40 % of dry weight of isolated algal cell walls and can easily be extracted using either hot water or an acid solution (Berteau and Mulloy 2003). Fucoidan is viscous in very low concentrations and susceptible to breakdown by diluted acids and bases. Fucoidans are produced as complex, heterogeneous polysaccharides, which contribute to intercellular mucilage. Their structural complexity varies in the degree of branching, substituents,sulfatation and type of linkages, the fine structure depending on the source of the polysaccharide. Although their composition varies with species and geographical origin, fucoidans always contain fucose and sulfate with small proportions of uronic acids, galactose, xylose, arabinose and mannose. Algal fucoidans have one of two types of homofucose backbone chains, with either repeated (1→3)-linked α-L-fucopyranosyl residues or alternating (1→3)- and (1→4)-linked α-L-fucopyranosyl residues Cumashi et al 2007). Sulfonato- and acetyl-groups as well as α-L-fucopyranosyl, α-D-glucuronopyranosyl and some other sugar residues may occur at O-2 and/or at O-4 of the α-L-fucose units of the backbone. Fucoidans with backbones of first type have been isolated from seaweeds *Saccharina latissima, Laminaria digitata, Chorda filum,* and *Cladosiphon okamuranus.* The second type of backbone was found in fucoidans from *Fucus evanescense, Fucus distichus,* and *Ascophyllum nodosum.*

A Tasmanian company, Marinova Pty Ltd, is able to supply commercial volumes of fucoidan extract and their derivates, formulated to purity levels of up to 95 %. All fucoidans of the species *Undaria* sp., *Lessonia* sp., *Macrocystis* sp., *Cladosiphon* sp., *Durvillea* sp., *Laminaria* sp., *Ecklonia* sp., *Fucus* sp., *Sargassum* sp., *Ascophyllum* sp., and *Alaria* sp. are Halal and Kosher certified. Marinova has isolated fucoidans from a range of species (species-specific), and can provide characterised fractions for either investigational research or as ingredients for nutraceutical and cosmetic applications. Different therapeutic profiles are primarily due to the molecular structure. The company has developed the Maritech™ coldwater extraction process, which maintains the integrity of fucoidans, and produces nature-equivalent high molecular weight molecules with optimal bioactivity. Solvent-based extraction, which is commonly used, causes degradation of fucoidans, and limits the activity of these molecules in biological assays.

Sector	% Demand	Sector Growth	Ton	Value US$
Textile printing and technical grades	41.5	Flat	11,000	140.0m
Food and Pharma	49	60%	13,000	150.0m
Animal feed	3.8	-75%	1,000	18.0m
PGA	5.7	-25%	1,500	10.0m
Total	100		26,500	$318m

Table 4. Alginate markets (2009) – Tonnage, value and sector growth over last decade (Data; Bixler and Porse, 2010)

Although the major physiological purposes of fucans in the algae are not thoroughly understood, they are known to possess numerous biological properties with potential human health applications (Berteau and Mulloy 2003). The list of bioactivity of fucoidan for human health is long. Fucoidan found in seaweed such as *Undaria* and *Laminaria* shows anti-coagulant, anti-viral and anti-cancer properties; Chevolot et al. 1999; Zhuang et al. 1995).

Fucoidan preparations have been proposed as an alternative to the injectable anti-coagulant heparin, because fucoidan originates from plant matter and is less likely to contain infectious agents such as viruses (Berteau and Mulloy 2003). No toxicological changes were observed when 300 mg/kg body weight per day fucoidan was administered to rats, however, significantly prolonged blood-clotting times were observed when concentrations were increased three-fold (Li et al. 2005). The biological activity (e.g. antioxidant and anti-coagulant) of sulphated polysaccharides is not only related to molecular weight and sulphated ester content (role in the charge of the molecule), as previously determined, but also to glucuronic acid and fucose content, together with the position of the sulphate groups on the sugar resides (Berteau and Mulloy 2003; Li et al. 2005; Zhao et al. 2008). A large molecular weight is required to achieve anti-coagulant activity, as fucoidan needs a long sugar chain in order to be able to bind the thrombin (coagulation protein in the blood stream). Some researchers have measured fucoidan's molecular weight at approximately 100 kDa. while others have observed a molecular weight of 1600 kDa (Rioux et al. 2007). The native fucoidan from *Lessonia vadosa* with a molecular weight of 320 kDa showed good anti-coagulant activity compared to a smaller depolymerised fraction with a molecular weight of 32 kDa, which presented weaker anti-coagulant activity (Li et al. 2008a). Some structural features of fucoidan are most likely required for certain specific activities.

Fucoidan stimulates the immuno system in several ways, and the numerous important biological effects of fucoidans are related to their ability to modify cell surface properties (Usov et al. 2001). Oral intake of the fucoidans present in dietary brown seaweed might take the protective effects through direct inhibition of viral replication and stimulation of the immune system (innate and adaptive) functions (Hayashi et al. 2008). Fucoidan has been found to restore the immune functions of immune-suppressed mice, act as an immunomodulator directly on macrophage, T lymphocyte, B cell, natural killer cells (NK cell; Wang et al. 1994), promote the recovery of immunologic function in irradiated rats (Wu et al. 2003), induce the production of interleukin (IL-1) and interferon-γ (IFN-γ) *in vitro*, and promote the primary antibody response in sheep red blood cells *in vivo* (Yang et al. 1995).

The mechanism of anti-viral activities of fucoidan is to inhibit viral sorption so as to inhibit viral-induced syncytium formation (Mandal et al. 2007). Sulphate is necessary for the anti-viral activity and sulphate located at C-4 of (1-3)-linked fucopyranosyl units appears to be very important for the anti-herpetic activity of fucoidan (Mandal et al. 2007). Some anti-viral properties of sulphated fucans have also been characterised, for example inhibition of infection of human immunodeficiency virus (HIV), Herpes Simplex Virus (HSV) (Iqbal et al. 2000; Mandal et al. 2007; Witvrouw and DeClercq 1997), poliovirus III, adenovirus III, ECH06 virus, coxsackie B3 virus, coxsackie A16 (Li et al. 1995), cytomegalovirus and bovine viral diarrhea virus (Iqbal et al. 2000).

Fucoidan is known to have anti-tumour effects but its mode of action is not fully understood. A study done by Alekseyenko et al. 2007 demonstrated that when 10 mg/kg of fucoidan was administered in mice with transplanted Lewis lung adenocarcinoma, it produced moderate anti-tumour and anti-metastatic effects (Li et al. 2008b). These polyanionic polysaccharides have anti-angiogenesis, antiproliferation for tumour cells, they inhibit tumour growth and reduce tumour size (Ellouali et al. 1993; Li et al. 2008a), inhibit tumour cell adhesion to various substrata (Liu et al. 2005), and have direct anti-cancer effects on human HS-Sultan cells through caspase and ERK pathways (Aisa et al. 2005).

Besides directly inhibiting the growth of tumour cells, fucoidans can also restrain the development and diffusion of tumour cells through enhancing the body's immuno-modulatory activities, because fucoidan mediates tumour destruction through type 1 T-helper (Th1) cell and NK cell responses (Maruyama et al. 2007). In addition, at a dose of 25 mg/kg, fucoidan potentiated the toxic effect of cyclophosphamide used to treat various types of cancer and some auto-immune disorders (Alekseyenko et al. 2007).

Many studies suggest that fucoidan has potential for use as an anti-inflammatory agent. A study showed that fucoidan treatment led to less severe symptoms in the early stages of *Staphylococcus aureus*-triggered arthritis in mice, but delayed phagocyte recruitment and decreased clearance of the bacterium (Verdrengh et al. 2000). Additionally, injection of fucoidan into sensitized mice before hapten challenge can reduce contact hypersensitivity reactions (Nasu et al. 1997). Furthermore, recruitment of leukocytes into cerebrospinal fluid in a meningitis model is reduced by fucoidan (Granert et al. 1999) as is IL-1 (interleukin-1) production in a similar model (Ostergaard et al. 2000).

Fucoidan can act as a ligand for either L- or P- selectins, both of which interact with the sulphated oligosaccharides and this interaction has physiological consequences that could be therapeutically beneficial (Omata et al. 1997). Selectins are a group of lectins (sugar-binding proteins) that interact with oligosaccharides clustered on cell surfaces during the margination and rolling of leukocytes prior to firm adhesion, extravasation and migration to sites of infection (Lasky 1995).

In addition, fucoidan is an excellent natural antioxidant and presents significant antioxidant activity in experiments *in vitro*. In recent years, sulphated polysaccharides from the marine algae *Porphyra haitanesis* (Zhang et al. 2003), *Ulva pertusa* (Qi et al. 2005b; Qi et al. 2005a), *Fucus vesiculosus* (Ruperez et al. 2002), *Laminaria japonica* (Xue et al. 2004) and *Ecklonia kurome* (Hu et al. 2001) have been demonstrated to possess antioxidant activity. There are few reports however detailing the relationship between structure and antioxidant activity of sulphated polysaccharides from marine algae. Fucan showed low antioxidant activity relative to fucoidan (Rocha de Souza et al. 2007) and as mentioned previously, the ratio of sulphate content/fucose and the molecular weight were effective indicators to antioxidant activity of the samples (Wang et al. 2008). Fucoidan may have potential for preventing free radical mediated diseases such as Alzheimer's and the aging process. Previously, fucoidan was extracted from *L. japonica*, a commercially important algae species in China. Three sulphated polysaccharide fractions were successfully isolated through anion-exchange column chromatography and had their antioxidant activities investigated employing various established *in vitro* systems, including superoxide and hydroxyl radical scavenging activity, chelating ability, and reducing power (Wang et al. 2008). All fractions were more effective than the unprocessed fucoidan. Two galactose-rich fractions had the most potent scavenging activity against superoxide (generated in the PMS-NADH system) and hydroxyl radicals with EC_{50} values of 1.7 µg mL^{-1} and 1.42 mg mL^{-1}, respectively. One of these fractions also showed the strongest ferrous ion chelating ability at 0.76 mg mL^{-1} (Wang et al. 2008). Additionally, fucoidan (homofucan) from F. vesiculosus and fucans (heterofucans) from Padina gymnospora had an inhibitory effect on the formation of hydroxyl radical and superoxide radical (Rocha de Souza, et al. 2007). Healing of dermal wounds with macromolecular agents such as natural polymers is a growing area of research interest in pharmaceutical biotechnology. Fucoidan has been shown to modulate the effects of a variety of growth factors through mechanisms thought to be similar to the action of heparin. Fucoidan-chitosan films can promote re-epithelization and contraction of the wound area. Moreover, fucoidan-chitosan films may be suitable for use in hydrogel formulations for the treatment of dermal burns (Sezer, et al. 2007).

4.6. Mannitol

Mannitol is an important sugar alcohol which is present in many species of brown algae, especially in *Laminaria* and *Ecklonia*. The mannitol content is subject to wide seasonal fluctuations and varies with environment. Mannitol is the sugar alcohol corresponding to mannose. It usually constitutes less than 10 % of the dry weight in both *Ascophyllum nodosum* and *L. hyperborea* stipe. In autumn fronds of *L. hyperborea*, however, the content may be as high as 25 % of the dry weight .

Applications of mannitol are extremely diverse. It is used in pharmaceuticals, in making chewing gum, in the paint and varnish industry, in leather and paper manufacture, in the plastics industry and in the production of explosives. The US, the UK, France and Japan are the main centres of production. Mannitol can be used in a variety of foods, candies and chocolate-flavoured compound coatings because it can replace sucrose to make sugar-free compound coatings. Sugar-free chocolates are especially popular for people with diabetes, a growing problem in modern society. It is used as a flavour enhancer because of its sweet and pleasantly cool taste. Mannitol can be used to maintain the proper moisture level in foods so as to increase shelf-life and stability, because it is non-hygroscopic and chemically inert. Mannitol is the preferred excipient for chewable tablets due to its favourable feel in the mouth. It is non-carcinogenic and can be used in pediatric and geriatric food products, as it will not contribute to tooth decay (Nabors 2004).

4.7. Laminarin

Laminaran is a glucan, built up from (1→3)- and (1→6)-β-glucose residues. It is a linear polysaccharide, with a β(1→3):β(1→6) ratio around 3:1. Laminarin is found in the fronds of Laminaria/Saccharina and, to a lesser extent, in Ascophyllum and Fucus species and Undaria. The content varies seasonally and with habitat and can reach up to 32 % of the dry weight. Laminaran does not gel nor form any viscous solution and its main potential appears to lie in medical and pharmaceutical uses.

Commercial applications of the extract have so far been limited, although some progress has been made in France as an anti-viral in agricultural applications (Goemar 2010) or as dietary fiber (Deville et al. 2004). Especially the use of laminarin as substratum for prebiotic bacteria seems to have a good commercial application (Deville et al. 2004). Laminarin does not gel or form any viscous solution, and its main potential appears to lie in medical and pharmaceutical uses. It has been shown to be a safe surgical dusting powder, and may have value as a tumour-inhibiting agent and, in the form of a sulphate ester, as an anti-coagulant (Miao et al. 1999). The presence of anti-coagulant activity in brown algae was first reported in 1941, when *Laminaria* showed anti-coagulant properties with its active compound being located in the holdfasts (Shanmugam and Mody 2000). There are about 60 brown algal species identified to have blood anti-coagulant properties. Laminarin only shows anti-coagulant activity after structural modifications such as sulphation, reduction or oxidation. The anti-coagulant activity is improved chemically by increasing the degree of sulphation (Shanmugam and Mody 2000).

Preparations containing 1→3:1→6-β-D-glucans, laminarin, and fucoidan are manufactured by the health industry and marketed for their beneficial properties on the immune system. The producers of these tablets cite numerous papers discussing the biological activity of these glucans.

Laminarin provides protection against infection by bacterial pathogens, and protection against severe irradiation, it boosts the immune system by increasing the B cells and helper T cells, reduces cholesterol levels in serum and lowers systolic blood pressure, among other effects (Table 4c; Hoffman et al. 1995) lower the levels of total cholesterol, free cholesterol, triglyceride and phospholipid in the liver (Miao et al. 1999; Renn et al. 1994a; 1994b). The hypocholesterolemic and hypolipidemic responses are noted to be due to reduced cholesterol absorption in the gut (Kiriyama et al. 1969; Lamela et al. 1989; Panlasigui et al. 2003). This is often coupled with an increase in the faecal cholesterol content and a hypoglycaemic response (Dumelod et al. 1999; Ito and Tsuchida 1972; Nishide et al. 1993). A high level of low density lipid (LDL) cholesterol can lead to plaque forming and clog arteries and lead to cardiovascular diseases and heart attacks or strokes, a major cause of disease in the US. Laminarin as a potential cancer therapeutic is under intensive investigation (Miao et al. 1999).

4.8. Ulvan

The name ulvan is derived from the original terms ulvin and ulvacin introduced by Kylin in reference to different fractions of *Ulva lactuca* water-soluble sulphated polysaccharides. It is now being used to refer to polysaccharides from members of the Ulvales, mainly, *Ulva* sp. Ulvans are highly charged sulphated polyelectrolytes, composed mainly of rhamnose, uronic acid and xylose as the main monomer sugars, and containing a common constituting disaccharide; the aldobiuronic acid, [→4)-D-glucuronic acid-(1→4)- L-rhamnose3-sulfate-(1→]. Iduronic acid is also a constituent sugar. The average molecular weight of ulvans ranges from 189 to 8,200 kDa (Lahaye 1998). The cell-wall polysaccharides of ulvales represent 38 % to 54 % of the dry algal matter (Lahaye, 1998). Two major kinds have been identified: water soluble ulvan and insoluble cellulose-like material.

The mechanism of gel formation is unique among polysaccharide hydrogels. It is very complex and not yet fully understood. The viscosity of ulvan solutions as isolated polysaccharides has been compared to that of arabic gum. Whether ulvans present other functional properties of this gum remains to be established. The gelling properties of ulvans are affected by boric acid, divalent cations and pH. They are thermoreversible without thermal hysteresis. The gelling properties can be of interest for applications where gel formation needs to be precisely controlled (by pH or temperature), like the release of trapped molecules or particles under specific conditions (Percival and McDowell 1990; Lahaye et al. 1998). As already mentioned, highly absorbent, biodegradable hydrocolloid wound dressings limit wound secretions and minimise bacterial contamination. Polysaccharide fibres trapped in a wound are readily biodegraded. In the context of BSE (mad cow disease) and other prion contamination diseases, macromolecular materials from algal biomasses such as ulvans can constitute an effective and low-cost alternative to meat-derived products, because their rheological and gelling properties make them suitable as a substitute for gelatin and related compounds (Choi et al. 1999).

Ulvans are a source of sugars for the synthesis of fine chemicals. In particular, they are a potential source of iduronic acid, the only occurrence to date of this rare uronic acid in plants (Lahaye and Ray 1996). Iduronic acid is used in the synthesis of heparin fragment analogues with anti-thrombotic activities, and obtaining it requires a lengthy synthetic procedure that could be more cost-effectively replaced by a natural source (Lahaye 1998). Oligosaccharides from *Ulva* could be used as reference compounds for analyzing biologically active domains of glycosaminoglycans (GAG) like heparin. The use of oligo-ulvans as anti-coagulant agents could be expected since other, rarer, sulphated polysaccharides, like dermatan sulphate or fucan in brown algae, have shown this anti-thrombinic activity. Regular oligomers from ulvans may provide better-tolerated anti-thrombinic drugs (Paradossi et al. 2002).

Rhamnose, a major component of ulvans, is a rare sugar, used as a precursor for the synthesis of aroma compounds. Combinatorial libraries in glycopeptide mimetics are another example of the use of L-rhamnose in the pharmaceutical industry. The production of rhamnose from *Monostroma*, a Japanese species of Codiales, has been patented. Rare sulphated sugars such as rhamnose 3-sulphate and xylose 2-sulphate are also of interest (Lahaye and Robic 2007).

Other potential applications of ulvan oligomers and polymers are related to their biological properties. Recent studies have demonstrated that ulvans and their oligosaccharides were able to modify the adhesion and proliferation of normal and tumoural human colonic cells as well as the expression of transforming growth factors (TGF) and surface glycosyl markers related to cellular differentiation (Lahaye and Robic 2007) . Earlier work demonstrated strain-specific anti-influenza activities of ulvan from *U. lactuca* and the use of rhamnan, rhamnose and oligomers from desulphated *Monostroma* ulvans has been patented for the treatment of gastric ulcers (Fujiwara-Arasaki et al. 1984; Nagaoka et al. 2003).

Oligomers from seaweed species such as *Laminaria* sp. or *Fucus serratus* have also been studied as plant elicitors. These are natural compounds which stimulate the natural defences of plants. Several products derived from brown algae are already marketed worldwide. The success is because of their size and availability rather than their chemical composition. *Ulva* cell walls bind heavy metals and ulvans are the main contributors with 2.8 to 3.77 meq g^{-1} polysaccharide. The ion-exchange property of ulvans explains why they have been chosen as bioindicators for monitoring heavy metal pollution in coastal waters (Nagaoka et al. 1999).

4.9. Porphyran and Xylans

In red algae, the fibrillar network is made of low crystalline cellulose, mannan or xylan and represents only about 10% of the cell wall weight. It can also contain minor amounts of sulphated glucans, mannoglycans and complex galactans. Most of our current knowledge of red algal cell wall polysaccharides is on the gelling and thickening water soluble galactans, agars and carrageenans, used in various applications. Unlike most red seaweed generally studied, *Palmaria palmata* does not produce matrix galactans, but instead (10/4)- and (10/3)-linked b-D-xylan together with a minor amount of fibrillar cellulose and b-(10/4)-Dxylan. Xylans can be 35 % of the dry weight of *Palmaria* (Lahaye et al. 2003). Xylans have not yet been of economic interest and only few applications are known. Species of *Porphyra* contain a sulphated polysaccharide called porphyran; a complex galactan. Porphyran is dietary fiber of good quality, and chemically resembles agar. A powder consisting of 20 % nori mixed with a basic diet given orally to rats prevented 1,2-dimethylhydrazine-induced intestinal carcinogenesis. Porphyran showed appreciable anti-tumour activity against Meth-A fibrosarcoma. In addition, it can significantly lower the artificially enhanced level of hypertension and blood-cholesterol in rats (Noda 1993).

4.10. Digestibility of polysaccharides

The majority of edible seaweed fibres are soluble anionic polysaccharides which are little-degraded or not fermented by the human colonic micro flora (Lahaye and Kaeffer 1997). The amount of dietary fibres in marine algae not digested by the human digestive tract is higher than that of other food materials (Murata and Nakazoe 2001). Most of the total algal fibres disappeared after 24 h (range 60 % - 76 %) in *in vitro* fermentation of e.g. *L. digitata* and *U. pinnatifida* using human faecal flora. However, unlike the reference substrate (sugar beet fibres), the algal fibres were not completely metabolized to short chain fatty acids (SCFA; range 47 % - 62 %). Among the purified algal fibres, disappearance of laminarins was approximately 90 % and metabolism to SCFA was approximately 85 %, in close agreement with the fermentation pattern of reference fibres. Sulphated fucans were not degraded. Sodium alginates (Na-alginates) exhibited a fermentation pattern quite similar to that of the whole algal fibres, with a more pronounced discrepancy between disappearance and production of SCFA: disappearance was approximately 83 % but metabolism was only approximately 57 %. Laminarin seemed to be a modulator of the intestinal metabolism by its effects on mucus composition, intestinal pH and short chain fatty acid production, especially butyrate. The characteristic fermentation pattern of the total fibres from the brown algae investigated was attributed to the peculiar fermentation of alginates (Michel et al. 1996; Deville et al. 2007).

Phycolloids are more or less degraded following adaptation of the human micro flora, but none of the seaweed polysaccharides have been shown to be metabolized, although some may be partly absorbed. Nothing is known about the fate of other algal polysaccharides in the human digestive tract, except that they cannot be digested by human endogenous enzymes. Results of fermentation *in vitro* with human faecal bacteria, indicate that brown seaweed fibres exhibit an original fermentation pathway (Mabeau and Fleurence 1993). Carrageenan is a good source of soluble fibre (Burtin 2003). Rats excrete carrageenan quantitatively in the faeces, if it is administered in the diet at levels of 2 % - 20 % and it therefore has no direct nutritive value (Hawkins and Yaphe 1965). Weight gain was significantly reduced, especially at higher levels. Furthermore, food efficiency showed interference with utilization of other nutrients in the diet. Only 10 % - 15 % appeared digestible from faecal examination (Hawkins and Yaphe 1965). An experimental feeding with *L digitata* seaweed extract in pig resulted in a higher production of butyric acid in the caecum and colon compared to the control group (Reilly et al. 2008). Butyrate is a beneficial metabolite for intestinal bacteria, because it is quickly metabolised by colonoyctes and accounts for about 70 % of total energy consumption of the colon (Reilly et al. 2008). Therefore, it is desirable to promote butyrate production in the colon by laminarin fermentation.

The particular chemical structure of ulvan (and of *Ulva*) is responsible for the resistance of this polysaccharide to colonic bacterial fermentation. Consumption of dietary fibres from *Ulva* sp. could be expected to act mainly as bullring agents with little effect on nutrient metabolism due to colonic bacterial fermentation products (short-chain fatty acids; Bobin-Dubigeon et al. 1997). All soluble fibre fractions of *P palmata* consisted of linear beta-1,4/beta1,3 mixed linked xylans containing similar amounts of 1,4 linkages (70.5 % - 80.2 %). The insoluble fibres contained essentially 1,4 linked xylans with some 1,3 linked xylose and a small amount of 1,4-linked glucose (cellulose). Soluble fibres were fermented within 6 hours by human faecal bacteria into short chain fatty acids (Lahaye et al. 1993).

4.11. Commercial products, patents and applications

Due to their plethora of bioactive molecules, marine macroalgae have great potential for further development as products in the nutraceutical, functional food, and pharmaceutical markets. Patent activity in this area has increased and several novel products based on macroalgae have entered the market in recent years. In respect of carbohydrates for example, the Kabushiki Kaisha Yakult Honsha company, Japan has patented a polysaccharide derivative (which contains fucoidan and rhamnan or rhamnan sulphate polysaccharides), extracted from the marine brown macroalgae, such as, *Cladosiphon okamuranus*, *Chordaricles nemacystus*, *Hydrilla* sp., *Fucus* sp., and a green alga *Monostroma nitidum*. The purpose of this compound is as a therapeutic agent for the prevention and treatment of gastric ulcers (specifically inhibiting the adhesion of *Helicobacter pylori* and administered as tablets, granules, powders or capsules (Nagaoka et al., 2003). Furthermore, Takara Shuzo Company, Kyoto, Japan has developed a medicinal composition exemplified by viscous polysaccharides isolated from red algae (specifically: *Gelidium amansil*, *G. japonicum*, *G. pacificum*, *G. subcostatum*, *Pteocladia tennis* and *Hypneaceae* species consisting of at least one 3,6- anhydrogalactopyranose. This compound is proposed for the treatment or prevention of diabetes, rheumatism, cancer and contains various inhibitory factors (Enoki et al., 2003). Sulphated fucans from *Fuscus vesiculosis* and *Ascophyllum nodosum* have been patented as anticoagulant substances (Smit, et al., 2004). In practical gastroenterology, mixtures of alginic acid and alginates with antacids are used to prevent gastro-esophageal reflux and to cure epigastric burning (Klinkenberg- Knol et al., 1995; Zeitoun et al., 1998). Indeed, in several countries such as the US, Germany, Japan, Belgium and Canada the use of alginic acid and its derivatives for the treatment of gastritis and gastroduodenal ulcers, as well as the use of alginates as atiulcer remedies, is protected by patents (Bogentoff et al, 1981; Borgo et al., 1984; Reckitt et al., 1987; Sheth et al., 1967). The medical application of pure fucoidan fraction has also been patented by the French research institute IFREMER-CNRS (PATENT WO/32099, however the extract of brown seaweeds (containing fucoidan fractions) can still be applied in cosmetology as fibroblast proliferation activators in the context of treatments aimed at aesthetics, for example of antiwrinkle treatments or of prevention of skin ageing without patent infringement.

Company	Compound	Activity / disorder	Development
Various	Heparin and derivatives	Anti-coagulants	Since 1940's
Astellas	Auranofin (Ridaura)	Anti-rheumatic	1983
GSK	Zanamivir (Relenza)	Anti-influenza	1992
Johnson & Johnson	Topiramate (Topamax)	Anti-epileptic	1987
Bayer	Acarbose (Glucobay) (Pseudo-oligosaccharide)	Type II diabetes (a-glucosidase, a-amylase inhibitor)	1990
Ortho-McNeil Janssen Pharmaceutical	Elmiron (Pentosan polysulfate)	Cystitis (for CJD)	1996
Alfa Wassermann	Sulodexide (Vessel™)	cardiovascular indications	Marketed since 1980's
Hunter Fleming (now Newron)	HF0420 – low MWt oligosaccharide	neuroprotective	Phase I
Progen (Australia)	PI-88 (Phosphomannopentaose sulphate, Heparan sulfate mimetics)	Anti-angiogenic/anti-metastatic. (hepatocellular carcinoma)	Phase III
	PG500 series (Heparan sulfate mimetics)	Anti-angiogenic/anti-metastatic.	Preclinical
Endotis Pharma	EP42675 (org) EP224283 (org)	Anticoagulant Neutralizable antithrombotic	Phase I Preclinical
	EP37	venous and arterial thromboses	Preclinical
	EP8000 programme	Anti-angiogenesis, anti-tumour growth /metastasis	Preclinical
Biotec Pharmacpn (Norway)	SBG (Soluble beta glucan – beta-1,3/1,6 glucan)	Immune stimulation Anti-cancer	GRAS nutraceutical

Table 5. Examples of carbohydrate-based drugs in use or development

The macroalgae polysaccharides described in this chapter show that these can be an interesting natural source of potential functional ingredients. As the content of proteins, carbohydrates, lipids and fiber can be influenced by the growing parameters (water, temperature, salinity, light and nutrients), macroalgae can be considered bioreactors that may be able to provide different polysaccharides at different quantities (Rui et. Al., 1990). From the descriptions above it is clear that macroalgae polysaccharides possess a multitude of bioactivities that might have antioxidant, antibacterial, antiviral, anticarcinogenic, anticoagulant and other bioactive properties for use as functional foods, pharmaceuticals and cosmeceuticals. The use of carbohydrate-based drugs is in its infancy, although there are several well-known examples (Table 5). Heparin is the key example of a major medicinally used carbohydrate based molecule. Low molecular weight heparins (eg. Certoparin, Dalteparin) and various derivatives (Fondaparinux – fully synthetic) have been developed to improve efficacy and half-life, and some are now being trialed for non-thrombotic/vascular applications (eg. Certoparin for inflammatory aspects of Alzheimers disease). It is only comparatively recently that the anti-inflammatory properties of heparin have been discovered. The presence of abundant 3-O-sulfated glucosamines in Heparan sulfates from human follicular fluid has suggested that some biological activities could be mimicked by other sulfated polysaccharides or derivatives (de Agostini et al., 2008). Low molecular weight fucoidan (LMWF) has been used to demonstrate this (Colliec et al., 1991, Millet et al., 1999 and Durand et al., 2008).

The drugs listed in Table 5 have overcome some of the perceived limitations of sugar-based molecules in terms of delivery, synthesis and immunogenicity. Although many still require intravenous delivery, several are available orally (e.g. Pentosan polysulfate), and further research is targeting to the improvement of oral availability by reducing compounds to their smallest active components, or by combining with other molecules (e.g. sulodexide). Improved synthesis has meant that some compounds can be synthesised (e.g. Fondaparinux), and small active components can be selected or modified to improve efficacy. No major problems have been reported with immunogenicity either in animal trials or as approved drugs. Sulphated polysaccharides may also be good anticoagulants through non-antithrombin mediated mechanisms, such as inhibition of thrombin mediated by heparin cofactor II (HC2). Like heparin, sulphated polysaccharides modulate cell growth-related activities, with inhibitory or promoting effects, depending on their carbohydrate backbones, degrees of sulphation, and distribution of sulphate groups. These activities can be usefully modulated through modification of sulphation patterns (Casu et al., 2002; Naggi et al., 2005). Several biological activities of sulphated polysaccharides are also molecular weight-dependent and the average length of their chains should be carefully controlled in order to maximize the desired activity. In some systems, opposite activities can be achieved with chains of different lengths of the same sulphated polysaccharide. Typically, relatively small oligosaccharides of heparin and heparinoids bind to growth factors and act as inhibitors of angiogenesis. On the other hand, the corresponding highermolecular weight species induce growth factors activation through oligomerization and binding with their receptors, thus promoting cell growth signalling (Casu and Lindahl, 2002; Goodner et al., 2008). Also inhibiton of heparanase and heparanase-related biological activities such as angiogenesis and metastasis by heparin species (Naggi et al., 2005) and sulfated hyaluronates (Naggi et al., 2005) are dependent on molecular weights and sulfation patterns. Similarly, sulphated polysaccharides are expected to be exploitable as drugs either in their natural form or with depolymerization and/or chemical derivatisation. Macroalgal suphated polysaccharides need to be fully characterised and the biological activities associated with specific structures determined so that their development can progress.

5. Brown algae – Growth, cultivation and productivity

Due to the relatively high carbohydrate content in percentage of the dryweight (Table 1) brown macroalgae in particular species of the Laminariaceae have come under the spotlight for mass-cultivation. Specifically in respect of carbohydrate production for fermentation purposes for ethanol and or biodegradable plastics production or other medical and food applications. Brown macroalgae exploitation in Europe is currently restricted to manual and mechanised harvesting of natural stocks, although several EU projects explore mass cultivation in European waters. The majority of Asian seaweed resources are cultivated. The most common system in Europe to obtain seaweed biomass is by harvesting natural stocks in coastal areas with rocky shores and a tidal system. The natural population of seaweed is a significant resource. Depending on water temperature, some groups will dominate, like brown seaweeds in cold waters and reds in warmer waters. In Europe the main harvesting countries are Norway and France. Between 120,000 and 130,000 tonnes of Laminaria are harvested annually in Norway. The standing stock is estimated to be 10 million tonnes (Vae and Ask, 2011). France harvests about 30,000 – 50,000 tonnes annually, mainly Laminaria species for hydrocolloid production. The green alga Ulva, is commonly encountered in estuaries and inshore coastal areas. When mass proliferation occurs they are known as 'green tides.' They tend to develop at more and more locations along European coasts, due to eutrophication, and are used to a small extent as fertilisers but not yet used for industrial applications. Nevertheless they may form an interesting source and feedstock for carbohydrates (Kraan & Guiry, 2006).

In Europe, knowledge of seaweed cultivation is scattered across several R&D groups and a few industrial groups. The amount of cultivated seaweed is very low, mainly very small companies with local facilities for cultivating high value species. Existing industries having large scale cultivation are located in Asian countries (China, Philippines, Korea, Indonesia, and Japan) and in Chile. The main obstacle in European countries will be labour cost. Development of mechanized seaweed cultivation will be required in Europe to achieve cost objectives. In Ireland also the aquaculture sector is gradually building know-how and basic infrastructure for Laminaria cultivation. Technologies to cultivate Laminaria are well known. For instance, the FAO publish a guide to Laminaria culture which is very detailed (Chen, 2005). The main producers of Laminaria are located in China, Korea and Japan, where preservation of natural stocks is not always sustainably managed. The main reason for an increased harvest is increased productivity due to selection of the best performing strains, improved crop-care, less variability, fertilizing techniques and faster harvesting. This strategy will also be required to achieve the low material cost with a high carbohydrate content needed for biofuel applications.

Brown seaweeds from the *Laminariaceae* or *Fucaceae* families have a generative cycle including *microstages*, i.e. zoospores and gametophytes, before fertilization results into new sporophytes. Gametophytes are independently living organisms which can be grown and propagated vegetatively. The interesting thing is that the gametophytes can be grown in biofermentors in a controlled way like microalgae or bacteria. This procedure is used for example for production of fucoxanthin from *Undaria pinnatifida* cultivated under microscopic gametophyte form. Because cloning of these gametophyte structures is possible, genetically uniform strains can be obtained quickly. These clones can be used also for breeding purposes towards varieties of the seaweed in their mature stage. Selection and breeding is an essential step to reach a uniform crop and to optimize yield of the targeted compounds. Little is known about the composition of these microstages, but based on their survival rate in non favorable (a)biotic conditions interesting levels of bioactive polysaccharides may be obtained. Cultivation is the most efficient solution to guarantee consistent contents in bioactive compounds, whereas seaweeds harvested from wild resources undergo uncontrolled composition variations due to changes in growing environment. If highly valuables molecules such as fucoidans are aimed as products, seaweed aquaculture is therefore a very interesting option. The current cultivation methods are still based on Asian techniques although currently programs are initiated in the EU to develop open sea based seaweed cultivation technology. Although seaweeds are known for their richness in bioactive substances like polysaccharides, proteins, lipids, vitamins and polyphenols and have been shown to have a wide range of potential cosmetic, pharmaceutical and medical applications, their economic potential is still insufficiently developed.

5.1. Eutrophication and suitable carbohydrate feed stocks

Inputs of biodegradable organic matter and inputs deriving from fertilizer run-off together with run off or dilution from finfish and shellfish rearing in near-shore waters and land based activities have many effects on the quality of coastal inshore waters and are a primary cause of eutrophication due to increased availability of nutrients (EPA, 2003). In estuaries and shallow coastal bays, this can lead to the proliferation of vast green algal mats, known as 'green tides'(Fletcher 1996). Kelp farms (inshore and near-shore) are able to act as bio-filters and are able to remove nitrates and phosphates from the surrounding eutrophic inshore waters. This allows for increased production of farmed seaweed as demonstrated by Chopin et al.,2001, 2008b Eutrophe waters are high in ammonia and phosphorous which can be stripped from the water by seaweed at rates varying from 60% up to 90% of the nutrient input. Macroalgae are able to take up nitrogen from seawater with rates to allow for a biomass increase of ca. 10% day-1. It is well documented that in tank systems the green alga Ulva is able to remove 90% of the nitrogenous compounds and the red alga Gracilaria up to 95% of dissolved ammonium from fish effluent (Neori et al., 2000). By cultivating and harvesting macroalgae as biofilters integrated with other shellfish or fish production systems, nutrient polution from these aquaculture systems could be alleviated through a process called IMTA (Integrated Multitrophic Aquaculture) while increasing production and carrying capacity (Chopin et al.,2001, 2008; Troell et al., 2009). Production of macroalgae in near-shore sea cultivation can be harvested for the bioethanol market to produce a value added marketable product acting as both an economic incentive and environmental incentive.

5.2. Effect of enrichment on carbohydrates

High ammonia and nitrate concentrations will alter the proximate composition in the macroalgae and cause a shift with higher protein and generally lower carbohydrate levels such as starch or dietary fiber as demonstrated by Rosenberg and Ramus, (1982); Pinchetti et al., (1998). How exactly it effects the carbohydrate composition is not known but might be an interesting way to manipulate carbohydrate content and composition in algae. In the red carrageenophyte Kappaphycus alvarezii the effect of ammonium addition in an otherwise nitrogen starved environment had a profound impact on the carrageenan content and showed increased gel strength of the carrageenan with an increase of ammonia (Rui et. al., 1990).

6. From a hydrocarbon society to carbohydrate society

Global demand for bio-fuels continues unabated. Rising concerns over environmental pollution and global warming, has encouraged the movement to alternate fuels, the world ethanol market is projected to reach 100 billion litres this year. Bioethanol is currently produced from carbohydrates from and-based crops such as corn and sugar cane. A continued use of these crops drives the food versus fuel debate. An alternate feed-stock which is abundant and carbohydrate-rich is necessary. The production of such crop needs to be sustainable and reduce competition with production of food, feed and industrial crops on agricultural inputs (pesticides, fertiliser, land, water). Macroalgae, in specific brown seaweeds could meet these challenges, being an abundant and carbon neutral renewable resource with potential to reduce green-house gas (GHG) emissions and the man-made impact on climate change.

Macroalgae are fast growing marine plants that can grow to considerable size (up to tens of meters in length in the case of Pacific kelp species), although Atlantic species would be smaller at ~ 3 m length (Lüning 1990). Growth rates of marine macroalgae far exceed those of terrestrial biomass. The large brown algae of kelp forests found on rocky shores inhabit an environment of vigorous water movement and turbulent diffusion. This allows very high levels of nutrient uptake, photosynthesis and growth. Highest productivity of kelp forests is found along the North American Pacific coast, which out-performs that of the most productive terrestrial systems (Velimirov et al. 1977). Laminaria-dominated communities of the European coasts have an annual productivity of approximately 2 kg carbon per m^2, which is still higher than, for example, temperate tree plantations or grasslands with a productivity of generally less than 1 kg carbon per m^2 (Thomas 2002). Production figures have been reported in the range of 3.3 – 11.3 kg dry weight m^{-2} yr^{-1} for non-cultured and up to 13.1 kg dry weight m^{-2} over 7 month for cultured brown algae compared with 6.1 – 9.5 kg fresh weight m^{-2} yr^{-1} for sugar cane, a most productive land plant. In addition marine biomass does not require fertilisation as currents and water exchange provide a continuous flow of a base level of nitrates and phosphates and large scale cultures may be useful in alleviating increased nitrogen levels in inshore waters. Due to the absence of lignin and a low content of cellulose, brown macroalgae carbohydrates may be easily convertible in biological processes compared to land plants.

Seaweeds are already farmed on a massive scale in Asia and substantial quantities are also harvested from natural populations. Recent research has shown the potential for large scale culture of macroalgae in Atlantic waters (Germany; Buck and Buchholz 2004: Canada; Chopin et al. 2008: France; Kaas 2006: Ireland; Kraan et al. 2000: Isle of Man, UK; Kain et al. 1990, Spain; Peteiro et. al., 2006). The challenges now lie in further developing cost-effective methodologies to grow, harvest transport and process large quantities of macroalgae.

A large body of research on fermentation of seaweeds into methane exists starting in the early 1980s and is extensively reviewed in Kelly and Dworjanyn (2008).

The only commercially available biofuels today are first generation biofuels, mainly bioethanol and biodiesel, produced from e.g. sugar cane and corn, and rapeseed, respectively. However, continued use of these crops will drive the food versus fuels debate even more as demand for ethanol increases. Not only does the large-scale production of corn and sugar cane damage the environment by the use of pesticides, it uses two other valuable resources: arable land and enormous quantities of water. For instance, the production of corn in the USA uses over 3 trillion litres of water a year in 2007 (Chiu et al. 2009).

Increased demand and the competition with food production has called for the development of second generation biofuels, based on utilization of lignocellulosic biomass, such as wood and agricultural waste. Second generation biofuels do not compete with food as a feedstock, but they compete for land and fresh water resources. Therefore the challenge is to find a feedstock which is abundant and carbohydrate-rich. This crop must be sustainable, use no agricultural inputs (pesticides, fertiliser, land, water), and must not be part of the human or animal food chain. Such a feedstock and an alternative to terrestrial biomass are marine macroalgae or seaweeds. Macroalgae and aquatic biomass are emerging as one of the most promising potential sources for biofuels production.

6.1. Suitable species and production

Several species of macroalgae accumulate high levels of carbohydrates, which are suitable as substrate for microbial conversion processes, e.g. for production of bioethanol, biobutanol as biofuels or other desirable chemicals with an attractive high product value. Green algae species such as the *Ulva* sp Linnaeus with high levels of the polysaccharide Ulvan (Lahaye and Ray 1996; Lahaye 1998) have been used in ethanol and methane production production (Morand et al. 1991, Adams et al. 2009). Brown macroalgae in particular kelp contain 50-60% carbohydrates of the dryweight and cultivation techniques have been firmly established for the last 50 years. Moreover, kelp is cultivated in large quantities up to 15.5 million wet tonnes in the Far East (FAO 2010).

Five Atlantic kelp species are suitable for cultivation and have a high carbohydrate level, i.e., *Saccorhiza polyschides* (Lightfoot) Batters; *Alaria esculenta* (L.) Greville; *Laminaria hyperborea* (Gunnerus) Foslie; *Laminaria digitata* (Hudson) J.V. Lamouroux; and *Saccharina latissima* (Linnaeus) C.E. Lane, C. Mayes, Druehl & G.W. Saunders (Werner and Kraan 2004). They differ in various aspects, such as morphology, ecophysiology and longevity. *Laminaria digitata* and *Laminaria hyperborea* are the only species that form extended monospecific kelp beds.

The biomass productivity of macroalgae ranges converted to carbon is about 1 to 3.4 kg carbon m^{-2} year^{-1} (Gao and McKinley 1994; Mann 1982; Mohammed and Fredriksen 2004). Seaweed communities of the North Atlantic coasts have an annual productivity of approximately 2 kg C per m^2, which is far higher than, for example, temperate tree plantations or grasslands with a productivity of generally less than 1 kg C m^{-2}, year^{-1} (Mann 1982; Chapman 1987; Thomas 2002; Lüning and Pang 2003; Mohammed and Fredrikson 2004), and 2.8 times higher than for sugar cane (Gao and McKinley 1994). Macroalgae can be cultivated in the open sea (Zhang et al. 2008; Bartsch et al. 2008; Kelly and Dworjanyn 2008). Ocean farming of seaweed does not depend on fresh water and does not occupy land areas (Yarish and Pereira 2008). Sustainable utilization of algal biomass - a largely unexplored feed stock resource can be a complement to terrestrial biomass for the future global energy and carbon security and thereby also strengthen the maritime economies.

Ocean farming of seaweed has the potential to produce in the order of 40 ton dry weight biomass per hectare per annum. An area of 2500 km2, the size of Luxembourg, would be able to provide 10 million ton dry biomass, representing 5.6-5.8 million ton carbohydrates. With current 90% enzymatic conversion into ethanol (Wargacki et al., 2012) this would yield close to 2 billion litres of bioethanol. This is about 2-3 % of the world's global bioethanol production (F.O. Licht 2009); however, it would cover about 50% of the EU's ethanol demand (Annon 2008b). The use of algal biomass has the potential to not only replace fossil resources, and thereby mitigate climate change, but also aid in the recycling potential of nitrates and phosphates in near and inshore waters.

6.2. Processing and fermentation of macroalgal biomass

The water content in macroalgae is higher than in terrestrial biomass (80-85 %), making seaweeds more suited for microbial conversion than for direct combustion or thermo-chemical conversion processes, which is an alternative for land-based biomass (Horn et al. 2000a; Ross et al. 2007). Seaweed carbohydrates may be used as substrates for microbial production of a wide range of fuels and chemicals. Ethanol production from hexose sugars such as glucose, sucrose, laminarin etc. derived from e.g. corn stover or sugar cane, is a well-known process. However, hexose-based polysaccharides constitute only about 30-40% of the carbohydrates in kelp. The remaining fraction is composed of C-5 sugars that until now have not been applied as substrates for industrial microbial production processes. However, recent breakthroughs have been made in C5 sugar fermentation technology allowing up to 90% of the available carbohydrates to be fermented (Wargacki, et al., 2012).

6.3. Pre-treatment of the seaweed biomass and hydrolysis of the polysaccharides of brown seaweed biomass

Fresh harvested brown seaweed contains about 15-20 % carbohydrates of the total wet weight, which equals about 200 g carbohydrates per kg wet weight, which is an appropriate substrate concentration for microbial conversion processes (Horn et al. 2000a). Lack of lignin in seaweeds implies that the harsh pre-treatment applied for release of fermentable sugars from lignocellulosic biomass is not required. Laminaran and mannitol can easily be extracted by water (Horn et al. 2000b). Alginates (consisting of polymer blocks of Uronic and guluronic acid) is present in the macroalgae biomass at 30-40% (Honya et al. 1993; McHugh 2003). Sodium alginate can be removed from the initial extraction solution by adding a calcium salt. This causes calcium alginate to form with a fibrous texture; it does not dissolve in water and can be separated from it. The separated calcium alginate is suspended in water and acid is added to convert it into alginic acid. This fibrous alginic acid is easily separated, placed in a planetary type mixer with alcohol, and sodium carbonate is gradually added to the paste until all the alginic acid is converted to sodium alginate. The paste of sodium alginate is sometimes extruded into pellets that are then dried and milled (McHugh 2003) . Due to the high viscosity, dilution with large amounts of water is required. The process is operating at alginate concentrations in the order of ~2 %. Such a dilution cannot be applied to processes aimed at use of alginate as fermentation substrate. Preferably, no water should be added, as it will increase the downstream processing costs (McHugh 2003).

Bioethanol production from cellulosic materials can be achieved by running simultaneously enzymatic hydrolysis and fermentation. For the wet seaweed biomass it is not as easy due to the alginates which are harder to release from the biomass causing enzymatic degradation of un-treated biomass and will be rate-limiting step if combined with fermentation. Hydrolysis should therefore be a part of the biomass pre-treatment. This hydrolysis can be carried out mechanically through grinding and emulsifying equipment, chemically using acid or alkali, or enzymatic. Several methods for partial or complete degradation of alginate as an integrated part of the mechanical pre-treatment of the biomass are known. Chemical hydrolysis should follow existing technologies , e.g. by modification and adaptation of methods used for acid and alkali pre-treatment of wood biomass (e.g. Ballesteros 2001; Klinke et al. 2001) and by combination of acid- or alkali with steam treatment. Other studies with macroalgae demonstrated the need for pre-treatment at 65°C, pH 2 for 1 h prior to fermentation (Horn, 2000a; 200b, Percival and McDowell, 1967). This in contrast with Adams et al. (2009) who found that these pre-treatments are not required for the fermentations with *Saccharina latissima* conducted, with higher ethanol yields being achieved in untreated fermentations than in those with altered pH or temperature pre-reatments. This result was seen in fresh and defrosted macroalgae samples using *Saccharomyces cerevisiae* and 1 unit of the enzyme laminarinase per kg of defrosted macroalgae. Nevertheless, the easiest and environmental friendliest way of pre-treatment of algal biomass is through a combination of mechanical and enzymatic hydrolysis (Doubet and Quatrano 1982).

6.4. Ethanol and butanol from brown seaweeds

Ethanol production from fermentation is the most obvious one as it has a direct application in the replacement of fossil fuels. However other products such as butanol and itaconic acid can be produced as well which can substitute and/or replace similar products produced from fossil resources. There are many microorganisms in the marine environment that can degrade and utilize algal carbohydrates as a carbon source for energy. Often these microorganisms are associated with the seaweeds being present on the blade surface or in tissue as many kelp species produce exo-polysaccharides as mucus layer or shed entire skin . This would imply that these organisms possess the necessary enzymes for cleavage of the algal polysaccharides. However, compounds such as ethanol and butanol are produced by anaerobic fermentation that require the presence of specific metabolic pathways generating these compounds as end products, e.g. yeast for ethanol production and *Clostridia* for butanol production. Limited information is available on the efficiency of these processes with seaweed carbohydrates (Horn et al. 2000a; 2000b), although several brake troughs have recently been made in respect of ethanol production from brown seaweeds (Wargacki, et al., 2012).

6.4.1. Ethanol

The potential of ethanol production from seaweeds can be calculated and is based on the following assumptions: A carbohydrate content of 60 % of dry weight and a 90 % conversion ratio to ethanol. Through fermentation one gram of sugar can yield 0.4 g ethanol. This will yield 0.22 kg or 0.27 l ethanol from 1 kg dry weight seaweed biomass, corresponding to approximately 0.05 l ethanol per kg wet weight.

Bacteria can be metabolize uronic acids to pyruvate and glyceraldehyde-3-P, which may then be fermented to ethanol by the glycolytic pathway (van Maris et al. 2006). In anaerobic fermentation processes, as ethanol and butanol production, oxygen is not available for removal of excess hydrogen generated. This implies that the conversion reaction from sub-strate to product must be red-ox balanced. Ethanol-production from hexose sugars is red-ox balanced, while production from pentoses or mannitol generates excess hydrogen. In many bacteria but not in yeast the enzyme transhydrogenases, solves this problem. Yeasts can avoid the problem by receiving a small, controlled supply of oxygen. However, oxygen leads to complete oxidation of the substrate to CO_2 and water, and reduced ethanol yields. Another strategy is introduction of transhydrogenase into strains that lack this, through genetic engineering (Fortman et al. 2008; Lee et al. 2008). Prospecting macroalgae (seaweeds) as feedstocks for bioconversion into biofuels and commodity chemical compounds is limited primarily by the availability of tractable microorganisms that can metabolize alginate polysaccharides. Wargacki, et al., (2012) present the discovery of a 36–kilo–base pair DNA fragment from Vibrio splendidus encoding enzymes for alginate transport and metabolism. The genomic integration of this ensemble, together with an engineered system for extracellular alginate depolymerization, generated a microbial platform that can simultaneously degrade,uptake, and metabolize alginate. They further engineered for ethanol synthesis, this platform enables bioethanol production directly from macroalgae via a consolidated process.

6.4.2. Butanol

Butanol is an alternative to ethanol with a higher energy content (butanol 29.2 MJ/l, ethanol 19.6 MJ/l), compared to gasoline (32 MJ/l). It can be used to supplement both gasoline and diesel fuels and can be handled by existing infrastructures (Fortman et al. 2008). Butanol is an important industrial chemical and is currently produced via petrochemical processes. In the last century butanol was produced through bacterial fermentation of starch rich compounds using *Clostridia* strains (Zverlov et al. 2006), which can use hexose as well as pentose sugers.

7. Outlook

Macroalgae are an interesting source for a myriad of different bioactive polysaccharides ranging from industrial applications to novel food applications. They possess many different interesting and often exotic polysaccharides that are currently explored for their functional properties in food and biomedicine. However, a far larger application would be the use of carbohydrates from cultivated seaweeds for alternative fuel sources. It is the exploitation of nature's energy cycle, photosynthesis and the resulting plant biomass that can accelerate this application. Society has to make a transition from a hydrocarbon to a carbohydrate economy, with the accrued benefits of carbon neutral biofuel, plastics and medecine. Macroalgae are efficient solar energy converters, and can create large amounts of biomass in a short-term, however, marine biomass is often an overlooked source, and potentially represents a significant source of carbohydrates as a renewable energy source.

Author details

Stefan Kraan
Ocean Harvest Technology Ltd., N17 Business Park, Milltown, Co. Galway, Ireland

8. References

Adams, J.M., Gallagher, J.A., Donnison, I.S. (2009). Fermentation study on *Saccharina latissima* for bioethanol production considering variable pre-treatments. *J. Appl. Phycol.*, 21(5), 569-574

Aisa Y, Miyakawa Y, Nakazato T, Shibata H, Saito K, Ikeda Y, Kizaki M (2005) Fucoidan induces apoptosis of human HS-Sultan cells accompanied by activation of caspase-3 and down-regulation of ERK pathways. *Am J Hematol* 78:7-14

Alekseyenko TV, Zhanayeva SY, Venediktova AA, Zvyagintseva TN, Kuznetsova TA, Besednova NN et al. (2007).Antitumor and antimetastatic activity of fucoidan, a sulfated polysaccharide isolated from the Okhotsk sea Fucus evanescens brown alga. *Bulletin of Experimental Biology and Medicine* 143(6), 730-732.

Amano, H., Kakinuma, M., Coury, D. A., Ohno, H., Hara, T. (2005). Effect of a seaweed mixture on serum lipid level and platelet aggregation in rats. *Fisheries Science*, 71(1160-1166.

Annon   (2008b).   Http://www.euractiv.com/29/images/UEPA%20Balanced%20Approach _tcm29-172883.pdf. Accessed 15 April 2012

Arasaki, S. & Arasaki, T. (1983). Low calorie, high nutrition vegetables from the sea. To help you look and feel better. Japan Publications, Tokyo, 196 pp.

Athukorala, Y., Lee, K. W., Kim, S. K., Jeon, Y. J. (2007). Anticoagulant activity of marine green and brown algae collected from Jeju Island in Korea. *Bioresource Technology*, 98(9), 1711-1716.

Ballesteros, I. Oliva, J.M., Negro, M.J., Manzanares, P., Ballesteros, M. (2002). Enzymic hydrolysis of steam exploded herbaceous agricultural waste (*Brassica carinata*) at different particle sizes. *Process Biochem.* 38(2), 187–92

Bartsch, I., Wiencke, C., Bischof, K., Buchholz, C.M., Buck, B.H., Eggert. A., Feuerpeeil, P., Hanelt, D., Jacobsen, S., Karez, R., Karsten. U., Molis, M., Roleda, M.Y., Schubert, H., Schumann, R., Valentin, K., Weinberger, F., Wiese, J. (2008). The genus *Laminaria sensu lato*: Recent insights and developments. *Eur. J. Phycol.*, 43, 1-86

Berteau, O. & Mulloy, B. (2003). Sulfated fucans, fresh perspectives: structures, functions, and biological properties of sulfated fucans and an overview of enzymes active toward this class of polysaccharide. *Glycobiology*, 13(6), 29R-40R.

Bixler, H.J. and Porse, H. (2010). A decade of change in the seaweed hydrocolloids industry. *Journal of Applied Phycology*, 23, 321-335

Bobin-Dubigeon, C., Lahaye, M., Barry, J. L. (1997). Human colonic bacterial degradability of dietary fibres from sea-lettuce (*Ulva* sp). *Journal of the Science of Food and Agriculture*, 73(2), 149-159.

Bogentoff, C. B. (1981). 2722484).

Borgo, E. (1984). 1176984).

Buck, B. H. and Buchholz, C.M. (2004). The offshore-ring: A new system design for the open ocean aquaculture of macroalgae. *J. Appl. Phycol.* 16, 355–368

Buck, C. B., Thompson, C. D., Roberts, J. N., Muller, M., Lowy, D. R., Schiller, J. T. (2006). Carrageenan is a potent inhibitor of papillomavirus infection. *Plos Pathogens*, 2(7), 671-680.

Burtin, P. (2003). Nutritional value of seaweeds. *Electronic Journal of Environmental, Agricultural and Food Chemistry*, 2(4), 498-503.

Caceres, P. J., Carlucci, M. J., Damonte, E. B., Matsuhiro, B., Zuniga, E. A. (2000). Carrageenans from chilean samples of *Stenogramme interrupta* (Phyllophoraceae): structural analysis and biological activity. *Phytochemistry*, 53(1), 81-86.

Carlucci, M. J., Pujol, C. A., Ciancia, M., Noseda, M. D., Matulewicz, M. C., Damonte, E. B. et al. (1997). Antiherpetic and anticoagulant properties of carrageenans from the red seaweed *Gigartina skottsbergii* and their cyclized derivatives: Correlation between structure and biological activity. *International Journal of Biological Macromolecules*, 20(2), 97-105.

Carthew, P. (2002). Safety of carrageenan in foods. *Environmental Health Perspectives*, 110(4), A176-A176.

Casu B, Lindahl U (2001). Structure and biological interactions of heparin and heparan sulfate. Adv. Carbohydr. Chem. Biochem. 57:159-206.

Casu B, Naggi A, Torri G (2002). Selective chemical modification as a strategy to study structure-activity relationships of glycosaminoglycans. Sem. Thromb. Hemost 28:335-342.

Cérantola, S., Breton, F., Ar Gall, E., Deslandes, E. (2006). Co-occurrence and antioxidant activities of fucol and fucophlorethol classes of polymeric phenols in *Fucus spiralis*. *Botanica Marina*, 49(4), 347-351.

Cha, S. H., Ahn, G. N., Heo, S. J., Kim, K. N., Lee, K. W., Song, C. B. et al. (2006). Screening of extracts from marine green and brown algae in Jeju for potential marine angiotensin-I converting enzyme (ACE) inhibitory activity. *Journal of the Korean Society of Food Science and Nutrition*, 35(307-314.

Chapman, A. R. O. (1987). The wild harvest and culture of *Laminaria longicruris* de la Pylaie in Eastern

Canada. In M.S. Doty, T.F. Caddy & B. Santelices (Eds.). *Case studies of seven commercial seaweed resources*. FAO Fisheries Technical Paper, 181: 193-238

Chapman, V. J. (1970). Seaweeds and their uses, 2nd edn. Methuen,London, 304 pp

Chen, H.-M., Zheng, L., Yan, X.-J. (2005). The preparation and bioactivity research of agaro-oligosaccharides. *Food Technology and Biotechnology*, 43(1), 29-36.

Chevolot, L., Foucault, A., Chaubet, F., Kervarec, N., Sinquin, C., Fisher, A. M. et al. (1999). Further data on the structure of brown seaweed fucans: relationships with anticoagulant activity. *Carbohydrate Research*, 319(1-4), 154-165.

Chiu, Y. W., Walsh, B. Suh, S. (2009). Water Embodied in Bioethanol in the United States. *Environ. Sci.Technol.*, 43, 2688–2692

Choi, Y. S., Hong, S. R., Lee, Y. M., Song, K. W., Park, M. H., Nam, Y. S. (1999). Study on gelatin-containing artificial skin: I. Preparation and characteristics of novel gelatin-alginate sponge. *Biomaterials*, 20(5), 409-417.

Chopin, T. et al. (2008). Mussels and kelps help salmon farmers reduce pollution. SeaWeb Vol 1, No2, 3 July Département d'Ecologie Côtière, B.P. 70, 29280 Plouzané http://www.ifremer.fr/envlit/documentation/documents.htm. Accessed 10 May 2010

Chopin, T., Buschmann, A.H., Halling, C., Troell, M., Kautsky, N., Neori, A., Kraemer, G.P., Zertuche-Gonzalez, J.A., Yarish, C., Neefus, C. (2001). Integrating seaweeds into marine aquaculture systems: a key towards sustainability. *Journal of Phycology* 37:975-986

Chopin, T., Robinson, S.M.C., Troell, M., Neori, A., Buschmann, A.H., and Fang, J., (2008b). Multitrophic integration for sustainable marine aquaculture: 2463-2475. *In: The Encyclopedia of Ecology, Ecological Engineering (Vol. 3)*. S.E. Jørgensen and B.D. Fath (Eds.). Elsevier, Oxford

Chopin, T., Sharp, G., Belyea, E., Semple, R., Jones, D. (1999). Open-water aquaculture of the red alga *Chondrus crispus* in Prince Edward Island, Canada. *Hydrobiologia*, 398/399, 417-425.

Colliec S, Fischer AM, Tapon-Bretaudière J, Boisson C, Durand P, Jozefonvicz J. (1991). Anticoagulant properties of a fucoidan fraction. Thromb Res 64(2),143-154

Critchley, A. T., R. D. Gillespie & K. W. G. Rotmann, 1998. The seaweed resources of South Africa. In Ohno, M., A. T. Critchley, D. Largo & R. D. Gillespie (eds), Seaweed Resources of the World. JICA, Japan: 413–427.

Critchley, A., Ohno, M. Largo, D. (2006). *The seaweed resources of the world*. A CD-rom project Eds. Alan Critchley , Masao Ohno and Danilo Largo, ETI, Amsterdam The Netherlands

Cumashi A, Ushakova NA, Preobrazhenskaya ME, D'Incecco A, Piccoli A, Totani L, Tinari N, Morozevich GE,

Berman AE, Bilan MI, Usov AI, Ustyuzhanina NE, Grachev AA, Sanderson CJ, Kelly M, Rabinovich GA, Iacobelli S, Nifantiev NE (2007). A comparative study of the anti-inflammatory, anticoagulant, antiangiogenic, and antiadhesive activities of nine different fucoidans from brown seaweeds. Glycobiology 17:541-552.

Dannhardt, G. & Kiefer, W. (2001). Cyclooxygenase inhibitors - current status and future prospects. *European Journal of Medicinal Chemistry*, 36(2), 109-126.

de Agostini AI, Dong JC, de Vantéry Arrighi C, Ramus MA, Dentand-Quadri I, Thalmann S, Ventura P, Ibecheole V, Monge F, Fischer AM, HajMohammadi S, Shworak NW, Zhang L, Zhang Z, Linhardt RJJ (2008). Human follicular fluid heparan sulfate contains abundant 3-O-sulfated chains with anticoagulant activity. Biol Chem., 283(42):28115-24.

Deville, C., Damas, J., Forget, P., Dandrifosse, G., Peulen, O. (2004). Laminarin in the dietary fibre concept. *Journal of the Science of Food and Agriculture*, 84(9), 1030-1038.

Deville, C., Gharbi, M., Dandrifosse, G., Peulen, O. (2007). Study on the effects of laminarin, a polysaccharide from seaweed, on gut characteristics. *Journal of the Science of Food and Agriculture*, 87(9), 1717-1725.

Doubet, R.S., Quatrano, R.S. (1982). Isolation of marine bacteria capable of producing specific lyases for alginate degradation. *Applied and Environmental Microbiology*, 44, 3754-756

Doyle, J. W., Roth, T. P., Smith, R. M., Li, Y. Q., Dunn, R. M. (1996). Effect of calcium alginate on cellular wound healing processes modeled in vitro. *Journal of Biomedical Materials Research*, 32(4), 561-568.

Dumelod, B. D., Ramirez, R. P. B., Tiangson, C. L. P., Barrios, E. B., Panlasigui, L. N. (1999). Carbohydrate availability of arroz caldo with lambda-carrageenan. *International Journal of Food Sciences and Nutrition*, 50(4), 283-289.

Durand E, Helley D, Al Haj Zen A, Dujols C, Bruneval P, Colliec-Jouault S, Fischer A-M, Lafont A (2008). Effect of Low molecular weight Fucoidan and Low molecular weight Heparin in a Rabbit Model of Arterial Thrombosis.J. Vascular Res 45:529-537.

Earons, G. (1994) Littoral Seaweed Resource Management. A Report Prepared for the Minch Project By Environment & Resource Technology Ltd Prepared for the Web by Gavin Earons, IT Unit, Comhairle nan Eilean Siar, 37p.

Ellouali, M., Boissonvidal, C., Durand, P., Jozefonvicz, J. (1993). Antitumor-activity of low-molecular-weight fucans extracted from brown seaweed *Ascophyllum nodosum*. *Anticancer Research*, 13(6A), 2011-2019.

Enoki, T., Sagawa, H., Tominaga, T., Nishiyama, E., Komyama, N., Sakai, T. et al. (2003). Drugs, foods or drinks with the use of algae-derived physiologically active substances. US Patent 0105029 A1).

EPA (2003). National Environmental Monitoring Programme for Transitional, Coastal and Marine Waters, Environmental Protection Agency.

F.O. Licht (2009). World Ethanol and Biofuels Report , vol. 7, no. 18, p. 365.

FAO (2012). FAO-statistics. Fishery Statistical Collections Global Aquaculture Production. http://www.fao.org/fishery/statistics/global-aquaculture-production/en. Accessed 15 April 2012.

Fernandez, L. E., Valiente, O. G., Mainardi, V., Bello, J. L., Velez, H., Rosado, A. (1989). Isolation and characterization of an antitumor active agar-type polysaccharide of *Gracilaria dominguensis*. *Carbohydrate Research*, 190(1), 77-83.

Fletcher, R. L. (1996). The occurrence of "green tides" - a review. *Marine benthic vegetation: recent changes and the effects of eutrophication*. W. S. A. P. H. Nienhuis. Berlin, Springer: 7-43.

Fleury, N. & Lahaye, M. (1991). Chemical and physicochemical characterization of fibers from *Laminaria digitata* (Kombu Breton) - A physiological approach. *Journal of the Science of Food and Agriculture*, 55(3), 389-400.

Fortman, J.L., Chhabra, S., Mukhopadhyay, A., Chou, H., Lee, T.S., Steen, E., Keasling, J.D. (2008). Biofuels alternatives to ethanol: pumping the microbial well. *Trends Biotechnol* , 26:375-381.

Fujiwara-Arasaki, T., Mino, N., Kuroda, M. (1984). The protein value in human nutrition of edible marine algae in Japan. *Hydrobiologia*, 116/117(513-516.

Gao, K., McKinley, K.R. (1994). Use of macroalgae for marine biomass production and CO2 remediation: a review. *J. Appl.Phycol.* 6, 45–60.

Ghosh, T., Chattopadhyay, K., Marschall, M., Karmakar, P., Mandal, P., Ray, B. (2009). Focus on antivirally active sulfated polysaccharides: From structure-activity analysis to clinical evaluation. *Glycobiology*, 19(1), 2-15.

Glyantsev, S., Annaev, A., Savvina, T., V (1993). Morphological basis for selecting composition and structure of biologically active compounds based on sodium alginate for wound treatment. *Byulleten' Eksperimental'noi Biologii i Meditsiny*, 115(1), 65-67.

Goemar (2010). Goemar. *http://www goemar fr.*

Granert, C., Raud, J., Waage, A., Lindquist, L. (1999). Effects of polysaccharide fucoidin on cerebrospinal fluid interleukin-1 and tumor necrosis factor alpha in pneumococcal meningitis in the rabbit. *Infection and Immunity*, 67(5), 2071-2074.

Haug, A. & Jensen, A. (1954). Seasonal variation in the chemical composition of *Alaria esculenta, Laminaria saccharina, Laminaria hyperborea* and *Laminaria digitata* from Northern Norway. 4,1-14.

Hawkins, W. W. & Leonard, V. G. (1963). Antithrombic Activity of Carrageenin in Human Blood. *Canadian Journal of Biochemistry and Physiology*, 41(5), 1325-&.

Hawkins, W. W. & Yaphe, W. (1965). Carrageenan as a dietary constituent for rat - faecal excretion nitrogen absorption and growth. *Canadian Journal of Biochemistry*, 43(4), 479-&.

Hawkins, W. W., Leonard, V. G., Maxwell, J. E., Rastogi, K. S. (1962). A study of the prolonged intake of small amounts of EDTA on the utilization of low dietary levels of calcium and iron by the rat. *Canadian Journal of Biochemistry and Physiology*, 40(391-.

Hayashi, L., Yokoya, N. S., Ostini, S., Pereira, R. T. L., Braga, E. S., Oliveira, E. C. (2008). Nutrients removed by *Kappaphycus alvarezii* (Rhodophyta, Solieriaceae) in integrated cultivation with fishes in re-circulating water. *Aquaculture*, 277, 185-191

Hennequart, F. (2007). Seaweed applications in human health/nutrition: The example of algal extracts as functional ingredients in novel beverages. 4th International Symposium Health and Sea, Granville, France.
http://www.sante-mer.com/pdf_dwnld/O17_Presentation_H_HENNEQUART.pdf, accessed 14 April 2012

Hoffman, R., Paper, D. H., Donaldson, J., Alban, S., Franz, G. (1995). Characterization of a laminarin sulfate which inhibits basic fibroblast growth-factor binding and endothelial-cell proliferation. *Journal of Cell Science*, 108(3591-3598.

Holdt, S and Kraan, S. (2011). Bioactive compounds in seaweed; functional food applications and legislation. Review Article. *Journal of Applied Phycology*, 23, 543-597

Honya, M., Kinoshita, T., Ishikawa, M., Mori, H., Nisizawa, K. (1993). Monthly determination of alginate M/G ratio, mannitol and minerals in cultivated *Laminaria japonica*. *Nippon Suisan Gakkaishi* 59(2), 295-299.

Horn, S.J., Aasen, I.M., Østgaard, K. (2000a). Production of ethanol from mannitol by *Zymobacter palmae*. *J. Ind. Microbiol. Biotechnol.*, 24, 51-57.

Horn, S.J., Aasen, I.M., Østgaard, K. (2000b). Ethanol production from seaweed extract. *J. Ind. Microbiol. Biotechnol.*, 25, 249-254

Hu JF, Geng MY, Zhang JT, Jiang HD (2001) An in vitro study of the structure–activity relationships of sulfated polysaccharide from brown algae to its antioxidant effect. J Asian Nat Prod Res 3:353–358

Iqbal, M., Flick-Smith, H., McCauley, J. W. (2000). Interactions of bovine viral diarrhoea virus glycoprotein E-rns with cell surface glycosaminoglycans. *Journal of General Virology*, 81(451-459.

Ito, K. & Tsuchida, Y. (1972). The effect of algal polysaccharides on depressing of plasma cholesterol level in rats. *Proceeding of The Seventh International Seaweed Symposium*, 451-455.

Jenkins, D. J. A., Wolever, T. M. S., Leeds, A. R., Gassull, M. A., Haisman, P., Dilawari, J. et al. (1978). Dietary fibers, fiber analogs, and glucose-tolerance - importance of viscosity. *British Medical Journal*, 1(6124), 1392-1394.

Jensen, A. & Haug, A. (1956). Geographical and seasonal variation in the chemical composition of *Laminaria hyperborea* and *Laminaria digitata* from the Norwegian coast.

Norwegian Institute of Seaweed Research, Akademisk Trykningssentral, Blindern, Oslo. Report 14, pp 1–8

Jensen, A. (1956). Component sugars of some common brown algae. Norwegian Institute of Seaweed Research, Akademisk Trykningssentral, Blindern, Oslo. Report 9, pp 1–8

Jeon Y-J, Athukorala Y, Lee J (2005) Characterization of agarose product from agar using DMSO. Algae 20:61–67

Kaas, R. (2006). The seaweed resources of France. In: Critchly, A.T. Ohno, M. Largo D.B. *The seaweed resources of the world.* Interactive CD ROM, ETI, The Netherlands.

Kain, J.M., Holt T.J., Dawes C.P. (1990). European Laminariales and their cultivation. In Yarish C., Penniman C.A., Van Patten P. (eds), Economically Important Marine Plants of the Atlantic: Their Biology and Cultivation. Connecticut Sea Grant College Program, Groton: 95-111.

Katayama, S., Ohshita, J., Sugaya, K., Hirano, M., Momose, Y., Yamamura, S. (1998). New medicinal treatment for severe gingivostomatitis. *International Journal of Molecular Medicine*, 2(6), 675-679.

Kelly M, Dworjanyn, S. (2008). The potential of marine biomass for anaerobic biogas production. The Crown Estate, 103 pages, London.

Khotimchenko, Y. S., Kovalev, V. V., Savchenko, O. V., Ziganshina, O. A. (2001). Physical-chemical properties, physiological activity, and usage of alginates, the polysaccharides of brown algae. *Russian Journal of Marine Biology*, 27(1), 53-64.

Kim, I. H. & Lee, J. H. (2008). Antimicrobial activities against methicillin-resistant *Staphylococcus aureus* from macroalgae. *Journal of Industrial and Engineering Chemistry*, 14(5), 568-572.

Kimura Y, Watanabe K, Okuda H (1996) Effects of soluble sodium alginate on cholesterol excretion and glucose tolerance in rats. *J. Ethnopharmacol* 54:47–54

Kindness, G., Williamson, F. B., Long, W. F. (1979). Effect of polyanetholesulphonic acid and xylan sulphate on antithrombin III activity. *Biochemical and Biophysical Research Communications*, 13(88), 1062-1068.

Kiriyama, S., Okazaki, Y., Yoshida, A. (1969). Hypocholesterolemic effect of polysaccharides and polysaccharide-rich foodstuffs in cholesterol-fed rats. *Journal of Nutrition*, 97(3), 382-&.

Klinke, H.B., Thomsen, A., Ahring, B.K. (2001) Potential inhibitors from wet oxidation of wheat straw and their effect on growth and ethanol production by *Thermoanaerobacter mathranii. Appl. Microbiol. Biotechnol.* 57,631–638.

Klinkenberg-Knol EC, Festen HP, Meuwissen SG (1995) Pharmacological management of gastro-oesophageal reflux disease. Drugs 49:495–710

Kraan, S., Guiry, M.D. (2006). The seaweed resources of Ireland. In: *Seaweed resources of the world*, A CD-rom project Eds. Alan Critchley , Masao Ohno and Danilo Largo, ETI, Amsterdam The Netherlands.

Kraan, S., Verges Tramullas, A., Guiry, M. D. (2000). The edible brown seaweed *Alaria esculenta* (Phaeophyceae, Laminariales): hybridisation, growth and genetic comparisons of six Irish populations. *Journal of Applied Phycology*, 12, 577-583

Kumar, C. S., Ganesan, P., Bhaskar, N. (2008a). In vitro antioxidant activities of three selected brown seaweeds of India. *Food Chemistry*, 107(2), 707-713.

Kylin, H. (1913). Zur Biochemie der Meeresalgen. *Hoppe-Seyler's Zeitschrift für physiologische Chemie*, 83(171-197.

Lahaye, M. & Kaeffer, B. (1997). Seaweed dietary fibres: structure, physico-chemical and biological properties relevant to intestinal physiology. *Sciences des Aliments*, 17(6), 563-584.

Lahaye, M. & Ray, B. (1996). Cell-wall polysaccharides from the marine green alga *Ulva rigida* (Ulvales, Chlorophyta) - NMR analysis of ulvan oligosaccharides. *Carbohydrate Research*, 283,161-173.

Lahaye, M. & Robic, A. (2007). Structure and functional properties of Ulvan, a polysaccharide from green seaweeds. *Biomacromolecules*, 8(6), 1765-1774.

Lahaye, M. (1991). Marine-algae as sources of fibers - determination of soluble and insoluble dietary fiber contents in some sea vegetables. *Journal of the Science of Food and Agriculture*, 54(4), 587-594.

Lahaye, M. (1998). NMR spectroscopic characterisation of oligosaccharides from two *Ulva rigida* ulvan samples (Ulvales, Chlorophyta) degraded by a lyase. *Carbohydrate Research*, 314(1-2), 1-12.

Lahaye, M., Michel, C., Barry, J. L. (1993). Chemical, physicochemical and in vitro fermentation characteristics of dietary-fibers from *Palmaria palmata* (L) Kuntze. *Food Chemistry*, 47(1), 29-36.

Lamela, M., Anca, J., Villar, R., Otero, J., Calleja, J. M. (1989). Hypoglycemic activity of several seaweed extracts. *Journal of Ethnopharmacology*, 27(1-2), 35-43.

Lasky, L. A. (1995). Selectin-carbohydrate interactions and the initiation of the inflammatory response. *Annual Review of Biochemistry*, 64(113-139.

Lee, J. B., Hayashi, K., Hashimoto, M., Nakano, T., Hayashi, T. (2004). Novel antiviral fucoidan from sporophyll of *Undaria pinnatifida* (Mekabu). *Chemical & Pharmaceutical Bulletin*, 52(9), 1091-1094.

Lee, S.K., Chou, H., Ham, T.S., Lee, T.S., Keasling, J.D. (2008). Metabolic engineering of microorganisms for biofuels production: from bugs to synthetic biology to fuels. *Current Opinion in Biotechnology*, 19,556–563.

Li F, Tian TC, Shi YC (1995) Study on anti virus effect of fucoidan in vitro. J Norman Bethune Univ Med Sci 21:255–257

Li, B., Lu, F., Wei, X. J., Zhao, R. X. (2008a). Fucoidan: Structure and bioactivity. *Molecules*, 13(8), 1671-1695.

Li, N., Zhang, Q. B., Song, J. M. (2005). Toxicological evaluation of fucoidan extracted from *Laminaria japonica* in Wistar rats. *Food and Chemical Toxicology*, 43(3), 421-426.

Li, Y., Qian, Z. J., Le, Q. T., Kim, M. M., Kim, S. K. (2008b). Bioactive phloroglucinol derivatives isolated from an edible marine brown alga, Ecklonia cava. *Journal of Biotechnology*, 136(Supplement 1), S578-S578.

Liu, R. M., Bignon, J., Haroun-Bouhedja, F., Bittoun, P., Vassy, J., Fermandjian, S. et al. (2005). Inhibitory effect of fucoidan on the adhesion of adenocarcinoma cells to fibronectin. *Anticancer Research*, 25(3B), 2129-2133.

Luescher-Mattli, M. (2003). Algae, a possible source for new drugs in the treatment of HIV and other viral diseases. *Current Medicinal Chemistry*, 2(219-225.

Lüning, K. (1990). *Seaweeds, Their environment, Biogeography and Ecophysiology.* Eds. C. Yarish and H. Kirkman, John Wiley & Sons, Inc., New York, 527 pp.

Lüning, K., Pang, S. (2003). Mass cultivation of seaweeds: current aspects and approaches. *J. Appl. Phycol.,* 15, 115–119.

Mabeau, S. & Fleurence, J. (1993). Seaweed in food products: biochemical and nutritional aspects. *Trends in Food Science & Technology*, 4(4), 103-107.

Mandal, P., Mateu, C. G., Chattopadhyay, K., Pujol, C. A., Damonte, E. B., Ray, B. (2007). Structural features and antiviral activity of sulphated fucans from the brown seaweed *Cystoseira indica. Antiviral Chemistry and Chemotherapy*, 18(153-162.

Mann, K.L. (1982). *Ecology of coastal waters: a systems approach.* University of California Press: Berkeley, CA.

Maruyama, H., Tanaka, M., Hashimoto, M., Inoue, M., Sasahara, T. (2007). The suppressive effect of Mekabu fucoidan on an attachment of *Cryptosporidium parvum* oocysts to the intestinal epithelial cells in neonatal mice. *Life Sciences*, 80(8), 775-781.

McHugh, D. J. (1987). Production and utilization of products from commercial seaweeds. 288, 1-189. FAO Fisheries Technical Paper No. 288, pp 1–189

Miao, H. Q., Elkin, M., Aingorn, E., Ishai-Michaeli, R., Stein, C. A., Vlodavsky, I. (1999). Inhibition of heparanase activity and tumor metastasis by laminarin sulfate and synthetic phosphorothioate oligodeoxynucleotides. *International Journal of Cancer*, 83(3), 424-431.

Michel, C., Lahaye, M., Bonnet, C., Mabeau, S., Barry, J. L. (1996). *In vitro* fermentation by human faecal bacteria of total and purified dietary fibres from brown seaweeds. *British Journal of Nutrition*, 75(2), 263-280.

Millet J, Colliec-Jouault S, Mauray S, Theveniaux J, Sternberg C, Boisson Vidal C, Fischer AM (1999). Antithrombotic and Anticoagulant Activities of a Low Molecular Weight Fucoidan by Subcutaneous Route. Thromb Haemostasis 81:391-395.

Mohammed, A. I., Fredriksen, S. (2004). Production, respiration and exudation of dissolved organic matter by the kelp *Laminaria hyperborea* along the west coast of Norway. *J. Mar. Biol. Ass. U.K.*, 84, 887-894.

Morand, P., Carpentier, B., Charlier, R.H., Mazé, J., Orlandini, M., Plunkett, B.A., de Waart, J. (1991). Bioconversion of seaweeds. In: Seaweed resources in Europe, uses and potential. Eds. Guiry, M.D. and Blunden, G. John Wiley &Sons, Chichester 432 pp.

Morrissey, J., Kraan, S., Guiry, M. D. (2001). A guide to commercially important seaweeds on the Irish coast. 1-66.

Mouradi-Givernaud A, Givernaud T, Morvan H, Cosson J (1992) Agar from Gelidium latifolium (Rhodophyceae, Gelidiales)—biochemical composition and seasonal variations. Bot Mar 35:153–159

Murata, M. & Nakazoe, J. (2001). Production and use of marine algae in Japan. *Jarq-Japan Agricultural Research Quarterly*, 35(4), 281-290.

Nabors, L. O. B. (2004). Alternative sweeteners. *Agro Food Industry Hi-Tech*, 15(4), 39-41.

Nagaoka, M., Shibata, H., Kimura, I., Hashimoto, S. (2003). Oligosaccharide derivatives and process for producing the same. United States Patent 6,645,940).

Nagaoka, M., Shibata, H., Kimura-Takagi, I., Hashimoto, S., Kimura, K., Makino, T. et al. (1999). Structural study of fucoidan from *Cladosiphon okamuranus* Tokida. *Glycoconjugate Journal*, 16(1), 19-26.

Naggi A, Casu B, Perez M, Torri, G, Cassinelli, G, Penco, S, Pisano, C, Giannini, G, Ishai-Michaeli R, Vlodavsky I (2005). Modulation of heparanase-inhibiting activity of heparin through selective desulfation, graded N-acetylation, and glycol-splitting. J. Biol. Chem 280:12103-12113.

Nasu, T., Fukuda, Y., Nagahira, K., Kawashima, H., Noguchi, C., Nakanishi, T. (1997). Fucoidin, a potent inhibitor of L-selectin function, reduces contact hypersensitivity reaction in mice. *Immunology Letters*, 59(1), 47-51.

Neori, A., Shpigel, M., Ben-Ezra, D. (2000). A sustainable integrated system for culture of fish, seaweed and abalone. *Aquaculture* 186, 279–291.

Nishide, E., Anzai, H., Uchida, N. (1993). Effects of alginates on the ingestion and excretion of cholesterol in the rat. *Journal of Applied Phycology*, 5(2), 207-211.

Noa, M., Mas, R., Carbajal, D., Valdes, S. (2000). Effect of D-002 on acetic acid-induced colitis in rats at single and repeated doses. *Pharmacological Research*, 41(4), 391-395.

Noda, H. (1993). Health benefits and nutritional properties of nori. *Journal of Applied Phycology*, 5(2), 255-258.

Omata, M., Matsui, N., Inomata, N., Ohno, T. (1997). Protective effects of polysaccharide fucoidin on myocardial ischemia-reperfusion injury in rats. *Journal of Cardiovascular Pharmacology*, 30(6), 717-724.

Ostergaard, C., Yieng-Kow, R. V., Benfield, T., Frimodt-Moller, N., Espersen, F., Lundgren, J. D. (2000). Inhibition of leukocyte entry into the brain by the selectin blocker fucoidin decreases interleukin-1 (IL-1) levels but increases IL-8 levels in cerebrospinal fluid during experimental pneumococcal meningitis in rabbits. *Infection and Immunity*, 68(6), 3153-3157.

Panlasigui, L. N., Baello, O. Q., Dimatangal, J. M., Dumelod, B. D. (2003). Blood cholesterol and lipid-lowering effects of carrageenan on human volunteers. *Asia Pacific Journal of Clinical Nutrition*, 12(2), 209-214.

Paradossi G, Cavalieri F, Chiessi E (2002) A conformational study on the algal polysaccharide ulvan. Macromolecules 35:6404–6411

Peteiro, C. Salinas, j. M & Fuertes, C. (2006). Cultivation of the autoctonous seaweed *Laminaria saccharina* off the galician coast (nw spain): Production and features of the sporophytes for An annual and biennial harvest. Thalassas, 22 (1): 45-53

Percival E, McDowell RH (1990) Algal polysaccharides. In: Dey PM (ed) Methods in plant biochemistry, volume 2: carbohydrates. Academic, London, pp 523–547

Pinchetti, J.L.G., del Campo Fernández, E., Moreno Díez, P., and and García Reina, G. (1998). Nitrogen availability influences the biochemical composition and photosynthesis of tank-cultivated Ulva rigida (Chlorophyta). Journal of Applied Phycology, 10, 383-389.

Qi, H. M., Zhang, Q. B., Zhao, T. T., Chen, R., Zhang, H., Niu, X. Z. et al. (2005a). Antioxidant activity of different sulfate content derivatives of polysaccharide extracted from Ulva pertusa (Chlorophyta) in vitro. International Journal of Biological Macromolecules, 37(4), 195-199.

Qi, H. M., Zhao, T. T., Zhang, Q. B., Li, Z., Zhao, Z. Q., Xing, R. (2005b). Antioxidant activity of different molecular weight sulfated polysaccharides from Ulva pertusa Kjellm (Chlorophyta). Journal of Applied Phycology, 17(6), 527-534.

Rasmussen, R. S. & Morrissey, M. T. (2007). Marine biotechnology for production of food ingredients. 237-292.

Reckitt and Colman Products Ltd (1987). 858 003.

Reilly, P., O'Doherty, J. V., Pierce, K. M., Callan, J. J., O'Sullivan, J. T., Sweeney, T. (2008). The effects of seaweed extract inclusion on gut morphology, selected intestinal microbiota, nutrient digestibility, volatile fatty acid concentrations and the immune status of the weaned pig. Animal, 2(10), 1465-1473.

Renn, D. W., Noda, H., Amano, H., Nishino, T., Nishizana, K. (1994a). Antihypertensive and antihyperlipidemic effects of funoran. Fisheries Science, 60(423-427.

Renn, D. W., Noda, H., Amano, H., Nishino, T., Nishizana, K. (1994b). Study on hypertensive and antihyperlipidemic effect of marine algae. Fisheries Science, 60(83-88.

Reynolds JEF, Prasad AB (1982) Martindale the extra pharmacopoeia. 28th Pharmaceutical Press, London, 735 pp

Rioux, L. E., Turgeon, S. L., Beaulieu, M. (2007). Characterization of polysaccharides extracted from brown seaweeds. Carbohydrate Polymers, 69(3), 530-537.

Rocha de Souza M, Marques C, Guerra Dore C, Ferreira da Silva F, Oliveira Rocha H, Leite E (2007) Antioxidant activities of sulfated polysaccharides from brown and red seaweeds. J Appl Phycol 19:153–160

Rodrigueza, M. R. C. & Montaño, M. N. E. (2007). Bioremediation potential of three carrageenophytes cultivated in tanks with seawater from fish farms. Journal of Applied Phycology, 19(6), 755-762.

Rosenberg, C. and Ramus, J. (1982). Ecological growth strategies in the seaweeds Gracilaria foliifera (Rhodophyceae) and Ulva sp. (Chlorophyceae): Soluble nitrogen and reserve carbohydrates. Marine Biology, 251-259

Ross, A., Jones, J.M., Kubacki, M.L., Bridgeman, T.G. (2008). Classification of macroalgae as fuel and its thermochemical behaviour. Bioresource Technology, 6494-6504.

Rui, L., Jiajun, L. and Chaoyuan, W. (1990). Effect of ammonium on growth and carrageenan content in Kappaphycus alvarezii (Gigartinales, Rhodophyta). Hydrobiologia 204-205, 499-503.

Rupérez, P. & Saura-Calixto, F. (2001). Dietary fibre and physicochemical properties of edible Spanish seaweeds. *European Food Research and Technology*, 212(3), 349-354.

Ruperez, P., Ahrazem, O., Leal, J. A. (2002). Potential antioxidant capacity of sulfated polysaccharides from the edible marine brown seaweed *Fucus vesiculosus*. *Journal of Agricultural and Food Chemistry*, 50(4), 840-845.

Savitskaya, I. M. (1986). Trial of a local hemostatic with alginate base. *Klin Khirurgiya*, 3(39-40.

Segal, H. C., Hunt, B. J., Gilding, K. (1998). The effects of alginate and non-alginate wound dressings on blood coagulation and platelet activation. *Journal of Biomaterials Applications*, 12(3), 249-257.

Sen, A. K., Das, A. K., Banerji, N., Siddhanta, A. K., Mody, K. H., Ramavat, B. K. et al. (1994). A new sulfated polysaccharide with potent blood anti-coagulant activity from the red seaweed *Grateloupia Indica*. *International Journal of Biological Macromolecules*, 16(5), 279-280.

Sezer, A.D.; Hatipoğlu, F.; Cevher, E.; Oğurtan, Z.; Baş, A.L. & Akbuğa, J. (2007). Chitosan films containing fucoidan as a wound dressing for dermal burn healing: Preparation and in vitro/in vivo evaluation, AAPS PharmSciTech, 8, 1-8.

Shanmugam, M. & Mody, K. H. (2000). Heparinoid-active sulphated polysaccharides from marine algae as potential blood anticoagulant agents. *Current Science*, 79(12), 1672-1683.

Sheth, B. B. (1967). 3 326755.

Skoler-Karpoff, S., Ramjee, G., Ahmed, K., Altini, L., Plagianos, M. G., Friedland, B. et al. (2008). Efficacy of Carraguard for prevention of HIV infection in women in South Africa: a randomised, double-blind, placebo-controlled trial. *The Lancet*, 372(9654), 1977-1987.

Smit, A. J. (2004). Medicinal and pharmaceutical uses of seaweed natural products: A review. *Journal of Applied Phycology*, 16(4), 245-262.

Spieler, R. (2002). Seaweed compound's anti-HIV efficacy will be tested in southern Africa. *Lancet*, 359(9318), 1675-1675.

Terada, A., Hara, H., Mitsuoka, T. (1995). Effect of dietary alginate on the faecal microbiota and faecal metabolic activity in humans. *Microbial Ecology in Health and Disease*, 8(6), 259-266.

Thomas, D.N. (2002). *Seaweeds*, Natural History Museum, London

Tobacman, J. K. (2001). Review of harmful gastrointestinal effects of carrageenan in animal experiments. *Environmental Health Perspectives*, 109(10), 983-994.

Torsdottir, I., Alpsten, M., Holm, G., Sandberg, A. S., Tolli, J. (1991). A small dose of soluble alginate-fiber affects postprandial glycemia and gastric-emptying in humans with diabetes. *Journal of Nutrition*, 121(6), 795-799.

Troell, M., Joyce, A., Chopin, T., Neori, A., Buschmann, A.H., Fang, J-G (2009). Ecological engineering in aquaculture — Potential for integrated multi-trophic aquaculture (IMTA) in marine offshore systems. *Aquaculture* 297, 1-9.

Tseng, C. K. (2001). Algal biotechnology industries and research activities in China. *Journal of Applied Phycology*, 13(4), 375-380.

Urbano, M. G. & Goñi, I. (2002). Bioavailability of nutrients in rats fed on edible seaweeds, Nori (*Porphyra tenera*) and Wakame (*Undaria pinnatifida*), as a source of dietary fibre. *Food Chemistry*, 76(3), 281-286.

Usov, A. I., Smirnova, G. P., Klochkova, N. G. (2001). Polysaccharides of algae: 55. Polysaccharide composition of several brown algae from Kamchatka. *Russian Journal of Bioorganic Chemistry*, 27(6), 395-399.

van de Velde, F., Lourenco, N. D., Pinheiro, H. M., Bakker, M. (2002). Carrageenan: A food-grade and biocompatible support for immobilisation techniques. *Advanced Synthesis & Catalysis*, 344(8), 815-835.

van Maris, A.J.A., Abbot, D.A., Bellisimi, E., van den Brink, J., Kuyper, M., Luttik, M.A.H., Wisselink, H.W., Scheffers, W.A., van Dijken, J.P., Pronk, J.T. (2006). Alcoholic fermentation of carbon sources in biomass hydrolysates by *Saccharomyces cerevisiae*: current status. *Antonie van Leeuwenhoek*, 90, 391-418.

Vaugelade, P., Hoebler, C., Bernard, F., Guillon, F., Lahaye, M., Duee, P. H. et al. (2000). Non-starch polysaccharides extracted from seaweed can modulate intestinal absorption of glucose and insulin response in the pig. *Reproduction Nutrition Development*, 40(1), 33-47.

Vea, J. and Ask, E.,(2011). Creating a sustainable commercial harvest of Laminaria hyperborea, in Norway. Journal of Applied Phycology 23, 489-494

Velimirov, B, Field, J.G., Griffiths, C.L., Zoutendyk, P. (1977). The ecology of kelp bed communities in the Benguela upwelling system. Analysis of biomass and spatial distribution. *Helgoländer wissenschaftliche Meeresuntersuchungen* 30,495-518.

Verdrengh, M., Erlandsson-Harris, H., Tarkowski, A. (2000). Role of selectins in experimental *Staphylococcus aureus*-induced arthritis. *European Journal of Immunology*, 30(6), 1606-1613.

Vlieghe, P., Clerc, T., Pannecouque, C., Witvrouw, M., De Clercq, E., Salles, J. P. et al. (2002). Synthesis of new covalently bound kappa-carrageenan-AZT conjugates with improved anti-HIV activities. *Journal of Medicinal Chemistry*, 45(6), 1275-1283.

Wang, J., Zhang, Q. B., Zhang, Z. S., Li, Z. (2008). Antioxidant activity of sulfated polysaccharide fractions extracted from *Laminaria japonica*. *International Journal of Biological Macromolecules*, 42(2), 127-132.

Wang, W. T., Zhou, J. H., Xing, S. T. (1994). Immunomodulating action of marine algae sulfated polysaccharides on normal and immunosuppressed mice. *Chinese Journal of Pharm Toxicol*, 8(199-202.

Wang, Y., Han, F., Hu, B., Li, J. B., Yu, W. G. (2006). In vivo prebiotic properties of alginate oligosaccharides prepared through enzymatic hydrolysis of alginate. *Nutrition Research*, 26(11), 597-603.

Wargacki,A.J., Leonard, E., Nyan Win, M., Regitsky,D.D., Santos, C.N.S., Kim,P.B., Cooper, S.R., Raisner, R.M., Herman, A., Sivitz, A.B., Lakshmanaswamy, A., Kashiyama,Y.,

Baker, D., Yoshikuni, Y. (2012). An Engineered Microbial Platform for Direct Biofuel Production from Brown Macroalgae. *Science* 335, 308.

Washington, N. & Denton, G. (1995). Effect of alginate and alginate-cimetidine combination therapy on stimulated postprandial gastro-oesophageal reflux. *Journal of Pharmacy and Pharmacology*, 47(11), 879-882.

Werner, A., Kraan, S. (2004). Review of the potential mechanisation of kelp harvesting in Ireland. Marine Environment & Health Series, No.17 pp 1-52, Marine Institute, Galway.

Witvrouw, M. & DeClercq, E. (1997). Sulfated polysaccharides extracted from sea algae as potential antiviral drugs. *General Pharmacology-the Vascular System*, 29(4), 497-511.

Wu, X. W., Yang, M. L., Huang, X. L., Yan, J., Luo, Q. (2003). Effect of fucoidan on splenic lymphocyte apoptosis induced by radiation. *Chin J radiol Med Prot*, 23(430-.

Xue, C. H., Chen, L., Li, Z. J., Cai, Y. P., Lin, H., Fang, Y. (2004). Antioxidative activities of low molecular fucoidans from kelp *Laminaria japonica*. 139-145.

Yan, X. J., Zheng, L., Chen, H. M., Lin, W., Zhang, W. W. (2004). Enriched accumulation and biotransformation of selenium in the edible seaweed *Laminaria japonica*. *Journal of Agricultural and Food Chemistry*, 52(21), 6460-6464.

Yang XL, Sun JY, Xu HN (1995). An experimental study on immunoregulatory effect of fucoidan. Chin J Mar Drugs

Yarish, C., Pereira, R. (2008). Mass production of marine macroalgae. In Sven Erik Jørgensen and Brian D. Fath (Editor-in-Chief), *Ecological Engineering*. Vol. [3] of Encyclopedia of Ecology, 5 vols. pp. [2236-2247] Oxford, Elsevier.

Ye, H., Wang, K., Zhou, C., Liu, J., Zeng, X. (2008). Purification, antitumor and antioxidant activities in vitro of polysaccharides from the brown seaweed Sargassum pallidum. *Food Chemistry*, 111(2), 428-432.

Zee S (1991) Body weight loss with the aid of alginic acid. Arc Intern Med 3-4:113–114

Zeitlin, L., Whaley, K. J., Hegarty, T. A., Moench, T. R., Cone, R. A. (1997). Tests of vaginal microbicides in the mouse genital herpes model. *Contraception*, 56(5), 329-335.

Zeitoun P, Salmon L, Bouche O, Jolly, D, Thiefin G (1998) Outcome of erosive/ulcerative reflux oesophagitis in 181 consecutive patients 5 years after diagnosis. Ital J Gastroenterol Hepatol 30:470–474

Zhang, Q. B., Li, N., Zhou, G. F., Lu, X. L., Xu, Z. H., Li, Z. (2003). In vivo antioxidant activity of polysaccharide fraction from *Porphyra haitanesis* (Rhodephyta) in aging mice. *Pharmacological Research*, 48(2), 151-155.

Zhang, Q.S., Qu, S.C., Cong, Y.Z., Luo, S.J., Tang, X.-X. (2008). High throughput culture and gametogenesis induction of *Laminaria japonica* gametophyte clones. *J. Appl. Phycol.* 20, 205-211.

Zhao, X., Xue, C. H., Li, B. F. (2008). Study of antioxidant activities of sulfated polysaccharides from Laminaria japonica. *Journal of Applied Phycology*, 20(4), 431-436.

Zhou, Y., Yang, H. S., Hu, H. Y., Liu, Y., Mao, Y. Z., Zhou, H. et al. (2006b). Bioremediation potential of the macroalga *Gracilaria lemaneiformis* (Rhodophyta) integrated into fed fish culture in coastal waters of north China. *Aquaculture*, 252(2-4), 264-276.

Zhuang, C., Itoh, H., Mizuno, T., Ito, H. (1995). Antitumor active fucoidan from the brown seaweed, Umitoranoo (*Sargassum-thunbergii*). *Bioscience Biotechnology and Biochemistry*, 59(4), 563-567.

Zverlov, V.V., Berezina, O.V., Velikodvirskaya, G.A., Schwarz, W. H. (2006). Bacterial acetone and butanol production by industrial fermentation in the Soviet Union: use of hydrolyzed agricultural waste for biorefinery. *Applied Microbiology and Biotechnology* 71, 587-597.

Permissions

The contributors of this book come from diverse backgrounds, making this book a truly international effort. This book will bring forth new frontiers with its revolutionizing research information and detailed analysis of the nascent developments around the world.

We would like to thank Chuan-Fa Chang, for lending his expertise to make the book truly unique. He has played a crucial role in the development of this book. Without his invaluable contribution this book wouldn't have been possible. He has made vital efforts to compile up to date information on the varied aspects of this subject to make this book a valuable addition to the collection of many professionals and students.

This book was conceptualized with the vision of imparting up-to-date information and advanced data in this field. To ensure the same, a matchless editorial board was set up. Every individual on the board went through rigorous rounds of assessment to prove their worth. After which they invested a large part of their time researching and compiling the most relevant data for our readers. Conferences and sessions were held from time to time between the editorial board and the contributing authors to present the data in the most comprehensible form. The editorial team has worked tirelessly to provide valuable and valid information to help people across the globe.

Every chapter published in this book has been scrutinized by our experts. Their significance has been extensively debated. The topics covered herein carry significant findings which will fuel the growth of the discipline. They may even be implemented as practical applications or may be referred to as a beginning point for another development. Chapters in this book were first published by InTech; hereby published with permission under the Creative Commons Attribution License or equivalent.

The editorial board has been involved in producing this book since its inception. They have spent rigorous hours researching and exploring the diverse topics which have resulted in the successful publishing of this book. They have passed on their knowledge of decades through this book. To expedite this challenging task, the publisher supported the team at every step. A small team of assistant editors was also appointed to further simplify the editing procedure and attain best results for the readers.

Our editorial team has been hand-picked from every corner of the world. Their multi-ethnicity adds dynamic inputs to the discussions which result in innovative outcomes. These outcomes are then further discussed with the researchers and contributors who give their valuable feedback and opinion regarding the same. The feedback is then collaborated with the researches and they are edited in a comprehensive manner to aid the understanding of the subject.

Apart from the editorial board, the designing team has also invested a significant amount of their time in understanding the subject and creating the most relevant covers. They scrutinized every image to scout for the most suitable representation of the subject and create an appropriate cover for the book.

The publishing team has been involved in this book since its early stages. They were actively engaged in every process, be it collecting the data, connecting with the contributors or procuring relevant information. The team has been an ardent support to the editorial, designing and production team. Their endless efforts to recruit the best for this project, has resulted in the accomplishment of this book. They are a veteran in the field of academics and their pool of knowledge is as vast as their experience in printing. Their expertise and guidance has proved useful at every step. Their uncompromising quality standards have made this book an exceptional effort. Their encouragement from time to time has been an inspiration for everyone.

The publisher and the editorial board hope that this book will prove to be a valuable piece of knowledge for researchers, students, practitioners and scholars across the globe.

List of Contributors

Suman Mishra and John Monro
Food Industry Science Centre, The New Zealand Institute for Plant & Food Research Limited, Palmerston North, New Zealand

Allan Hardacre
Institute of Food Nutrition and Human Health, Massey University, Palmerston North, New Zealand

Katarzyna Śliżewska
Institute of Fermentation Technology and Microbiology, Faculty of Biotechnology and Food Sciences, Technical University of Lodz, Lodz, Poland

Janusz Kapuśniak, Renata Barczyńska and Kamila Jochym
Institute of Chemistry, Environmental Protection and Biotechnology, Jan Dlugosz University in Czestochowa, Czestochowa, Poland

Alejandro Reyes-DelaTorre, María Teresa Peña-Rangel and Juan Rafael Riesgo-Escovar
Developmental Neurobiology and Neurophysiology Dept., Instituto de Neurobiología, Universidad Nacional Autónoma de México Querétaro, México

Anna Novials and Serafín Murillo
Department of Endocrinology and Nutrition, Hospital Clínic de Barcelona, Barcelona, Spain
Institut d'Investigacions Biomèdiques August Pi i Sunyer (IDIBAPS), Barcelona, Spain
Spanish Biomedical Research Centre in Diabetes and Associated Metabolic Disorders (CIBERDEM), Spain

Monia El Bour
National Institute of Sea Sciences and Technologies, Salammbô, Tunisia

Iwona Morkunas and Magda Formela
Department of Plant Physiology, Faculty of Horticulture and Landscape Architecture, Poznań University of Life Sciences, Poznań, Poland

Sławomir Borek and Lech Ratajczak
Department of Plant Physiology, Faculty of Biology, Adam Mickiewicz University, Poznań, Poland

José Luis da Silva Nunes
BADESUL Desenvolvimento, Brasil

Paulo Vitor Dutra de Souza, Gilmar Arduino Bettio Marodin and Jorge Ernesto de Araújo Mariath
Universidade Federal do Rio Grande do Sul, Brasil

José Carlos Fachinello
Universidade Federal de Pelotas, Brasil

Ali Ashraf Jafari
Plant Breeding, Gene Bank Division, Research Institute of Forests and Rangelands, Tehran, Iran

Luis Henrique Souza Guimarães
Faculdade de Filosofia, Ciências e Letras de Ribeirão Preto, University of São Paulo, Brasil

Amira El-Fallal, Mohammed Abou Dobara, Ahmed El-Sayed and Noha Omar
Faculty of Science, Damietta Branch, Egypt

Edson Holanda Teixeira, Francisco Vassiliepe Sousa Arruda and Bruno Rocha da Silva
LIBS, Integrated Laboratory of Biomolecules, Faculty of Medicine of Sobral, Federal University of Ceará, Fortaleza, CE, Brazil

Kyria Santiago do Nascimento, Victor Alves Carneiro and Benildo Sousa Cavada
BioMol, Laboratory of Biologically Active Molecules, Department of Biochemistry and Molecular Biology, Federal University of Ceará, Fortaleza, CE, Brazil

Celso Shiniti Nagano and Alexandre Holanda Sampaio
BioMar, Laboratory of Marine Biotechnology, Department of Fishing Engineering, Federal University of Ceará, Fortaleza, CE, Brazil

Stefan Kraan
Ocean Harvest Technology Ltd., N17 Business Park, Milltown, Co. Galway, Ireland